Maria do Carmo Nicoletti and Lakhmi C. Jain (Eds.)

Computational Intelligence Techniques for Bioprocess Modelling, Supervision and Control

Studies in Computational Intelligence, Volume 218

Editor-in-Chief
Prof. Janusz Kacprzyk
Systems Research Institute
Polish Academy of Sciences
ul. Newelska 6
01-447 Warsaw
Poland
E-mail: kacprzyk@ibspan.waw.pl

Further volumes of this series can be found on our homepage: springer.com

Vol. 197. Mauro Birattari
Tuning Metaheuristics, 2009
ISBN 978-3-642-00482-7

Vol. 198. Efrén Mezura-Montes (Ed.)
Constraint-Handling in Evolutionary Optimization, 2009
ISBN 978-3-642-00618-0

Vol. 199. Kazumi Nakamatsu, Gloria Phillips-Wren, Lakhmi C. Jain, and Robert J. Howlett (Eds.)
New Advances in Intelligent Decision Technologies, 2009
ISBN 978-3-642-00908-2

Vol. 200. Dimitri Plemenos and Georgios Miaoulis *Visual Complexity and Intelligent Computer Graphics Techniques Enhancements*, 2009
ISBN 978-3-642-01258-7

Vol. 201. Aboul-Ella Hassanien, Ajith Abraham, Athanasios V. Vasilakos, and Witold Pedrycz (Eds.)
Foundations of Computational Intelligence Volume 1, 2009
ISBN 978-3-642-01081-1

Vol. 202. Aboul-Ella Hassanien, Ajith Abraham, and Francisco Herrera (Eds.)
Foundations of Computational Intelligence Volume 2, 2009
ISBN 978-3-642-01532-8

Vol. 203. Ajith Abraham, Aboul-Ella Hassanien, Patrick Siarry, and Andries Engelbrecht (Eds.)
Foundations of Computational Intelligence Volume 3, 2009
ISBN 978-3-642-01084-2

Vol. 204. Ajith Abraham, Aboul-Ella Hassanien, and André Ponce de Leon F. de Carvalho (Eds.)
Foundations of Computational Intelligence Volume 4, 2009
ISBN 978-3-642-01087-3

Vol. 205. Ajith Abraham, Aboul-Ella Hassanien, and Václav Snášel (Eds.)
Foundations of Computational Intelligence Volume 5, 2009
ISBN 978-3-642-01535-9

Vol. 206. Ajith Abraham, Aboul-Ella Hassanien, André Ponce de Leon F. de Carvalho, and Václav Snášel (Eds.)
Foundations of Computational Intelligence Volume 6, 2009
ISBN 978-3-642-01090-3

Vol. 207. Santo Fortunato, Giuseppe Mangioni, Ronaldo Menezes, and Vincenzo Nicosia (Eds.)
Complex Networks, 2009
ISBN 978-3-642-01205-1

Vol. 208. Roger Lee, Gongzu Hu, and Huaikou Miao (Eds.)
Computer and Information Science 2009, 2009
ISBN 978-3-642-01208-2

Vol. 209. Roger Lee and Naohiro Ishii (Eds.)
Software Engineering, Artificial Intelligence, Networking and Parallel/Distributed Computing, 2009
ISBN 978-3-642-01202-0

Vol. 210. Andrew Lewis, Sanaz Mostaghim, and Marcus Randall (Eds.)
Biologically-Inspired Optimisation Methods, 2009
ISBN 978-3-642-01261-7

Vol. 211. Godfrey C. Onwubolu (Ed.)
Hybrid Self-Organizing Modeling Systems, 2009
ISBN 978-3-642-01529-8

Vol. 212. Viktor M. Kureychik, Sergey P. Malyukov, Vladimir V. Kureychik, and Alexander S. Malyoukov
Genetic Algorithms for Applied CAD Problems, 2009
ISBN 978-3-540-85280-3

Vol. 213. Stefano Cagnoni (Ed.)
Evolutionary Image Analysis and Signal Processing, 2009
ISBN 978-3-642-01635-6

Vol. 214. Been-Chian Chien and Tzung-Pei Hong (Eds.)
Opportunities and Challenges for Next-Generation Applied Intelligence, 2009
ISBN 978-3-540-92813-3

Vol. 215. Habib M. Ammari
Opportunities and Challenges of Connected k-Covered Wireless Sensor Networks, 2009
ISBN 978-3-642-01876-3

Vol. 216. Matthew E. Taylor
Transfer in Reinforcement Learning Domains, 2009
ISBN 978-3-642-01881-7

Vol. 217. Horia-Nicolai Teodorescu, Junzo Watada, and Lakhmi C. Jain (Eds.)
Intelligent Systems and Technologies, 2009
ISBN 978-3-642-01884-8

Vol. 218. Maria do Carmo Nicoletti and Lakhmi C. Jain (Eds.)
Computational Intelligence Techniques for Bioprocess Modelling, Supervision and Control, 2009
ISBN 978-3-642-01887-9

Maria do Carmo Nicoletti and Lakhmi C. Jain (Eds.)

Computational Intelligence Techniques for Bioprocess Modelling, Supervision and Control

Springer

Prof. Dr. Maria do Carmo Nicoletti
CS Department
Universidade Federal de S. Carlos
Rod. Washington Luiz, km 235
Caixa Postal 676
13565-905 S. Carlos - SP
Brazil
E-mail: carmo.nicoletti@pq.cnpq.br

Prof. Dr. Lakhmi Jain
PhD, ME, BE(Hons), Fellow IE(Aust)
Professor of Knowledge-Based Engineering
School of Electrical and Information Engineering
University of South Australia
Adelaide
Mawson Lakes Campus SA 5095
Australia
E-mail: Lakhmi.jain@unisa.edu.au

ISBN 978-3-642-42486-1 ISBN 978-3-642-01888-6 (eBook)

DOI 10.1007/978-3-642-01888-6

Studies in Computational Intelligence ISSN 1860949X

Softcover re-print of the Hardcover 1st edition 2009

Typeset & Cover Design: Scientific Publishing Services Pvt. Ltd., Chennai, India.

Printed in acid-free paper

9 8 7 6 5 4 3 2 1

springer.com

Preface

Computational Intelligence (CI) and Bioprocess are well-established research areas which have much to offer each other. Under the perspective of the CI area, Bioprocess can be considered a vast application area with a growing number of complex and challenging tasks to be dealt with, whose solutions can contribute to boosting the development of new intelligent techniques as well as to help the refinement and specialization of many of the already existing techniques. Under the perspective of the Bioprocess area, CI can be considered a useful repertoire of theories, methods and techniques that can contribute and offer interesting alternative approaches for solving many of its problems, particularly those hard to solve using conventional techniques.

Although throughout the past years CI and Bioprocess areas have accumulated substantial specific knowledge and progress has been quick and with a high degree of success, we believe there is still a long way to go in order to use the potentialities of the available CI techniques and knowledge at their full extent, as tools for supporting problem solving in bioprocesses. One of the reasons is the fact that both areas have progressed steadily and have been continuously accumulating and refining specific knowledge; another reason is the high level of technical expertise demanded by each of them.

The acquisition of technical skills, experience and good insights in either of the two areas is very demanding and a hard task to be accomplished by any professional. As often happens with interdisciplinary areas, it is difficult to find experts in one of the two areas willing to get deeply involved in the other area, to the point of learning its specialized vocabulary, becoming familiar with its terminology, acquiring sufficient technical language to master communication as well as to understand its main concepts and many of its specialized procedures, so to become autonomous and knowledgeable in both and be able to devise efficient CI methods customised to bioprocesses. Fortunately, in spite of their own technical vocabulary and language, both areas share a mathematical language, which can help to bridge the gap between the specialized technical languages employed by each of them.

We believe that without a deep understanding of the problem, a good insight into choosing the technique most suitable for solving the problem and a very good command of the chosen technique, its power and limitations, we will rarely find a reliable, appropriate and satisfactory solution to the problem at hand.

The eleven chapters in this book as well as the Appendix intend to help those interested in both CI techniques and Bioprocess become familiarized with the vocabulary, technical language and some of the main techniques and problems involved in both areas. The book covers the use of CI techniques in bioprocess related tasks namely modeling, supervision, monitoring and control, diagnostic, learning and optimization, with applications in several areas. Aimed at researchers, practitioners and graduate students, it may serve as a text for advanced courses in chemical engineering, bioinformatics and biotechnology and for computer scientists interested in bioprocesses. Chapters are self-contained and many of them, besides their focus on theoretical foundations, also include applications to real-world problems; they are briefly described next.

(1) Computational Intelligence Techniques as Tools for Bioprocess Modelling, Optimization, Supervision and Control

Presents an overview of recent and relevant works related to the use of CI techniques in Bioprocesses.

(2) Software Sensors and their Applications in Bioprocess

Clearly describes the motivations, design, implementation and use of software sensors and inferential estimation in bioprocesses, typically in fermentations. Research work on software sensor is reviewed and the associated techniques are introduced with examples and case studies.

(3) Monitoring of Bioprocesses: Mechanistic and Data-driven Approaches

Some state estimation techniques are theoretically discussed under two approaches, mechanistic and data-driven and real-world applications are presented.

(4) Novel Computational Methods for Modeling and Control in Chemical and Biochemical Process Systems

The chapter focuses on developing more efficient computational schemes for the modeling and control of chemical and biochemical processes. Artificial neural networks are introduced and successfully used on a couple of benchmark problems.

(5) Computational Intelligence Techniques for Supervision and Diagnosis of Biological Wastewater Treatment Systems

Surveys artificial intelligence as well as statistics techniques used for monitoring and controlling wastewater treatment systems. The chapter covers knowledge based systems, fuzzy logic, neural networks and multivariate statistical methods.

(6) Multiobjective Genetic Algorithms for the Optimization of Wastewater Treatment Processes

This chapter presents a methodology for combining multiobjective genetic algorithms with wastewater treatment plant (WWTP) models for the evaluation, optimization and comparison of WWTP control laws. The use of the methodology on a case study is described.

(7) Data Reconciliation Using Neural Networks for the Determination of K_La

The problem of estimating the oxygen mass transfer coefficient (K_La) in aerobic fermentation using data reconciliation is described in the chapter. Data reconciliation is implemented in two different ways: one by minimizing an objective function that takes into account measurements and four estimation methods, and the other by using a previously trained NN.

(8) A Computational Intelligent Based Approach for the Development of a Minimal Defined Medium Application to Human Interleukin-3 Production by *Streptomyces lividans 66*

The chapter describes an elaborated combination of different techniques aiming at identifying the composition of the optimum minimal medium for the production of rHuIL-3. The use of NN in combination with statistical techniques is part of the process.

(9) Bioprocess Modelling for Learning Model Predictive Control (L-MPC)

The chapter describes a data-driven modeling methodology for batch and fed batch processes. It also describes how developed models can be used for process monitoring, for ensuring process reproducibility through control and for optimizing process performance by enforcing learning from previous runs through a control methodology named Learning Model Predictive Control (L-MPC).

(10) Performance Monitoring and Batch to Batch Control of Biotechnological Processes

This chapter describes two approaches to ensuring the production quality of batch biotechnological processes.

(11) Modelling of Biotechnological Processes - An Approach Based on Artificial Neural Networks

The chapter describes a software tool named FerMoANN, suitable for modeling fermentation processes. The use of the tool in two fermentation processes is presented and discussed.

We wish to express our gratitude to the authors and reviewers for their contribution. We are also grateful to the editorial team of Springer-Verlag and SCI Data Processing Team for their assistance during the preparation of the manuscript. M.C. Nicoletti wishes to extend her thanks to Leonie C. Pearson for kindly reviewing many of her writings.

M.C. Nicoletti, Brazil
L.C. Jain, Australia

Contents

1

Computational Intelligence Techniques as Tools for Bioprocess Modelling, Optimization, Supervision and Control

M.C. Nicoletti[1], L.C. Jain[2], and R.C. Giordano[3]

[1] Computer Science Dept.,
[3] Chemical Engineering Dept.,
Universidade Federal de S. Carlos –UFSCar
Rod. Washington Luiz, km 235
Caixa Postal 676
13565-905 S. Carlos – SP, Brazil
carmo.nicoletti@pq.cnpq.br, roberto@power.ufscar.br
[2] School of Electrical and Information Engineering
University of South Australia
Adelaide, Australia
Lakhmi.jain@unisa.edu.au

Abstract. This is an introductory chapter that presents a general review of some Computational Intelligence (CI) techniques used today, both in the biotechnology industry and in academic research. Various applications in bioprocess-related tasks are presented and discussed. The aim of putting forth a surveying view of the main tendencies in this field is to provide a broad panorama of the research in the intersection between the two areas, to highlight the popularity of a few CI techniques in Bioprocess applications and to discuss the potential benefits that other not so explored CI techniques could offer.

1 Introduction

For the purpose of this introductory chapter we are calling Computational Intelligence (CI) the research area whose main focus is the investigation and use of techniques considered intelligent that can be automated by means of a computer program/system. They typically refer to search/optimization algorithms in general (including genetic algorithms, simulated annealing, particle swarm, etc.), neural networks, fuzzy logic and fuzzy inference, symbolic learning algorithms, clustering, feature subset selection and neuro-fuzzy systems. Although not conventionally considered computational intelligence techniques, many statistical methods are closely related to CI techniques and, as such, a few of them are dealt with in this book. It is also common to come across hybrid systems that implement a conveniently balanced combination of different CI (or CI/statistical) techniques aiming to compensate each other's drawbacks and whose design and development can be also considered as one of the goals of CI (see [1]).

Bioprocess modelling, optimization, supervision and control may be classified as a branch of Chemical (or, more recently, Biological or Bioprocess) Engineering. These

M. do Carmo Nicoletti, L.C. Jain (Eds.): Comp. Intel. Tech. for Bio. Mod., SCI 218, pp. 1–23.
springerlink.com

processes include upstream (feedstock preparation) and downstream (separation and purification of bioproducts) operations together, of course, with the core of the process: one or more bioreactors. These may be enzymatic reactors (with enzymes as catalysts) or reactors for cultivation of microorganisms or cells. These last bioreactors are usually called fermenters in industry, although often the cultivation occurs aerobically, and fermentation is an undesired side route. This technical denomination was probably inherited from one of the first bioprocesses in history, the production of alcoholic beverages. The wide realm of bioprocess operations turns impractical surveying all possibilities for application of CI techniques. Thus, this chapter is particularly (but not exclusively) focused on bioreactors for cultivation of microorganisms or cells, the so-called fermenters.

As stated in [2] within the context of bioreactor control,"Bioprocess control is defined as providing a near optimal environment for microorganisms to grow, multiply, and produce a desired product. This includes providing the right concentration of nutrients to the culture (e.g carbon, nitrogen, oxygen, phosphorous, sulfur, minerals), removing any toxic metabolic products (e.g CO_2), and controlling important internal cellular parameters (e.g temperature, pH)."

There are many state and input variables that characterize a bioreactor, its operation and control. Generally, they are grouped into physical operational parameters (refer to the operation of the bioreactor, e.g. agitation speed, agitation power, temperature, pressure, broth volume, mass, etc.); chemical ones (refer to the chemical environment inside the bioreactor, e.g. cultivation medium composition, oxygen and carbon dioxide concentrations, pH, amino acids, etc.), biochemical (intracellular) parameters (refer to the metabolic state of the cell during the cell growth, e.g. cell mass composition, enzymes, other proteins, etc.) and macro-biological parameters (refer to what is going on inside the bioreactor, e.g. contamination, degeneration, aggregation, mutation, etc.).

Since most of these processes are transient in nature, conventionally the modelling and control of bioprocesses is based on a set of differential equations corresponding to mass (and, more rarely, energy) balances. Finding an appropriate and reliable mathematical set of equations, as well as a reliable set of parameters, is a time-consuming task mainly due to the inherent complexity of such processes, such as their time-dependency, the non-linearity of the reactions and the uncertainties typical of biological reactions. In recent years, a growing number of works have investigated the use of computational intelligent techniques aiming at modelling and monitoring different aspects related to bioprocesses, in an attempt to provide alternative and more flexible strategies to implement these tasks.

This approach has received an important incentive ever since the proposition of the Process Analytical Technology initiative, PAT, by the FDA (US Food and Drug Administration) in 2003 [3]. This program was motivated by the observation that the rush for an approval by the FDA of new biotechnological processes was directing the industry towards inefficient technical solutions. Thus the standard requisites of this regulatory agency became more flexible in accepting applications of CI-related tools such as: software sensors (for instance, based on neural or neural-fuzzy networks), clustering and fault detection algorithms (fuzzy-based or based on multivariate statistical techniques such as Principal Component Analysis, PCA).

By browsing through the titles of the many research works employing CI techniques in bioprocess applications, there will be no doubt that the main CI technique is neural networks (NNs). A possible reason for that is the fact that, as commented in [4], neural networks have been used in a variety of applications where statistical methods are traditionally employed.

As a justification for the previous statement, the authors Warner and Misra say "These networks [multilayered feedforward neural networks, in this case] do not impose a functional relationship between the independent and dependent variables. Instead, the functional relationship is determined by the data in the process of finding values for the weights [supervised learning]. The advantage of this process is that the network is able to approximate any continuous function and we do not have to guess the function form. The disadvantage is that it is difficult to interpret the network. In linear regression models, we can interpret the coefficients in relation to the problem. Another disadvantage of the neural network is that convergence to a solution can be slow and depends on the network's initial conditions". As pointed out by Warner and Misra, however, as a conclusion from an example on real data, "...traditional statistical tools still have a role in analysis and that the use of any tool must be thought about carefully."

This chapter reviews the many different uses of the most popular CI techniques in bioprocesses, organized by the different roles they play in the process i.e., software sensors in Section 2, filters and smoothers in Section 3, support for the optimization of cultivation media in Section 4 and monitoring and controlling bioreactors in Section 5. Section 6 highlights a few other uses of CI techniques in relation to bioprocesses that do not completely conform to the previous ones and finally in Section 7 a few conclusions are presented based on both what was discussed throughout the chapter and some considerations about future perspectives in relation to works involving both areas, CI and Bioprocesses.

2 CI Techniques as Software Sensors

As stated by Chéruy in [5], "A bottleneck in bioprocess monitoring and control is often caused by the lack of reliable sensors, particularly for biological variables. That is why estimation techniques issued from control theory have been applied to on line estimate bioprocess variables. That induced the development of 'software sensors', which associate a sensor (hardware) and an estimation algorithm (software) in order to provide on line estimates of the unmeasureable variables and kinetic parameters."

Generally many of the software sensor proposals found in the literature can be roughly grouped in one of two conceptually different types, those called model-driven, that require phenomenological knowledge about the process they will be used with and those called data-driven, based on the knowledge inferred from collected data from similar processes. There are, however, hybrid proposals that cover a variety of combinations of the two approaches.

Harms and coworkers in [6] describe the many difficulties and problems related to bioprocess monitoring, in relation to the use and limitations of the different types of available sensors. The authors comment that "Although there is vigorous research being done on sensor development, only a small fraction of new sensors are suitable for use in fermentations. Bioprocesses are remarkably harsh environments for sensors, as the growing culture can infiltrate sensors and this invalidates their results... Another challenge is that bioprocesses may run for weeks, during which time the potential sensor must be able to perform without recalibration". Although contemplating the use of CI techniques in what they call virtual measurements, the authors are not too emphatic in their statement and remind that "... an artificial intelligence (technique) is only as good as its training set and the limits of any model need to be recognized". In spite of limitations and problems, the use of CI techniques for bioprocess monitoring has been particularly active in relation to the development of software sensors. Many works can be found in the literature that implement CI techniques (sometimes combined and sometimes on their own) to function as software sensors.

In the particular area of enzyme engineering and production, for example, Linko and coworkers in [7] present a detailed discussion of the state of the art use of neural network as software sensors in processes related to the production of α-amylase, β-galactosidase, glucoamylase, lipase and xylanase and conclude that "... well-trained neural networks can be employed as software sensors for the on-line estimation and prediction of key process parameters in enzyme production."

Bogaerts and Vande Wouwer in [8] review several software sensor design methods based on the theory of state observation that can be used with bioprocesses, such as extended Kalman filters, receding-horizon observers, asymptotic observers and hybrid observers. In [9], Lin and coworkers present a data-driven approach for soft sensor development based on a combination of PCA, regression techniques and dynamic least squares.

The development of a software sensor for the optimal control of a wastewater treatment process is described in [10]. The software sensor is a hybrid NN, which includes a PCA procedure for data preprocessing. The proposed hybrid NN, that infers wastewater quality parameters, can also be used for fault detection of processes, for the estimation of toxicity as well as the automation of other wastewater treatment processes. In their work the authors compared the performance of four different approaches for implementing software sensors: multivariate regression (11 inputs and 1 output); principal component regression (PCR) (11 inputs reduced to 5PCs), NN (11 input neurons, 1 output neuron and 1 hidden layer with 10 neurons, trained using the Quasi-Newton method) and the hybrid approach, PCA+NN (5 input neurons, 1 hidden layer with 5 neurons and 1 output neuron). The hybrid method PCA+NN exhibited the most accurate prediction among the four.

In their review on methods to estimate parameters and process variables that cannot be measured directly and the subsequent use of these measurements for bioprocess control, Komives and Parker [11] consider NNs as one of the techniques available under the umbrella of empirical models, to be used on its own, or as hybrid NNs (as used in [12]), where *a priori* knowledge about the process is inserted into the NN. The authors also place NNs in the group of the optimization algorithms, since they can be used in both estimation and process optimization, giving as an example the work described in [13] where Becker and coworkers used a feedforward neural network for the control and optimization of a pilot-scale beer fermentation.

In the work described in [14] for the on-line monitoring of yoghurt fermentation, for instance, data from an electronic nose (EN), a near-infrared spectrometer (NIRS) and standard bioreactor probes are fused using two NNs in sequence; the first for predicting quantitative process variables and the second for predicting qualitative process state variables describing critical process phases.

Desai and coworkers [15] used a machine learning based modelling formalism known as support regression vector (SRV) in software sensor applications; their approach was used in two simulated fed-batch processes, invertase and streptokinase production. In their paper the authors also provide a comparison between the SRV-based *versus* two NN-based (specifically, multilayer perceptron (MLP) and radial basis function network (RBFN)) software sensors and the results favored the former.

Traditional neural network (NN) training algorithms such as Backpropagation require the definition of the network architecture before training starts. Generally these methods work well only when the network architecture is appropriately chosen. The definition of neural network architecture suitable for a given application is a hard and time-consuming task and is a recurrent problem every time a NN is considered for solving a particular problem. A common practice adopted is to define several different architectures, train and evaluate each of them, retaining the one most appropriate for the problem.

There are two different approaches for automatically designing NN architectures, referred to as evolutionary and non-evolutionary. In the evolutionary approach, a NN can be evolved by means of an evolutionary technique, i.e. a population-based stochastic search strategy such as GA (see [16] [17]). In the non-evolutionary approach, the NN is built not as a result of an evolutionary process, but rather as the result of a specific algorithm designed to automatically construct it, as is the case of a constructive algorithm. Constructive neural network algorithms incrementally add hidden nodes and weights to the network during training until stopping criteria is satisfied. In so doing, the process of determining the architecture of the network and the learning process happen simultaneously (see [18] [19] [20] for a review on several constructive NNs).

In the literature there have not been many attempts of using constructive NN algorithms (CoNN) in applications involving bioprocesses. Probably one of the reasons is the fact that many CoNN algorithms are for classification tasks and bioprocess applications are typically regression problems (see [21] for a review of CoNN for regression tasks). One of the most well known CoNN algorithms for regression problems is the Cascade Correlation (CasCor) algorithm, proposed in [22]. The CasCor algorithm creates NNs with multiple hidden layers, where each hidden layer has one hidden neuron. The popularity of CasCor can be attested by the various ways this algorithm has inspired new variations.

A similar approach to CasCor called Constructive Backpropagation (CBP) was proposed in [23] and a recurrent extension to CasCor named RCC is described in [24]. An investigation into problems and improvements in relation to the basic CasCor can be found in [25]. CasCor has also inspired the proposal of the Fixed Cascade Error (FCE), described in [26] and [27].

Two NNs, a feedfoward trained using Levenberg-Marquardt (FFLM) and a CasCor have been used as software sensor for the on-line estimation of biomass concentration, in a fed-batch cultivation of baker´s yeast, as described in [28]. Experiment results indicated a moderately better performance from CasCor sensors over their FFLM counterparts. A very recent and broad review on soft sensors with application in many different process industry fields, like the chemical industry, bioprocess industry, steel industry etc. can be found in [29].

3 CI Techniques as Filters and Smoothers

A close issue to sensors and probes is related to the quality of the data they collect. Bioprocess monitoring and control are particularly dependent on on-line data obtained at discrete times, which usually are noisy. Noisy data can provoke a high level of disturbances in any control system, contributing to unreliable results. The way to deal with noisy data is to use conventional filters and/or smoothing methods (see [30] [31] [32]) for eliminating the noise present in the sampled data, before they are fed into the control system. However, as commented by Patnaik in [33], particularly in bioreactors, the use of traditional smoothing methods has a major drawback "... smoothing methods carry the risk of suppressing subtle but important changes and of eliminating genuine physiological effects along with spurious and erroneous data."

Aiming at more flexible methods that would prevent the previously mentioned drawback in bioprocess control and monitoring, the tendency in research is to explore unconventional techniques, which would be able to filter data and, at the same time, allow for some inherent disturbances to go through [34]. One such filtering method is based on NNs which, due to their plasticity and versatile nature, can accommodate intrinsic data disturbances and still produce good generalization results.

References [35], [36] and [37] describe a few works on the use of NN as filters in bioprocess applications. Particularly, Patnaik in [38] justifies the use of NN as filters by saying: "while static filters (such as Butterworth or the Kalman type) are commonly used to prune any inflow noise, it is difficult to program them to respond dynamically to disturbances with mixed time-dependent distributions. This can be done more readily by trained neural networks, such that the neurally filtered stream contains noise of a prescribed time-dependent mean and variance". The work described in [38] uses NN with different purposes, since it describes a hybrid model of a bioreactor, which combines a partial mathematical model with two small Elman NN for the intra-cellular variables, an autoassociative NN to filter the noise and a feedforward NN as the controller respectively. According to the author, the integrated system is operationally robust in the presence of disturbances and produces smoother profiles of the concentrations, improved gene expression and more β-galactosidase than a previous one-single Elman network system.

Patnaik in [39] describes the proposal of a hybrid filter for filtering the feed stream noise from oscillating cultures based on a combination of a neural network and an extended Kalman filter. The author justifies the hybrid approach by stating that "Prior removal of noise is done by filters based either on specific algorithms or on artificial intelligence. Neither method is perfect. Hybrid filters combine both methods and thereby capitalize on their strengths while minimizing their weaknesses". In the experiments five hybrid filters were implemented and compared, each combining an extended Kalman filter (EKF) with a different NN (i.e., backpropagation with momentum, backpropagation with adaptive learning, radial basis, auto-associative and Elman). The performance of all hybrid filters was superior to both, the EKF and the purely neural filters. It was also pointed out that although the auto-associative NN was not the quickest to learn, its performance comparatively to others, in relation to unseen disturbances, was the best. These results corroborate the results presented in [37], where three algorithmic filters and one auto-associative neural filter had their performance compared in continuous cultures of *Saccharomyces cerevisiae* with oscillatory responses. The results from experiments described in [40], also in relation to cultures of *Saccharomyces cerevisiae* with chaotic oscillations, support the hybrid filter (EKF combined with NN) as the best choice.

Patnaik in [41] describes a hybrid neural simulator (HNS) for improved filtering of noise from oscillating cultures that combines a hybrid neural filter (HNF) for filtering the noise in oscillating cultures, a hybrid description of the fermentation kinetics and macroscopic balance equations for the bioreactor. The HNF is a component of the HNS discussed in the previous paragraph.

In [34] a non-conventional approach to deal with online noise filtering of CO_2 mole fractions is proposed as a committee of constructive CasCor networks. A typical bioprocess signal in the case study was the 4-20mA analog output of an infrared CO_2 detector, installed in the gas outflow of a bioreactor during aerobic cultivations of *Bacillus megaterium.* A committee system composed by CasCor NNs, previously trained using real data, arbitrated the output results of two (or a few more) neural networks, promoting a more realistic and fair result in relation to a given input data.

4 CI Techniques for Optimization of the Preparation of Cultivation Media

An important factor for an efficient operation of a bioreactor is the definition of the culture medium composition, which traditionally is conducted using Design of Experiments (DOE) statistical techniques. This is a complicated problem since it should take into consideration a large number of medium components as well as the metabolic complexity of the microorganisms involved. A popular mathematical/statistical technique employed for fermentation media optimization is the RSM (Response Surface Methodology) (see [42]).

There have been a few attempts of using CI techniques, particularly genetic algorithms, as an alternative to traditional methods in relation to media composition. The problem of determining a suitable media composition can be approached as an optimization problem implemented as a stochastic search procedure. As mentioned by Weuster-Botz in [43], “these [procedures] require neither the unimodality of the response surface nor limitation in the number of medium components under consideration”. The stochastic search implemented by genetic algorithms is particularly convenient since it can explore large variable spaces.

Genetic algorithms have been used in the work described in [44], with three other optimization methods namely Simplex, Rosenbrock and iterative factorial experimental design (IFED), for the optimization of the media composition, taking into consideration the errors and inaccuracies present in the experimental data. The four methods were used for the optimization of a single parameter, the microorganism specific growth rate μ and different mathematical models that describe microorganism growth were used. According to the authors, “If we optimize a bioprocess on a shaker, where parallel experiments can be carried out, genetic algorithms are the preferred methods since they require the smallest number of experiments with high efficiency. For optimization of bioprocesses in a bioreactor where only consecutive experiments can be realized, Rosenbrock or Simplex methods are the methods of choice.” The handling of experimental errors is a major issue in media formulation and can interfere in the results, independently of the design method employed (see [45]).

Nagata and Chu combined neural networks and genetic algorithms aiming at the optimization of a fermentation medium, for the production of the enzyme hydantoinase by *Agrobacterium radiobacter*, as described in [46]. They used the experimental data reported in [47] for training two NNs; both with four input neurons, representing the concentrations of four medium components. In one of the NNs the output neuron represented hydantoinase and in the other, cell concentration. As usual, the topology of the NN was determined empirically, by trial and error, and both were three-layer NNs, with four neurons in the input layer, six neurons in the single hidden layer and one output neuron. GA was then used to optimize the functions represented by each of the two NNs, by finding the optimum settings (i.e., input values) for maximum enzyme and cell production.

Similarly to the work described in [41], Rao and coworkers proposed in [48] a hybrid system based on cooperation between a feedforward NN and genetic algorithm aiming to optimize the fermentation conditions to enhance the alkaline protease production by *Bacillus circulans.* Six different bacterial growth-associated factors were analyzed in terms of their individual influence on alkaline protease production. A fully connected three-layer NN with 6, 13 and 1 neurons respectively, learns (using Backpropagation) the nonlinear relationship between the fermentation factors and the enzyme yield; the function represented by the NN is then further optimized for alkaline protease production using genetic algorithm. According to the authors the hybrid system has provided a significant improvement in the alkaline protease yield. The number of hidden neurons was empirically determined.

The work described in [46] is again described in [49] for the optimization of fermentation medium for the production of a novel lipopeptide by *Bacillus subtilis MO-01*. The NN topology with 4 input neurons, 5 neurons in the single hidden layer and 1 neuron in the output layer used in the experiments was empirically determined. In order to train the NN the authors used a dataset comprising 30 experiment runs reported in [50]. The NN was trained using a MATLAB built-in function based on Bayesian regularization Backpropagation coupled with the Levenberg-Marquardt optimization algorithm (LMBP). The function represented by the NN was then optimized using GA, aiming at determining the process variable values that maximized the lipopeptide yield.

Franco-Lara and coworkers in [51] discuss the simultaneous maximization of the biomass concentration and conversion of pentafluoroacetophenon (PFAP) with *Synechococcus* PCC 7942 in a collaborative process involving NN and genetic algorithms. Two NNs were constructed one representing the relationship between the concentrations of the various medium components and the biomass concentration after 72hs and the other representing the relationship between the concentrations of the various medium components and the conversion of PFAP after 24hs. For defining the topology of the NNs the authors employed the De Jong function [52]; the NN that delivered the minimal mean squared error had 4 input neurons and 15 hidden neurons. The Bayesian regularization training algorithm was chosen after comparing four different algorithms namely Backpropagation, Levenberg-Marquard, Quasi-Newton and the Bayesian regularization algorithm for the De Jong function modelling. In the GA implementation, the objective functions were the simultaneous maximization of biomass concentration and PFAP. In [53] Link and Weuster-Botz describe a new software tool named GAME.opt that uses a GA for multi-objective experimental optimization, based on the Strength Pareto Evolutionary Algorithm (SPEA) [54].

Using the fermentative production of biopolymer scleroglucan as case study, Desai and coworkers conducted a comparison of NNs-GA and RSM for fermentation media optimization, described in [55]. The yield of scleroglucan was modelled and optimized as a function of four independent variables (sucrose, yeast extract, magnesium sulphate and dipotassium hydrogen phosphate) using both, NN-GA and RSM. The comparison between both was conducted focusing on three aspects namely, modelling, sensitivity analysis and optimization abilities. In the experiments conducted the NNs consistently outperformed RSM in all three aspects. The results support the combination NN-GA as a very promising alternative for media composition design.

Another popular evolutionary algorithm which has similarities with GA, known as particle swarm optimization (PSO) [56], has been used for medium optimization in the work described in [57]. The work aimed at exploring the applicability of evolutionary algorithms to fermentation medium optimization problems, particularly searching for possible advantages in using an alternative approach to statistically designed methods. The authors concluded that PSO algorithm was successful at providing a fermentation medium superior to control with respect to titre. Also, the population-based approach part of a PSO algorithm provided several different mediums with higher titre and lower viscosity than the control medium. This work differs from most works in CI in bioprocesses, in relation to the computational environment used. While most works are heavily dependent on a Matlab environment and a few of its toolboxes, in this work the algorithm was written in C and implemented on a HP-UX workstation. As commented by Alford in [2], "The particle swarm evolutionary technique has been tried on two existing fermentation processes at Lilly. In both fermentation processes, broth medium composition providing >50% yield improvement resulted, compared to the previous existing production medium."

5 CI Techniques in Parameter Estimation and Bioreactor Modelling and Control

Fed-batch cultivation is a very common type of biotechnological process which corresponds to a bioreactor to which substrate is continuously fed while the microorganisms and products remain in the fermenter until the end of the process.

The control of such a bioprocess can be approached in a few different ways. Depending on the adopted approach, CI techniques can be used with different goals and implemented with different purposes.

The classical approach to model-based bioreactor control relies on a series of mass balance equations. Depending on the granularity of the description, the classical approach can produce a very simplified or a very complex model. Since the use of CI techniques in classical control is tangential, the focus of this subsection is on hybrid control. This is a very extensive area, and in what follows, this review discusses some of the most recent works in the area.

Several references focus on reviews of one or a few CI techniques in bioprocess control in general or in some of its specific aspects, such as [2], [6], [11], [58] and [59]. Others, however, only mention them very briefly, such as [60] and [61].

A detailed report on a 2-year investigation on the use of NNs for process monitoring and control, where four main industrial applications were considered is described in [62]. Lennox and coauthors describe in detail the lessons they learned from the investigation, focusing on data issues as well as on NN issues. Their analysis, comments and conclusions are no doubt very useful and can be very helpful for those willing to consider the use of NN for solving such kind of problems.

Relevant issues to be taken into account when using NN in bioprocesses, as suggested in the review [63] are proper scaling of data, selection of an appropriate network structure – including suitable choices of input and output variables, and the purpose of the network. In their review the previous issues and different NNs were studied in relation to different microbiological systems.

Of vital importance in fed-batch fermentations is the estimation of the optimal feed rate of substrate; the estimation process can be characterized as a singular control problem [64] and, as such, it is a difficult optimization task to be solved. Supplementary substrate feeding control is a very sensitive process in the sense that it is quite hard to be tuned and has a deep impact on the final product; the underfeeding can induce microorganism starvation and the overfeeding can produce the formation of undesirable products as well as interfere in the product formation. The development of an adequate feeding strategy, consequently, is critical for successful fed-batch fermentation processes.

In the literature there have been many different approaches to solving the problem of estimating the optimal feeding strategies aiming at improving the outcome of a fed-batch bioreactor. Among the many proposals one can find those based on mathematical methods for representing the kinetics and mass balances of the fermentation process, such as dynamical programming (see [65]) and adaptive mesh selection (see [66]) as well as the data-driven methods, especially the CI based methods, such as NNs and GAs.

The popular study on search algorithms for biotechnical process optimization by Simutis & Lübbert, published in [67] emphasizes the importance of search algorithms (*versus* the classical optimization approach based on Pontgryagin´s maximum principle), showing that the use of random search procedures, which are comparatively much simpler, easy to understand and implement, can lead to essentially the same success as the classical procedure. In the study three random search methods were used, namely, chemotaxis algorithm, simulated annealing and evolutionary programming. The main conclusion from their study is advice to inexperienced users that "should take the evolutionary algorithm with an ANN with sigmoid basis functions in order to determine an optimized control profile of complicated processes."

Discussing how to increase the performance of models for process optimization and control, Simutis and coworkers in [68] conclude that "the performance of models can be increased by combining artificial neural nets, fuzzy rule systems and classical model approach in form of hybrid models. This proved to be better than using anyone of these techniques separately." The use of hybrid process models was also supported by the considerations and discussions presented in [69] as well.

Lee and coworkers present in [59] a research review paper on control of fed-batch fermentations describing the basic feed control strategies and the use of fuzzy representation and inference as well as NNs in bioprocess control and modelling.

Franco-Lara & Weuster-Botz in [70] describe a methodology for optimizing the feeding strategy, through a combination of genetic algorithms and neural networks that delivers smooth continuous nonlinear feed rate trajectories. The authors used GA as a search procedure for finding the weights and biases of the NN that optimize the objective function of the process. A three-layer NN architecture and the exponential sigmoid function as the activation function were chosen. The input layer was defined by one neuron and five bias components, the hidden layer with five neurons and one bias and the output layer with a single neuron. The genetic algorithm was in charge of searching through the space of weight-bias vectors, for an optimal 16-dimensional vector (10 dimensions referring to weights and 6 to biases).

Chaudhuri & Modak in [64] propose an algorithm for determining optimal substrate feeding policies for fed-batch fermentations based on feedforward NN. The problem was approached by dividing the entire operating period of the fermentation into N intervals, determining the feed rate for each of the N intervals and training a NN using process data available at the different sampling intervals. The work describes the use of their approach in two applications, secreted protein production and invertase production. Based on the obtained results, authors conclude that the NN model is capable of representing the essential features of the process kinetics and the model can be used for dynamic simulation of the process.

Petrova and coworkers in [71] and [72] describe the modelling of specific growth rate of *Saccharomyces cerevisiae* within the framework of the classical unstructured model of a fermentation process using what they call feedforward neural networks that take into account culture memory. In their proposal the so called culture memory is some sort of recurrence implemented as a feedforward architecture. From the experiments authors conclude that NN models that take into account the culture memory predict the process dynamics more accurately than NN models without memory groups.

A method known as neural network parameter function modelling, which combines material balances with the approximation capabilities of feed-forward neural networks was developed by Tholudur & Ramirez and is proposed in [73]. Paraphrasing their description, the process can be briefly stated as follows. In a set of differential equations that represent the dynamics of a process, there are parameter functions that are dependent on the state variables. The neural network parameter function technique consists of rewriting the differential equations so to represent the parameter functions in terms of the state variables of the system. A series of experiments is then conducted and, at every sampling point, the state variables are measured. Next, the state trajectories are calculated, smoothed and differentiated to produce the state derivative trajectories. Using the state and derivative trajectories, values for the parameter functions are calculated at each sampling point. Neural networks are then trained using the input-output data previously obtained and the trained NNs can then be reinserted into the original differential equations to provide a model that captures the process dynamics. The technique is useful due to the fact that it substitutes the need to assume specific functional forms for each of the system's parameter functions. As pointed out in [74], the method of NN parameter function modelling seems very effective in modelling protein production in biotechnology processes.

The technique just described was implemented as a dynamic hybrid model of a fermentation process by Laursen and coworkers, as described in [75], to produce a foreign protein which, generally, is a particularly difficult type of process to model and control. NNs were determined for five critical parameter functions of growth rate, glucose consumption rate, oxygen consumption rate, acetate production rate and protein production rate. The NNs were used in seventeen full simulations and, according to the authors, they predicted dynamic response data well.

The open-loop optimization problem of a fed-batch alcoholic fermentation by *Zymomonas mobilis* was solved by Henriques and coauthors in [76] using singular control theory with a hybrid neural model, developed using experimental data from batch and fed-batch fermentations. The optimal solution was implemented experimentally. This work may be classified within a strategy that combines a CI-based model with classical optimal control theory.

The same group in [77] uses a hybrid NN modelling methodology combining mass balance equations with functional link networks (FLNs) to represent kinetic rates. The authors assert that the simple structure of the FLNs allows a rapid estimation of network weights and, consequently, the use of the hybrid model in an adaptive form.

In the work described in [78] an *ad-hoc* recurrent NN architecture with 10 input neurons, one hidden layer with eight neurons and intra-connections between the four output neurons was developed for modelling the dynamics of fed-batch yeast fermentation. Four input neurons represent state variables (namely glucose, biomass, dissolved oxygen and ethanol) and another four represent the same variables that are the output of the NN and recurrently presented as input. The two other input neurons represent air flow and substrate feed rate. The number of neurons in the hidden layer was empirically determined.

The intra connections between output neurons are from the one particular neuron (dissolved oxygen concentration) to the other three neurons. This strategy supports the on-line adaptation (of weights) of the NN to changing conditions based only on the value of dissolved oxygen; the other state variables are not measured but predicted by the adapted network. Results show that using the strategy for the on-line updating of weights, it is possible to implement NN to control processes in a wide region outside its training domain (which is one of the drawbacks of using NNs for process control).

Four bioprocess related control problems found in the literature have been solved by the technique proposed and described by Sarkar & Modak in [79], where a multilayer feedforward NN was combined with simulated annealing (SA) to produce an efficient hybrid algorithm aiming at the determination of optimal control profiles. The case studies were: optimal production of secreted protein in a fed-batch bioreactor, induced foreign protein production by recombinant bacteria; isothermal continuous stirred tank reactor with complex reactions and biphasic growth of yeast and the results obtained using the NN-SA combination in the four cases agreed with those reported in the literature. In general lines, the SA searches for weight and bias values of the NN, directed by an objective function which is evaluated by solving the differential equations describing the process, using the control profile generated by the NN with the current weight and bias values.

Two of the problems considered in [79] that were solved by a hybrid NN-SA combination, namely the optimal production of secreted protein in a fed-batch bioreactor and the biphasic growth of yeast, had already been dealt with in a previous work by the same authors [80], where a GA was used for determining an optimal substrate feeding policy for fed-batch fermentation with a single control variable.

The work described in [81] deals with the optimization of fed-batch bioreactors using GA, considering multiple control variables; this is typical of industrial fermentation processes where microorganisms require more than one substrate for their growth and product formation [82]. One of the two case studies described in [81] identified as 'induced foreign protein production by recombinant bacteria', using two control variables, previously described, modelled and optimized by a hybrid NN-SA approach in [80] was revisited and a solution strategy based on a GA was proposed.

GA was also used as an optimization tool for protease production process in the proposal described in [83]. The main goal of the work was to develop a dynamic model of the bioprocess for optimization of protease production rate, reaction time and other important process parameters. Two approaches to modelling were tried, the basic model with Monod´s equation and a modified version, as described in [84]; the experimental data had a better fit with the modified model. A GA was then used for solving both problems the nonlinear estimation and the product maximization.

Simulated annealing was the technique used in [85] for determining optimal parameter values of three *Streptococcus pneumoniae* growth models (Monod, Moser and Tessier); the Tessier equation was identified as the kinetic expression which provided the best fit to the cultivation dataset used for parameter estimation. The model, comprised by the Tessier equation, estimated parameter values given by the SA and mass balance equations, was further validated by comparing the simulated results with data obtained from three cultivations carried out in similar conditions.

Aiming at finding the optimal control action during a fed-batch fermentation process for cloned invertase production in *S. Cerevisiae* yeast, the work by Valencia and coworkers in [86] describes a proposal for a controller that uses a fuzzy ARTMAP neural network [87] combined with dynamic programming.

Differential Evolution (DE) [88] [89] [90] [91] is a relatively recent heuristic evolutionary method for optimization of multidimensional multimodal functions that has been receiving substantial attention from the scientific community. The possible reasons for the DE growing in popularity are its robustness, easiness to implement and use as well as the small number of user defined parameter values required. An extended version of DE, named Hybrid Differential Evolution (HDE) was proposed by Chiou & Wang in [92] aiming at avoiding the premature convergence of the original method. The authors propose two new operations (migration and acceleration) aiming at the improvement of the convergence speed without decreasing the diversity among the individuals. In [93] the HDE was used to estimate the parameter values of the Monod model of a recombinant fermentation process. In [94] a multiobjective optimization approach is used to estimate the kinetic model parameters of batch and fed-batch fermentations for ethanol production using *Saccharomyces diastaticus* (LORRE 316). Both, batch and fed-batch data were simultaneously used to formulate the estimation problem, which could be characterized as a multiobjective optimization problem. The HDE was then used to solve the multiobjective parameter estimation problem to obtain a global Pareto solution. In [95] two kinetic models of batch fermentation for ethanol and glycerol production using the same microorganism were considered and the HDE was used to estimate the kinetic model parameters.

The use of an evolutionary algorithm for optimizing the feeding profile for a fed-batch bioreactor is described in [96]. Two different representation schemes were used for codifying a chromosome (each chromosome represents a feeding profile). In the first approach the profile was represented by a set of control variable values along with a number of time points. Depending on the time resolution this representation can result in a lengthy chromosome, leading to a noisy, highly fluctuating feeding profile. To circumvent this problem the resulting feeding profile was input to a filtering process. In the second approach the profile was represented by a functional (of time) chromosome, f(t), that incorporates previous knowledge about the process. The representation was based on the general feeding profile shape (see [97]) consisting of a maximum and a minimum flow rate and a singular arc – the exact sequence and flow rate values are dependent upon the type of fermentation and the initial conditions. The simple basis functions used, as considered in [67], were a polynomial function, a NN function or a RBF.

In [98] a dynamic neural model based on two recurrent neural networks, collaboratively working in sequence to represent the nonlinear relationship between feed rate and biomass product, is proposed. The first recurrent NN is responsible for estimating the concentration of dissolved oxygen, given the volume V and the feed rate F. The estimated dissolved oxygen, as well as V and F are then input to the second recurrent NN that predicts the biomass concentration. Both NNs have one hidden layer that uses sigmoid activation functions and the output layers in both use linear functions. The Levenberg-Marquardt Backpropagation (LMBP) was used as the training algorithm. The optimization of a fed-batch fermentation process modelled by the cascade NNs was then carried out by a GA searching for the best feed rate profiles. Experiments have shown that the cascade NN model was able to capture the dynamic nonlinear function embedded in the training data and could be used as the model of the bioprocess aiming at its optimization, which was conducted via a modified GA.

6 Miscellaneous

There have been many other uses of CI techniques, particularly NN and GAs, as tools for supporting bioprocesses. An on-line diagnosis system for fermentation processes is not only dependent on of a reliable diagnosis strategy but also on on-line measurements of several state variable values. Several of these values, however, are very rarely measured on-line; generally their measurements take place in laboratories. A combined approach to fault detection and diagnosis has been described in [99], using an extended Kalman filter and a NN classifier for fault diagnosis of glutamic acid fermentation process. The EKF is used for the on-line estimate of the process parameter values based on on-line measurements of related variables; the estimated values are then input to both, a NN, for detecting faults and to the controller itself (described in [100]).

NNs have also been used for fault diagnosis in a simulated continuous stirred tank reactor, in the work by Zhang, described in [101]. The proposed technique takes into account multiple NNs and combines their results in a process that can be characterized as information fusion that gives an overall diagnosis. Different combination schemes were investigated including averaging, majority voting and a proposed modified majority voting.

Shimizu and coworkers in [102] describe an autoassociative neural network for the on-line fault diagnosis for optimal rice α-amylase production process of a temperature-sensitive mutant of *Saccharomyces cerevisiae.* The diagnosis system was able to detect the on-line faults of a defective temperature sensor and the plasmid instability of the recombinant cells, allowing corrective actions to be taken. An autoassociative neural network was also used in the study described in [103], for the on-line detection of deviations from normal antibiotic production fermentation using conventional process variables. Aiming at obtaining a suitable NN for fault detection, the authors explored different issues such as the selection of process variables and the preprocessing of data.

The work by Calzone and coworkers described in [104] is quite unique since it is based on symbolic computation, implemented by their system BIOCHAM (Biochemical Abstract Machine) which allows discovering kinetic parameter values from a temporal logic specification with constraints on numerical concentrations. BIOCHAM offers a rule-based language for modelling biochemical systems, a non-deterministic Boolean simulator and a numerical simulator, an original query language and a machine learning system to infer interaction rules and kinetic parameters from observed temporal properties.

A knowledge based system written in Mathematica called BIOESTIM (BIOprocess ESTIMation) is presented in [5]; this software tool can generate the equations of a possible model-driven estimator, given that the user provides the system with knowledge about the process. A three-step interaction with the system consists of: (1) the user providing the system with process knowledge i.e. the dynamical balance model of the process; (2) BIOESTIM analyzing the structural properties of the process model taking into account the previous process knowledge and automatically generating the equations of possible indirect measurement or observer and (3) the system accepting experimental data (such as provided by a hardware sensor), to simulate the proposed estimator.

Due to the intrinsic time dependency associated with bioprocesses, the so called symbolic learning is not particularly suitable for the bioprocess related tasks described so far. However, due to some success in knowledge discovery in biochemical domains by ILP (Inductive Logic Programming) systems, this is a potential area still to be explored. In the work described in [105], for instance, the ILP system known as Progol was successfully used in a typical knowledge discovery task to obtain structural constraints associated with mutagenicity of molecules. The work described in [106] uses decision tree analysis based on *information gain* to reduce fermentation data through the identification of the significant inputs; according to authors the usefulness of the approach lies in its ability to treat categorical variables, which are typical of archived fermentation databases, as well as continuous variables.

7 Conclusions

In [107] the control of cultivation processes for recombinant protein production is carefully reviewed by Gnoth and coworkers, focusing on three main issues: (1) the necessary measurements to determine the actual state of the running process, and additional measurements which can be made in order to identify and validate the process models, (2) the controllers that can be employed to make sure that the process will run on predefined paths, and (3) the measures that detect and analyze failures during the process. Authors emphasize that "... the process operational strategy can only be as good as the knowledge about the process behavior or dynamics that can be made quantitatively available to process optimization studies. Therefore, quantitative models describing the process aspects relevant in keeping the process objectives under control, they may be mechanistic or data driven ones, are not simply a play with computers, but tools indispensable for optimizing the benefit/cost ratio." Another point that we would like to add to those already mentioned by Gnoth and coworkers, is the need for computer tools that have already been extensively tested and proven to be sound and reliable for dealing with real data in a real world bioprocess environment.

As can be inferred from the vast number of available works in the area, rather than being competitive, model-driven and data-driven approaches complement each other much more. They are suitable to cooperatively share the search for the solution of many bioprocess related tasks in a much more efficient way than either of them, on their own, would be able to.

On the one hand this chapter aims to give a review on literature dealing with the use of CI techniques in problems and tasks related to the Bioprocess area. On the other hand it intends to serve as evidence of the popularity of a few CI techniques and the many different roles they can play in bioprocesses. In the literature review we tried to cover the wide variety of recent works and organize them by grouping them into meaningful classes, based on the different functionalities the CI-techniques they implement can have. Particularly in relation to the use of CI techniques in bioprocess control, we intend next to propose a more systematic approach to reviewing the main techniques, by grouping their main uses into the following categories: (1) CI-based models with classical approaches to optimization; (2) Phenomenological-based models with CI-based optimization and (3) CI-based models with CI-based optimization.

Acronyms and Abbreviations

ART: Adaptive Resonance Theory
ASTM: American Society for Testing and Materials
BIOCHAM: Biochemical Abstract Machine
BIOESTIM: Bioprocess Estimation
CasCor: Cascade Correlation Algorithm
CBP: Constructive Backpropagation
CI: Computational Intelligence
CO_2: Carbon Dioxide
CoNN: Constructive Neural Network
DE: Differential Evolution
DOE: Design of Experiments
DPLS: Discriminant Partial Least Squares
EKF: Extended Kalman Filter
EN: Electronic Nose

FCE: Fixed Cascade Error
FDA: US Food and Drug Administration
FFLM: Feedforward Network trained using Levenberg-Marquardt
FLN: Functional Link Network
FUZZY ARTMAP: a particular supervised ART algorithm
GA: Genetic Algorithms
HDE: Hybrid Differential Evolution
HNS: Hybrid Neural Simulator
HNF: Hybrid Neural Filter
IFED: Iterative Factorial Experimental Design
ILP: Inductive Logic Programming
LMBP: Levenberg-Marquardt Backpropagation Algorithm
MLP: Multilayer Perceptron
NIRS: Near-Infrared Spectrometer
NLPCA: Non Linear Principal Component Analysis
NN: Neural Network
PAT: Process Analytical Initiative
PCA: Principal Component Analysis
PCR: Principal Component Regression
PFAP: Pentafluoroacetophenon
PSO: Particle Swarm Optimization
RCC: Recurrent Extension to CasCor
RBFN: Radial Basis Function Network
RSM: Response Surface Methodology
SA: Simulated Annealing
SPEA: Strength Pareto Evolutionary Algorithm
SRV: Support Regression Vector

References

1. Fulcher, J., Jain, L.C.: Computational Intelligence: A Compendium. Springer, Heidelberg (2008)
2. Alford, J.S.: Bioprocess control: advances and challenges. Computers & Chemical Engineering 30(10-12), 1464–1475 (2006)
3. U. S. Department of Health & Human Services, FDA (U. S. Food and Drug Administration), http://www.fda.gov/cder/OPS/PAT.htm (access, February 2009)
4. Warner, B., Misra, M.: Understanding neural networks as statistical tools. The American Statistician 50(4), 284–293 (1996)
5. Chéruy, A.: Software sensors in bioprocess engineering. Journal of Biotechnology 52(3), 193–199 (1997)
6. Harms, P., Kostov, Y., Rao, G.: Bioprocess monitoring. Current Opinion in Biotechnology 13(2), 124–127 (2002)
7. Linko, S., Zhu, Y.-H., Linko, P.: Applying neural networks as software sensors for enzyme engineering. Trends in Biotechnology 17(4), 155–162 (1999)
8. Bogaerts, P., Vande Wouwer, A.: Software sensors for bioprocesses. ISA Transactions 42, 547–558 (2003)

9. Lin, B., Recke, B., Knudsen, J.K.H., Jorgensen, S.B.: A systematic approach for soft sensor development. Computers & Chemical Engineering 31(5-6), 419–425 (2007)
10. Choi, D.-J., Park, H.Y.: A hybrid artificial neural network as a software sensor for optimal control of a wastewater treatment process. Water Research 35(16), 3959–3967 (2001)
11. Komives, C., Parker, R.S.: Bioreactor state estimation and control. Current Opinion in Biotechnology 14(5), 468–474 (2003)
12. Fellner, M., Delgado, A., Becker, T.: Functional nodes in dynamic neural networks for bioprocess modelling. Bioprocess and Biosystems Engineering 25(5), 263–270 (2003)
13. Becker, T., Enders, T., Delgado, A.: Dynamic neural networks as a tool for the online optimization of industrial fermentation. Bioprocess and Biosystems Engineering 24(6), 347–354 (2002)
14. Cimander, C., Carlsson, M., Mandenius, C.-F.: Sensor fusion for on-line monitoring of yoghurt fermentation. Journal of Biotechnology 99(3), 237–248 (2002)
15. Desai, K., Badhe, Y., Tambe, S.S., Kulkarni, B.D.: Soft-sensor development for fed-batch bioreactors using support vector regression. Biochemical Engineering Journal 27(3), 225–239 (2006)
16. Schaffer, J.D., Whitely, D., Eshelman, L.J.: Combinations of genetic algorithms and neural networks: a survey of the state of the art. In: Proceedings of the International Workshop of Genetic Algorithms and Neural Networks, pp. 1–37 (1992)
17. Yao, X.: Evolving neural networks. Proceedings of the IEEE 87(9), 1423–1447 (1999)
18. Campbell, C.: Constructive learning techniques for designing neural network systems. In: Leondes, C. (ed.) Neural Network Systems Technologies and Applications, vol. 2. Academic Press, San Diego (1997)
19. Muselli, M.: Sequential constructive techniques. In: Leondes, C. (ed.) Neural Network Systems Techniques and Applications, vol. 2, pp. 81–144. Academic, San Diego (1998)
20. Parekh, R.G., Yang, J., Honavar, V.: Constructive neural-network learning algorithms for pattern classification. IEEE Transactions on Neural Networks 11(2), 436–451 (2000)
21. Kwok, T.-Y., Yeung, D.-Y.: Constructive algorithms for structure learning in feedforward neural networks for regression problems. IEEE Transactions on Neural Networks 8(3), 630–645 (1999)
22. Fahlman, S., Lebiere, C.: The cascade correlation architecture. In: Touretzky, D.S. (ed.) Advances in Neural Information Processing Systems, vol. 2, pp. 524–532. Morgan Kaufman, Los Altos (1990)
23. Lehtokangas, M.: Modelling with constructive backpropagation. Neural Networks 12(4-5), 707–716 (1999)
24. Fahlman, S.: The recurrent cascade-correlation architecture. In: Advances in Neural Information Processing Systems, vol. 3, pp. 190–196. Morgan Kaufman, San Mateo (1991)
25. Prechelt, L.: Investigation of the CasCor family of learning algorithms. Neural Networks 10(5), 885–896 (1997)
26. Lahnajärvi, J.J.T., Lehtokangas, M.I., Saarinen, J.P.P.: Fixed cascade error – a novel constructive neural network for structure learning. In: Proceedings of the Artificial Neural Networks in Engineering Conference (ANNIE 1999), St. Louis, USA, pp. 25–30 (1999)
27. Lahnajärvi, J.J.T., Lehtokangas, M.I., Saarinen, J.P.P.: Evaluation of constructive neural networks with cascaded architectures. Neurocomputing 48(1), 573–607 (2002)
28. Vanek, M., Hrncirik, P., Vovsik, J., Nahlik, J.: On-line estimation of biomass concentration using a neural network and information about metabolic state. Bioprocess and Biosystems Engineering 27(1), 9–15 (2004)

29. Kadlec, P., Gabrys, B., Strandt, S.: Data-driven soft sensors in the process industry. Computers & Chemical Engineering (article in press) (2009)
30. Yin, L., Yang, R., Gabbouj, M., Neuvo, Y.: Weighted median filters: a tutorial. IEEE Transactions on circuits and systems – II: Analog and digital signal processing 43(3), 157–192 (1996)
31. Savitzky, A., Golay, M.J.E.: Smoothing and differentiation of data by simplified least squares procedures. Analytical Chemistry 36, 1627–1639 (1964)
32. Eilers, P.H.C.: A perfect smoother. Analytical Chemistry 75(14), 3631–3636 (2003)
33. Patnaik, P.R.: Further enhancement of fed-batch streptokinase yield in the presence of inflow noise by coupled neural networks - IMTECH communication no. 033/2000. Process Biochemistry 37(2), 145–151 (2001)
34. Giordano, R.C., Bertini Jr., J.R., Nicoletti, M.C., Giordano, R.L.C.: Online filtering of CO_2 signals from a bioreactor gas outflow using a committee of constructive neural networks. Bioprocess and Biosystems Engineering 31(2), 101–109 (2008)
35. Patnaik, P.R.: Coupling of a neural filter and a neural controller for improvement of fermentation performance. Biotechnology Techniques 13(11), 735–738 (1999)
36. Patnaik, P.R.: Improvement of the microbial production of Streptokinase by controlled filtering of process noise. Process Biochemistry 35(3-4), 309–315 (1999)
37. Patnaik, P.R.: On the performances of noise filters in the restoration of oscillatory behavior in continuous yeast cultures. Biotechnology Letters 25(9), 681–685 (2003)
38. Patnaik, P.R.: An integrated hybrid neural system for noise filtering, simulation and control of a fed-batch recombinant fermentation. Biochemical Engineering Journal 15(3), 165–175 (2003)
39. Patnaik, P.R.: Hybrid filtering of feed stream noise from oscillating yeast cultures by combined Kalman and neural network configurations. Bioprocess and Biosystems Engineering 30(3), 181–188 (2007)
40. Patnaik, P.R.: Hybrid filtering to rescue stable oscillations from noise-induced chaos in continuous cultures of budding yeast. FEMS Yeast Research 6(1), 129–138 (2006)
41. Patnaik, P.R.: A hybrid simulator for improved filtering of noise from oscillating microbial fermentations. Biochemical Engineering Journal 39(2), 389–396 (2008)
42. Myers, R.H., Montgomery, D.C., Anderson-Cook, C.M.: Response Surface Methodology: Process and Product Optimization Using Designed Experiments, 3rd edn. Wiley Series in Probability and Statistics. John Wiley & Sons, Chichester (2009)
43. Weuster-Botz, D.: Experimental design for fermentation media development: statistical design or global random search? Journal of Bioscience and Bioengineering 90(5), 473–483 (2000)
44. Milavec, P., Podgornik, A., Stravs, R., Koloini, T.: Effect of experimental error on the efficiency of different optimization methods for bioprocess media optimization. Bioprocess and Biosystems Engineering 25(2), 69–78 (2002)
45. Zuzek, M., Friedrich, J., Cestnik, B., Karalic, A., Cimerman, A.: Optimisation of fermentation medium by a modified method of genetic algorithms. Biotechnology Techniques 10(12), 991–996 (1996)
46. Nagata, Y., Chu, K.H.: Optimization of a fermentation medium using neural networks and genetic algorithms. Biotechnology Letters 25(21), 1837–1842 (2003)
47. Achary, A., Hariharan, K.A., Bandhyopadhyaya, S., Ramachandran, R., Jayaraman, K.: Application of numerical modeling for the development of optimized complex medium for D-hydantoinase production from Agrobacterium radiobacter NRRL-B-11291. Biotechnology and Bioengineering 55(1), 148–154 (1997)

48. Rao, C.S., Sathish, T., Mahalaxmi, M., Laxmi, G.S., Rao, R.S., Prakasham, R.S.: Modelling and optimization of fermentation factors for enhancement of alkaline protease production by isolated Bacillus circulans using feed-forward neural network and genetic algorithm. Journal of Applied Microbiology 104(3), 889–898 (2008)
49. Imandi, S.B., Karanam, S.K., Garapati, H.R.: Optimization of fermentation medium for the production of lipopeptide using artificial neural networks and genetic algorithms. International Journal of Natural and Engineering Sciences 2(2), 105–109 (2008)
50. Gu, X.B., Zheng, Z.M., Yu, H.Q., Wang, J., Liang, F.L., Liu, R.L.: Optimization of medium constituents for a novel lipopeptide production by Bacillus subtilis MO-01 by a response surface method. Process Biochemistry 40(10), 3196–3201 (2005)
51. Franco-Lara, E., Link, H., Weuster-Botz, D.: Evaluation of artificial neural networks for modelling and optimization of medium composition with a genetic algorithm. Process Biochemistry 41(10), 2200–2206 (2006)
52. De Jong, K.A.: An analysis of the behaviour of a class of genetic adaptive systems, Doctoral Thesis, Department of Computer and Communication Science, University of Michigan, Ann Arbor (1975)
53. Link, H., Weuster-Botz, D.: Genetic algorithm for multi-objective experimental optimization. Bioprocess and Biosystems Engineering 29(5-6), 385–390 (2006)
54. Zitzler, E., Thiele, L.: Multiobjective evolutionary algorithms: a comparative case study and the strength Pareto approach. IEEE Transactions on Evolutionary Computation 3(4), 257–271 (1999)
55. Desai, K.M., Survase, S.A., Saudagar, P.S., Lele, S.S., Singhal, R.S.: Comparison of artificial network (ANN) and response surface methodology (RSM) in fermentation media optimization: Case study of fermentative production of scleroglucan. Biochemical Engineering Journal 41(3), 266–273 (2008)
56. Kennedy, J., Eberhart, R.C.: Particle swarm optimization. In: Proceedings of the IEEE International Conference on Neural Networks, Perth, WA, Australia, vol. 4, pp. 1942–1948 (1995)
57. Cockshott, A.R., Hartman, B.E.: Improving the fermentation medium for Echinocandin B production part II: Particle swarm optimization. Process Biochemistry 36(7), 661–669 (2001)
58. Glassey, J., Ignova, M., Ward, A.C., Montague, G.A., Morris, A.J.: Bioprocess supervision: neural networks and knowledge based systems 52(3), 201–205 (1997)
59. Lee, J., Lee, S.Y., Park, S., Middleberg, A.P.J.: Control of fed-batch fermentations. Biotechnology Advances 17(1), 29–48 (1999)
60. Clementschitsch, F., Bayer, K.: Improvement of bioprocess monitoring: development of novel concepts. Microbial Cell Factories 5(19), 1–11 (2006)
61. Schügerl, K.: Progress in monitoring, modeling and control of bioprocesses during the last 20 years. Journal of Biotechnology 85(2), 149–173 (2001)
62. Lennox, B., Montague, G.A., Frith, A.M., Gent, C., Bevan, V.: Industrial application of neural networks – an investigation. Journal of Process Control 11(5), 497–507 (2001)
63. Karim, M.N., Yoshida, T., Rivera, S.L., Saucedo, V.M., Eikens, B., Oh, G.-S.: Global and local neural network models in biotechnology: Application to different cultivation processes. Journal of Fermentation and Bioengineering 83(1), 1–11 (1997)
64. Chaudhuri, B., Modak, J.M.: Optimization of fed-batch bioreactor using neural network model. Bioprocess Engineering 19(1), 71–79 (1998)
65. De Tremblay, M., Perrier, M., Chavarie, C., Archambault, J.: Optimization of fed-batch culture of hybridoma cells using dynamic programming: single and multi feed cases. Bioprocess and Biosystems Engineering 7(5), 229–234 (1992)

66. Waldraff, W., King, R., Gilles, D.D.: Optimal feeding strategies by adaptive mesh selection for fed-batch bioprocesses. Bioprocess Engineering 17(4), 221–227 (1997)
67. Simutis, R., Lübbert, A.: A comparative study on random search algorithms for biotechnical process optimization. Journal of Biotechnology 52(3), 245–256 (1997)
68. Simutis, R., Oliveira, R., Manikowski, M., Azevedo, S.F., Lübbert, A.: How to increase the performance of models for process optimization and control. Journal of Biotechnology 59(1-2), 73–89 (1997)
69. Galvanauskas, V., Simutis, R., Lübbert, A.: Hybrid process models for process optimization, monitoring and control. Bioprocess and Biosystems Engineering 26(6), 393–400 (2004)
70. Franco-Lara, E., Weuster-Botz, D.: Estimation of optimal feeding strategies for fed-batch bioprocesses. Bioprocess and Biosystems Engineering 27(4), 255–262 (2005)
71. Petrova, M., Koprinkova, P., Patarinska, T., Bliznakova, M.: Neural network modelling of fermentation process. Microorganisms cultivation model. Bioprocess Engineering 16(3), 145–149 (1997)
72. Petrova, M., Koprinkova, P., Patarinska, T., Bliznakova, M.: Neural network modelling of fermentation process. Bioprocess Engineering 18(4), 281–287 (1998)
73. Tholudur, A., Ramirez, W.F.: Optimization of fed-batch bioreactors using neural network parameter function models. Biotechnology Progress 12(3), 302–309 (1996)
74. Tholudur, A., Ramirez, W.F., McMillan, J.D.: Interpolated parameter functions for neural network models. Computers & Chemical Engineering 24(11), 2545–2553 (2000)
75. Laursen, S.Ö., Webb, D., Ramirez, W.F.: Dynamic hybrid neural network model of an industrial fed-batch fermentation process to produce foreign protein. Computers & Chemical Engineering 31(3), 163–170 (2007)
76. Henriques, A.W.S., Costa, A.C., Alves, T.L.M., Lima, E.L.: Optimization of fed-batch processes: challenges and solutions. Brazilian Journal of Chemical Engineering 16, 171–177 (1999)
77. Costa, A.C., Henriques, A.W.S., Alves, T.L.M., Maciel Filho, R., Lima, E.L.: A hybrid neural model for the optimization of fed-batch fermentations. Brazilian Journal of Chemical Engineering 16, 53–63 (1999)
78. Gadkar, K.G., Mehra, S., Gomes, J.: On-line adaptation of neural networks for bioprocess control. Computers & Chemical Engineering 29(5), 1047–1057 (2005)
79. Sarkar, D., Modak, J.M.: ANNSA: a hybrid artificial neural network/simulated annealing algorithm for optimal control problems. Chemical Engineering Science 58(14), 3131–3142 (2003)
80. Sarkar, D., Modak, J.M.: Optimisation of fed-batch bioreactors using genetic algorithms. Chemical Engineering Science 58(11), 2283–2296 (2003)
81. Sarkar, D., Modak, J.M.: Optimization of fed-batch bioreactors using genetic algorithm: multiple control variables. Computers & Chemical Engineering 28(5), 789–798 (2004)
82. Modak, J.M., Lim, H.C.: Optimal operation of fed-batch bioreactors with two control variables. The Chemical Engineering Journal 42, B15–B24 (1989)
83. Dutta, J.R., Dutta, P.K., Banerjee, R.: Modeling and optimization of protease production by a newly isolated Pseudomonas sp. using genetic algorithm. Process Biochemistry 40(2), 879–884 (2005)
84. Cruz, A.J.G., Silva, A.S., Araújo, M.L.G.C., Giordano, R.C., Hokka, C.O.: Modeling and optimization of the cephalosporin C production bioprocess in a fed batch bioreactor with invert sugar as substrate. Chemical Engineering Science 54(15-16), 3137–3142 (1999)

85. Montera, L., Horta, A.C.L., Zangirolami, T.C., Nicoletti, M.C., Carmo, T.S., Gonçalves, V.M.: A heuristic search for optimal parameter values of three biokinetic growth models for describing batch cultivations of Streptococcus pneumoniae in bioreactors. In: Nguyen, N.T., Borzemski, L., Grzech, A., Ali, M. (eds.) IEA/AIE 2008. LNCS (LNAI), vol. 5027, pp. 359–368. Springer, Heidelberg (2008)
86. Valencia, C., Espinosa, G., Giralt, J., Giralt, F.: Optimization of invertase production in a fed-batch bioreactor using simulation based dynamic programming coupled with a neural classifier. Computers & Chemical Engineering 31(9), 1131–1140 (2007)
87. Carpenter, G.A., Grossberg, S., Markuzon, N., Reynolds, J.H., Rosen, D.: Fuzzy ARTMAP: a neural network architecture for incremental supervised learning of analog muldimensional maps. IEEE Transactions on Neural Networks 3(5), 698–713 (1992)
88. Storn, R.M., Price, K.V.: Differential evolution - a simple and efficient heuristic for global optimization over continuous spaces. Journal of Global Optimization 11(4), 341–359 (1997)
89. Storn, R.M.: On the usage of differential evolution for function optimization. In: NAFIPS 1996, pp. 519–523 (1996)
90. Storn, R.M., Price, K.V.: Minimizing the real functions of the ICEC 1996 contest by differential evolution. In: IEEE Conference on Evolutionary Computation, Nagoya, Japan, pp. 842–844 (1996)
91. Price, K.V., Storn, R.M., Lampinen, J.A.: Differential Evolution: A Practical Approach to Global Optimization. Springer, Berlin (2005)
92. Chiou, J.-P., Wang, F.-S.: A hybrid method of differential evolution with application to optimal control problems of a bioprocess system. In: Proc. of The IEEE Conference on Evolutionary Computation, Anchorage, pp. 627–631 (1998)
93. Chiou, J., Wang, F.: Estimation of Monod model parameters by hybrid differential evolution. Bioprocess and Biosystems Engineering 24(2), 109–113 (2001)
94. Wang, F.-S., Sheu, J.-W.: Multiobjective parameter estimation problems of fermentation processes using a high ethanol tolerance yeast. Chemical Engineering Science 55(18), 3685–3695 (2000)
95. Wang, F.-S., Su, T.-L., Jang, H.-J.: Hybrid differential evolution for problems of kinetic parameter estimation and dynamic optimization of an ethanol fermentation process. Industrial and Engineering Chemistry Research 40, 2876–2885 (2001)
96. Ronen, M., Shabtai, Y., Guterman, H.: Optimization of feeding profile for a fed-batch bioreactor by an evolutionary algorithm. Journal of Biotechnology 97(3), 253–263 (2002)
97. Lim, H.C., Tayeb, Y.J., Modak, J.M., Bonte, P.: Computational algorithms for optimal feed rates for a class of fed-batch fermentation - numerical results for penicillin and cell-mass production. Biotechnology and Bioengineering 28(9), 1408–1420 (1986)
98. Chen, L., Nguang, S.K., Chen, X.D., Li, X.M.: Modelling and optimization of fed-batch fermentation processes using dynamic neural networks and genetic algorithms. Biochemical Engineering Journal 22(1), 51–61 (2004)
99. Liu, W.: An extended Kalman filter and neural network cascade fault diagnosis strategy for the glutamic acid fermentation process. Artificial Intelligence in Engineering 13(2), 131–140 (1999)
100. Liu, W., Tian, S.B.: Parameter estimation and optimal control of the batch glutamic acid fermentation process. In: Proceedings of the IEEE International Conference on Systems, Man, and Cybernetics, Beijing, vol. 1, pp. 314–319 (1988)
101. Zhang, J.: Improved on-line process fault diagnosis through information fusion in multiple neural networks. Computers & Chemical Engineering 30(3), 558–571 (2006)

102. Shimizu, H., Yasuoka, K., Uchiyama, K., Shioya, S.: On-line fault diagnosis for optimal rice α-amylase production process of a temperature-sensitive mutant of Saccharomyces cerevisiae by an autoassociative neural network. Journal of Fermentation and Bioengineering 83(5), 435–442 (1997)
103. Huang, J., Shimizu, H., Shioya, S.: Data preprocessing and output evaluation of an autoassociative neural network model for online fault detection in virginiamycin production. Journal of Bioscience and Bioengineering 94(1), 70–77 (2002)
104. Calzone, L., Chabrier-Rivier, N., Fages, F., Soliman, S.: A machine learning approach to biochemical reaction rules discovery. In: Francis, J., Doyle III (eds.) Proceedings of Foundations of Systems Biology and Engineering FOSBE 2005, Santa Barbara, pp. 375–379 (2005)
105. Muggleton, S., Srinivasan, A., King, R.D., Sternberg, M.J.E.: Biochemical knowledge discovery using inductive logic programming. In: Arikawa, S., Motoda, H. (eds.) DS 1998. LNCS (LNAI), vol. 1532, pp. 326–341. Springer, Heidelberg (1998)
106. Buck, K.K.S., Subramanian, V., Block, D.E.: Identification of critical batch operating parameters in fed-batch recombinant E. coli fermentations using decision tree analysis. Biotechnology Progress 18(6), 1366–1376 (2002)
107. Gnoth, S., Jenzsch, M., Simutis, R., Lübbert, A.: Control of cultivation processes for recombinant protein production: a review. Bioprocess and Biosystems Engineering 31(1), 21–39 (2008)

2

Software Sensors and Their Applications in Bioprocess

Hongwei Zhang

Faculty of Arts, Computing, Engineering and Science, Sheffield Hallam University, City Campus, Sheffield, UK S1 1WB
H.Zhang@shu.ac.uk

Abstract. Industrial bioprocesses present a very difficult challenge to control engineers. Problems associated with the nature of the organisms in the process and difficulties related to obtaining accurate information regarding the progression of the process make controlling and monitoring particularly challenging. The lack of suitable and robust on-line sensors for key variables such as biomass or product concentration has been considered as a serious obstruction for the implementation of control and optimization of bioprocesses. Considering biomass concentration alone, there are typically two methods available to measure this value – direct or indirect methods. To measure the biomass directly, several techniques have been applied: optical density measurements, capacity measurements, high performance liquid chromatography (HPLC), nuclear magnetic resonance (NMR), laser cytometry or biosensors. In addition to the high costs associated with these measuring devices, their reliability can be poor when applied to large-scale systems. It is still the case that most industrial bioprocess control policies are based upon the use of infrequent off-line assay information for process operator supervision. The low sampling frequency associated with such measurements and the inevitable delays in taking samples and performing laboratory tests inevitably compromises the quality of control that is possible using such measurements. As a result of this an alternative approach, that of indirect measurement has attracted a great deal of attention over the last 20 years or so. Indirect measurements of biomass are mathematical algorithms that can produce estimates of unmeasured biomass concentration using the continuously measured variables such as temperature, dissolved oxygen, pH and off-gas concentration. The method of estimating the quality related variables from measurements of secondary variables is referred to as 'Inferential Estimation' and these mathematical estimators are usually referred to as 'Software Sensors'. Improved control of the process can be achieved by measuring and setting up a feedback control system using these secondary variables. Such control strategies are referred to as 'Inferential Controllers'. Software sensors usually rely on a model to describe the process, thus different techniques have been proposed for on-line inferential estimation in bioprocesses just as different models exist. Among these applications, the majority have been based upon mechanistic, artificial neural network (ANN) or other empirical models. In this chapter, some of the important and recent research conducted on Software Sensors is reviewed and the associated techniques are introduced with examples and case studies.

M. do Carmo Nicoletti, L.C. Jain (Eds.): Comp. Intel. Tech. for Bio. Mod., SCI 218, pp. 25–56.
springerlink.com

1 Introduction

Industrial bioprocesses present a very difficult challenge to control engineers. Problems associated with the nature of the organisms in the process and difficulties related to obtaining accurate information regarding the progression of the process make controlling and monitoring particularly challenging.

The lack of suitable and robust on-line sensors for key primary process variables, such as metabolites, microorganism, or nutrient levels, has been considered as a serious obstruction for the implementation of control and optimization of bioprocesses [1]. Considering biomass concentration in a typical fermentation process, there are generally two methods available to measure this value – direct or indirect methods. To measure the biomass directly, several techniques have been applied: optical density measurements, capacity measurements, high performance liquid chromatography (HPLC), nuclear magnetic resonance (NMR), laser cytometry or biosensors [2]. In addition to the high costs associated with these measuring devices, their reliability can be poor when applied to large-scale systems [3]. It is still the case that most industrial bioprocess control policies are based upon the use of off-line assay information for process operator supervision [4]. Due to sterility or aseptic requirements, the frequency of sampling of the broth is restricted. The low sampling frequency associated with such measurements and the inevitable delays in taking samples and performing laboratory tests inevitably compromises the quality of control that is possible using such infrequent and irregular measurements. As a result of this an alternative approach, that of indirect measurement has attracted a great deal of attention over the last 20 years or so. Indirect measurements of biomass are mathematical algorithms that can produce estimates of unmeasured biomass or product concentration using the continuously measured secondary variables such as temperature, dissolved oxygen, pH and off-gas concentration. The method of estimating the quality related variables from measurements of secondary variables is referred to as 'Inferential Estimation' and these mathematical estimators are usually referred to as 'Software Sensors'. Improved control of the process can be achieved by measuring and setting up a feedback control system using these secondary variables. Such control strategies are referred to as 'Inferential Controllers' [5].

Inferential estimation usually relies on a model to describe the process, thus different techniques have been proposed for on-line inferential estimation in bioprocesses just as different models exist. Among these applications, the majority have been based upon mechanistic, neural network or other empirical models. A comprehensive review on the subject of software sensors has been presented by Montague [3]. De Assis and Filho [6] and Bogaerts and Vande Wouwer [7] also reviewed the software sensor development for online bioreactor state estimation. The following sections introduce and review some of the important and recent research conducted in this field with examples and case studies.

Fermentation processes are responsible for the production of many highly profitable substances such as enzymes, antibiotics and vitamins, therefore they have great importance in the biotechnology industry. In the reminder of this chapter, fermentation process is used as a typical bioprocess to demonstrate how software sensors can be designed and implemented in bioprocesses.

2 Principle of Software Sensors

A schematic diagram demonstrating the principle of software sensors is shown in Figure 1.

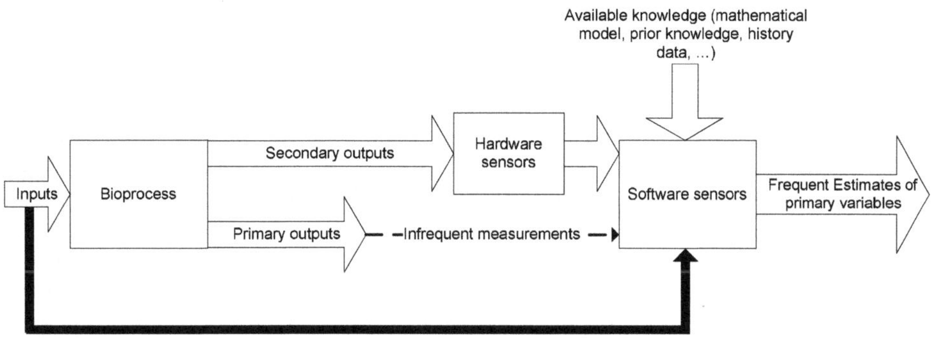

Fig. 1. Software sensor structure

As shown in Figure 1, secondary outputs of a bioprocess, for example temperature, gas-flow rates, oxygen and carbon dioxide concentration, dissolved oxygen tension (DOT) and pressures, are measured online using hardware sensors. These measurements and process inputs such as feed rates, together with infrequent measurements of primary outputs, are supplied to software sensors which utilize mathematical models and prior knowledge of the process to compute frequent estimates of quality variables. If sufficiently accurate, these estimates can then be used as feedback for automatic control and optimization, as well as for monitoring and assessing the health of the bioprocess concerned.

A critical element in the synthesis of software sensors is the available knowledge of the process. This knowledge is generally expressed through mathematical models. In the model, the relationship between a primary output and secondary outputs and inputs are described. The model can then be used to generate estimates of the difficult to measure primary output at the frequency at which the easily measured inputs and secondary variables are measured. According to the modeling paradigms that are employed, there are two types of software sensors being developed and successfully applied in bioprocess engineering.

2.1 Mechanistic Model-Based Software Sensors

Mechanistic models are based on fundamental physical principles such as conservation laws (e.g., mass and energy balances, and, if necessary, momentum and population balances). Depending on the end-use of the mechanistic model, there are two different approaches that can be applied to model bioprocesses: structured and unstructured models [8].

2.1.1 Structured Model

Models that include a fairly detailed description of the most important reactions inside the cell are known as structured metabolic models. This form of modeling requires knowledge about reaction kinetics, thermodynamic, transport and physical properties [9]. Although the structured models are often difficult to obtain, they represent better understanding of the bioprocess and thus have considerable practical value.

2.1.2 Unstructured Model

Models based on first principles such as mass and energy balances combined with Monod-type [10] expressions for kinetics are called unstructured models. When developing an unstructured model, modeling focused on the relations between state variables and control variables with the highest correlations. One of the disadvantages of unstructured models is their inability to represent the transient behavior in a precise manner. Unstructured models are simpler when compared to the structured model, yet are applicable to real fermentations provided consideration is made to their weaknesses.

The main objections to the use of unstructured models lie in the limited model validity and poor dynamics of empirical expressions that are built into the process model [11].

Among software sensors built upon mechanistic models, the most popular method is the Extended Kalman Filter (EKF). Section 3.2 details the construction and implementation of EKF as software sensors for bioprocesses.

2.2 Empirical Model-Based (Data-Driven) Software Sensors

Although various mechanistic models have been developed for bioprocesses over the years, when applied to highly nonlinear and multidimensional systems, they are often inaccurate because of the stiffness of differential equations. In this case, alternative approaches using empirical models can be more appropriate. Empirical models do not need a detailed process description, instead they require a large process database. Data-driven methods include kernel and spline smoothers, additive models, projection pursuit and artificial neural networks. Applications of ANN, fuzzy logic and multivariate statistical methods in developing empirical models have been frequently found in the literature.

2.2.1 ANN Model-Based Software Sensors

Being an established and powerful tool for modeling and pattern recognition, ANNs are used in various fields. Owing to the capability of ANNs to represent multivariable relationships, particularly those that occur in highly non-linear dynamic systems without any knowledge of the underlying process, they are widely used in modeling bioprocess. Many different types of ANNs architecture have been employed in fermentation modeling applications, including Feedforward Artificial Neural Networks (FANN), Recurrent Backpropagation Neural Networks (RBPN), Recurrent Tainable Neural Networks (RTNN), Control Affine Feedforward Neural Networks (CAFNN). ANNs have been proved to be very successful in many applications, however, they may lead to predictions which conflict with fundamental constraints represented by the conservation principles, particularly when outside the domain of training. ANNs are also an example of a highly complex and non-transparent model which makes them very difficult to analyze and assess.

Examples of ANN-based software sensors are discussed in Section 3.3.

2.2.2 Fuzzy Logic Model-Based Software Sensors

Since the introduction of fuzzy logic by Zadeh [12], it has been used comprehensively in the field of modeling, control and artificial intelligence. Fuzzy logic is capable of dealing with systems which do not have or require exact, quantitative information, thus, it is suitable for bioprocesses where real data are imprecise and often corrupted by noise, contaminants, etc. Fuzzy logic has been applied as a state estimator for fed-batch fermentation [13][14]. Fuzzy logic gains superiority over neural networks as system complexity increases because it allows for an easier design, system identification, and reassignment after a change [14]. Moreover, fuzzy rules can process numerical as well as symbolic data and offers some transparency. Although fuzzy logic model-based software sensors have found several successful applications, its suitability for different types of fermentation kinetics still needs to be further evaluated.

2.2.3 Multivariate Statistical Model-Based Software Sensors

When there is limited process understanding but a wealth of process data available from a range of operating conditions, another promising alternative for developing empirical models is to apply Multivariate Statistical Process Control (MSPC) techniques, such as Principal Component Analysis (PCA) and Partial Least Squares (PLS). PCA is able to reduce the dimensionality of a set of process data, which may contain many process variables. This reduced dimensional space identifies the underlying characteristics within a set of process data [15]. With PLS linear calibration, modeling of the multivariate input signals and the process variable is realized [16][17]. For fed-batch fermentation processes where time is also an important variable, data compression can be performed efficiently by Multiway Principal Components Analysis (MPCA) and Multiway Partial Least Squares (MPLS) reported by Nomikos and MacGregor [18][19]. MPCA and MPLS are extensions of PCA and PLS which transform the data collected from batch processes so that PCA and PLS models can be constructed from it. PCA and PLS models can be developed using nominal data which are collected during successful, high yield fermentation batches. The models then identify characteristics in the data sets that are synonymous with high yield production. Future batches can be compared with these models and any deviations from the nominal conditions indicate that the current batch is not consistent with high yield production. A principal advantage of MPCA and MPLS models is that the information contained in the original data sets is extracted in lower dimensional vectors and matrices, thus the data reduction achieved is significant [20], which allows comparisons and analysis to be made with greater ease than using the raw data. Although MPCA and MPLS are fast and efficient methods, one significant drawback is that the methods capture only the linear relationships between the principal components of the predictor (input) and response (output) variables. As a result, the models developed suffer from poor predictions for highly nonlinearly behaving fed-batch processes.

Section 3.4 introduces statistical model-based software sensors in great detail.

2.3 Hybrid Models

Since both mechanistic and empirical models have certain disadvantages, a natural form of fermentation process modeling has emerged that combines them in a hybrid formulation. Hybrid models combine the benefits of mechanistic and empirical modeling strategies to produce models that are often more accurate and extrapolate better than independent empirical and mechanistic models. The aim of such modeling is to make use of all the available information and knowledge of the process. There are several hybrid model structures that have been proposed in the literature, they all divide the process into subsystems and describe them by different kinds of representations. The most popular structure that has been applied uses a combination of mechanistic models and ANNs (Mechanistic-ANN) where mechanistic models are served as the foundation while the poorly known or unknown properties of the process, such as the reaction kinetics, are modeled with the aid of an ANN. The main disadvantage of the hybrid model is the relatively longer computation time for training the net. Genetic Programming (GP) and Genetic Algorithm (GA) are also found to be good choices to be integrated into hybrid models. McKay *et al.* [21] applied a hybrid technique of mechanistic modeling and GP to identify the fed-batch fermentation process defined in the problem statement of the Biotechnological Control Forum Modelling and Control Competition and proved the developed hybrid model was accurate and was extremely useful in both the optimization and control of this process. This model was also shown to be transparent which was a major advantage of the approach. There have been various alternative hybrid methods presented in the literature which include combining mechanistic models with adaptive regression splines and local stepwise regression [22] and the combination of PCA models with fuzzy logic reported by Jin, Wang and Chu [23].

3 Methods of Inferential Estimation

Method of inferential estimation is the 'brain' of a software sensor. This section reviews and introduces some of the most important methods of inferential estimation, with case studies to demonstrate how they can be implemented on bioprocesses.

3.1 Linear Estimator

The simplest software sensors are based on linear fixed parameter correlation models that relate bioprocess inputs to secondary and primary measurements. For example, during the anaerobic fermentation of sugars by yeasts, measurement of CO_2 evolution rate (CER) provides a good indicator of reactor ethanol productivity [24]. The gas evolution rate and ethanol produced in the reactor can be described by a linear correlation. Linear estimators are easy to construct and implement, but most bioprocesses are non-linear and this can compromise the performance of the linear software sensors.

To cope with the system's nonlinearity, Guilandoust *et al.* [25] presented an adaptive linear estimator. Lant, Tham and Montague [26] discussed this estimator further and successfully applied it to an industrial bioprocess plant. Golobic *et al.* [2] also developed an approach where the software sensors were based on a linear model while the parameters of the model were estimated using the recursive least squares method. The model was then upgraded using a time-varying forgetting factor mechanism to enable good tracking of the estimated parameters during rapid and large changes in the nonlinear and dynamic characteristics of the process. Although there have been successful applications of linear estimators, Montague [3] pointed out that rather than relying on a linear adaptive model to compensate for the inaccuracy in model structure, it is more appropriate to derive a model with a more representative nonlinear structure. One method of achieving this is through the use of extended Kalman filters.

3.2 Adaptive Observer and Extended Kalman Filter

Estimators based on mechanistic models have the potential to provide accurate predictions and many software sensors using adaptive observers and extended Kalman filters have been published in this field. Bastin and Dochain [27] proposed an adaptive non-linear observer for the estimation of specific growth rate and cell concentration. The method does not require an analytical description of the specific growth rate, instead this is simply considered to be a completely unknown time varying parameter. De Azevedo [28] used an asymptotic observer to estimate ammonium concentration based on the on-line measurement of the optical density in a fed-batch PHB (Poly-β-hydroxybuttyric acid) producing process. Nejjari *et al.* [29] chose a Joint Observer Estimator (JOE) after a comparative study was conducted using several estimation techniques. The joint observer/estimator is analogous to the extended Luenberger observer, using Kronecker's calculation. The estimator was shown to be able to reject disturbances, setpoints changes and ensure convergence with relatively short transient responses. Furthermore, stability of the software sensor was proved using the Lyapunov method. Boillereaux and Flaus [30] developed a so-called Moving Horizon State Estimation (MSHE) method to perform the state estimation based on the nonlinear least squares. Bogaerts and Hanus [31] presented a Full-Horizon Observer (FHO) which is based on the determination of the most likely initial concentrations of the culture. The method compared favorably with the Kalman filtering and Extended Kalman Filtering approaches.

3.2.1 The Extended Kalman Filter

The famous Kalman filter has become another popular approach used for inferential estimation and development of software sensors. The Kalman filter is the optimal state estimator for a linear system when a model for the system together with the knowledge of certain stochastic properties of measurement and disturbance noises are available. The EKF is an adaptation to the nonlinear case of the linear Kalman filter. The EKF method optimally tries to estimate the state of the system by assuming that, (1) the behavior of the system is described by a non-linear model; and (2) the mean and the covariance of the measurement errors are known. Many successful applications have been reported in the literature, using both simulation and experimental investigations.

Figure 2 shows the flow chart of the recursive state estimator algorithm [32].

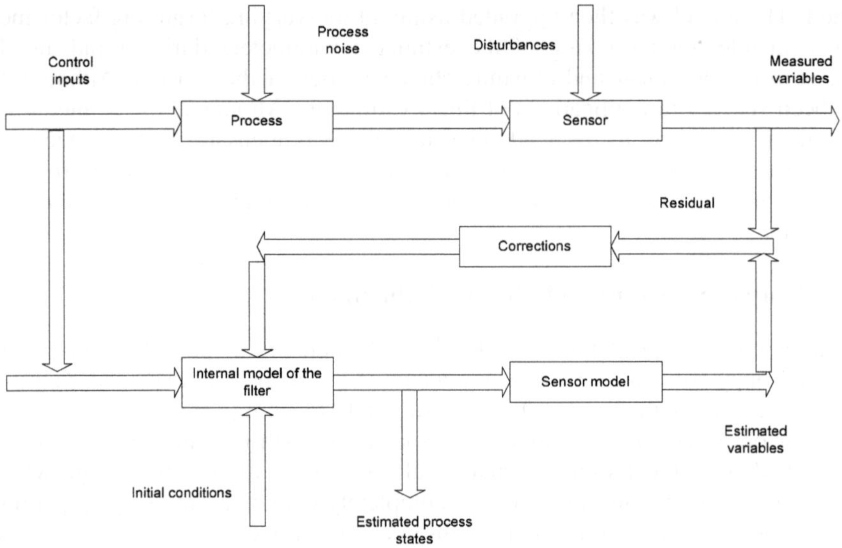

Fig. 2. Flow chart of the recursive state estimator [32]

It is assumed that the process has a state vector $x(t) \in \mathbb{R}^n$ and is described by the non-linear continuous process model

$$\mathbf{x} = \mathbf{f}(\mathbf{x}(t), \mathbf{u}(t)) + \mathbf{w}(t) \tag{1}$$

where $\mathbf{f}$ is the continuous nonlinear system function and $\mathbf{w}$ is the process noise vector which contains all the perturbations which act on the system and which are not described deterministically. These include un-modeled dynamics and noise corrupted (or unmeasured) input. The corresponding discrete-time process model is

$$\mathbf{x}_{k+1} = \overline{\mathbf{f}}(\mathbf{x}_k, \mathbf{u}_k) + \mathbf{w}_k \tag{2}$$

where $\overline{\mathbf{f}}$ is the discrete nonlinear system function.

It is assumed that the process noise is a Gaussian-distributed, zero-mean random variable, with covariance $\mathbf{Q}_k$ and is independent of the process noises or the states of the system which have occurred at any previous time:

$$p(\mathbf{w}_k) \sim N(0, \mathbf{Q}_k) \tag{3}$$

$$E\left[\mathbf{w}_k \mathbf{w}_i^T\right] = \begin{cases} \mathbf{Q}_k, & i = k \\ 0, & i \neq k \end{cases} \tag{4}$$

$$E\left[\mathbf{w}_k \mathbf{x}_i^T\right] = 0, \ \forall i, k \tag{5}$$

The measurements acquired at time k are aggregated into the p-dimensional observation vector $\mathbf{z}_k$, which is related to the state of the system by

$$\mathbf{z}_k = \bar{\mathbf{h}}(\mathbf{x}_k) + \mathbf{v}_k \tag{6}$$

where $\bar{\mathbf{h}}$ is the discrete measurement function. The vector $\mathbf{v}_k$ represents the measurement noise and is assumed to be a zero-mean Gaussian distributed random variable with variance $\mathbf{R}_k$:

$$p(\mathbf{v}_k) \sim N(0, \mathbf{R}_k) \tag{7}$$

The vector $\mathbf{v}_k$ is independent of the measurement noises at all previous time steps and is independent of the process noise:

$$E\left[\mathbf{v}_k \mathbf{v}_i^T\right] = \begin{cases} \mathbf{R}_k, & i = k \\ 0, & i \neq k \end{cases} \tag{8}$$

$$E\left[\mathbf{w}_k \mathbf{v}_i^T\right] = 0, \ \forall i, k \tag{9}$$

The *a priori* and *a posteriori* estimate errors are defined, respectively, as

$$\mathbf{e}_{k|k-1} = \mathbf{x}_k - \mathbf{x}_{k|k-1} \tag{10}$$

$$\mathbf{e}_{k|k} = \mathbf{x}_k - \mathbf{x}_{k|k} \tag{11}$$

The *a priori* estimate error covariance and the *a posteriori* estimate error covariance are then

$$\mathbf{P}_{k|k-1} = E\left[\mathbf{e}_{k|k-1} \mathbf{e}_{k|k-1}^T\right] \tag{12}$$

$$\mathbf{P}_{k|k} = E\left[\mathbf{e}_{k|k} \mathbf{e}_{k|k}^T\right] \tag{13}$$

The EKF is first initialized with $\mathbf{x}_{0|0} = \mathbf{x}_0$ and $\mathbf{P}_{0|0} = \mathbf{P}_0$ ($\mathbf{P}_0$ is positive definitive), and then it operates recursively performing a single cycle each time a new set of measurements becomes available. Each iteration propagates the estimate from the time when the last measurement was obtained to the current time. The equations for the Kalman filter fall into two groups: time update equations and measurement update equations.

The time update equations are responsible for projecting forward (in time) the current state and error covariance estimates to obtain the *a priori* estimates for the next time step. The measurement update equations are responsible for the feedback - i.e. for incorporating a new measurement into the *a priori* estimate to obtain an improved *a posteriori* estimate. The time update equations can also be thought of as *predictor equations*, while the measurement update equations can be thought of as *corrector equations*. Indeed the final estimation algorithm resembles that of a predictor-corrector algorithm for solving numerical problems.

3.2.1.1 Measurement Update. The *a posteriori* state estimate $\mathbf{x}_{k|k}$ is computed as a linear combination of an *a priori* estimate $\mathbf{x}_{k|k-1}$ and a weighted difference between an actual measurement $\mathbf{z}_k$ and a measurement prediction:

$$\hat{\mathbf{x}}_{k|k} = \hat{\mathbf{x}}_{k|k-1} + \mathbf{K}_k \left[\mathbf{z}_k - \overline{\mathbf{h}}\left(\hat{\mathbf{x}}_{k|k-1}\right)\right] \tag{14}$$

where

$$\mathbf{K}_k = \mathbf{P}_{k|k-1} \overline{\mathbf{H}}_k^T \left(\overline{\mathbf{H}}_k \mathbf{P}_{k|k-1} \overline{\mathbf{H}}_k^T + \mathbf{R}_k \right)^{-1} \tag{15}$$

$$\overline{\mathbf{H}}_k = \left(\frac{\partial \overline{\mathbf{h}}(\mathbf{x})}{\partial \mathbf{x}} \right)_{\mathbf{x}=\hat{\mathbf{x}}_{k|k-1}} \tag{16}$$

The matrix $\mathbf{K}_k$ is chosen to be the gain that minimizes the *a posteriori* error covariance (13). The covariance matrix is updated by

$$\mathbf{P}_{k|k} = \left(\mathbf{I} - \mathbf{K}_k \overline{\mathbf{H}}_k \right) \mathbf{P}_{k|k-1} \tag{17}$$

3.2.1.2 Time Update. The state and covariance matrix in the next sampling instant are estimated by

$$\hat{\mathbf{x}}_{k+1|k} = \overline{\mathbf{f}}\left(\hat{\mathbf{x}}_{k|k}, \mathbf{u}_k \right) \tag{18}$$

$$\mathbf{P}_{k+1|k} = \overline{\mathbf{A}}_k \mathbf{P}_{k|k} \overline{\mathbf{A}}_k^T + \mathbf{Q}_k \tag{19}$$

where

$$\overline{\mathbf{A}}_k = \left(\frac{\partial \overline{\mathbf{f}}(\mathbf{x}, \mathbf{u})}{\partial \mathbf{x}} \right)_{\mathbf{x}=\hat{\mathbf{x}}_{k|k}, \mathbf{u}=\mathbf{u}_k} \tag{20}$$

3.2.1.3 EKF Computation Procedure. The computation sequence of the EKF is described as follows:

(a) Save the actual (k) state process: $\left[\hat{\mathbf{x}}_{k|k}, \mathbf{P}_{k|k} \right]$ (For initialization, given $\left[\hat{\mathbf{x}}_0, \mathbf{P}_0 \right]$)

(b) Calculate the predicted state: $\hat{\mathbf{x}}_{k+1|k}$

(c) Calculate the covariance error: $\mathbf{P}_{k+1|k}$

(d) Calculate the filter gain: $\mathbf{K}_{k+1}$

(e) Calculate the state process (k+1): $\hat{\mathbf{x}}_{k+1|k+1}$

(f) Calculate the covariance error: $\mathbf{P}_{k+1|k+1}$

(g) Update the sample time k=k+1 and go back to step (a).

3.2.2 Case Study: Implementation of the EKF to Yeast Fermentation

In this case study, a fed-batch yeast fermentation is used as an example to show how EKF can be employed for software sensors developments based on mechanistic models. The commercial production of baker's yeast is an aerobic fed-batch process, however, some ethanol is initially produced and the process is controlled in such a way that this ethanol is later utilized as an energy source [33]. Figure 3 shows the variation in several process variables recorded during a typical fermentation involving baker's yeast production. It can be seen that, when the sugar concentration is higher than a critical value, labelled *Scrit* , ethanol is produced but, when the concentration declines below the critical value, ethanol is consumed resulting in an ethanol concentration peak. This process is best described by a model called the bottle-neck model, since it can be explained in terms of a critical maximum reaction rate, i.e. a bottle-neck. A simple fed-batch bottle-neck model which contains 5 mass balance equations and 8 kinetic equations is developed and used in the implementation of EKF software sensors. The biomass, substrate, ethanol and DOT concentrations as well as the fermenter volume are estimated by the EKF.

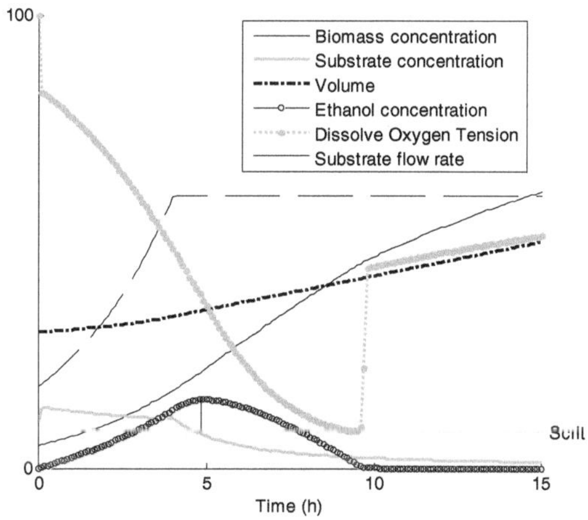

Fig. 3. A typical batch of a fed-batch yeast fermentation

3.2.2.1 Simulation Setup. The substrate concentration and the concentration of dissolved oxygen are chosen as observed variables. It is assumed that these variables are measured with a relatively infrequent sampling time at 0.15 h (i.e. 9min) per sample. In experimental works, these measurements are readily available through on-line sensors. The concentration of dissolved oxygen can be measured by a DO probe while a glucose sensor can provide online measurement of substrate (glucose) concentration in the liquid phase. The substrate feed rate is used as an input to the simulation with a sample time of 0.01h. In all the simulations, random deviations of white noise are assumed to exist in the observed variables.

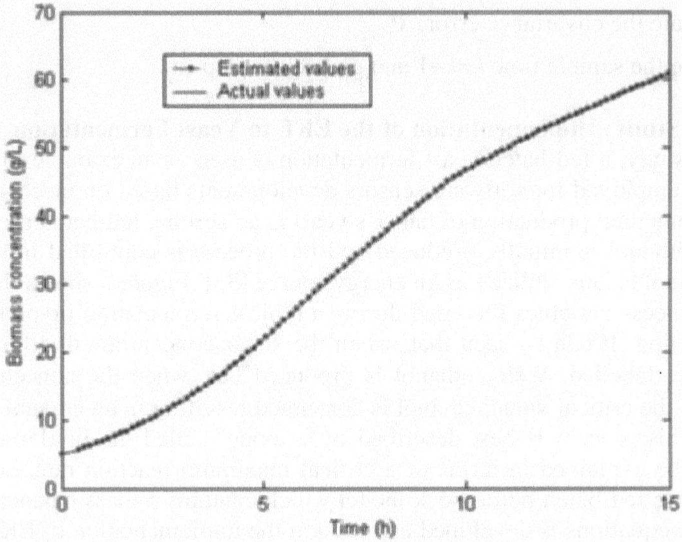

Fig. 4. Biomass concentration estimation

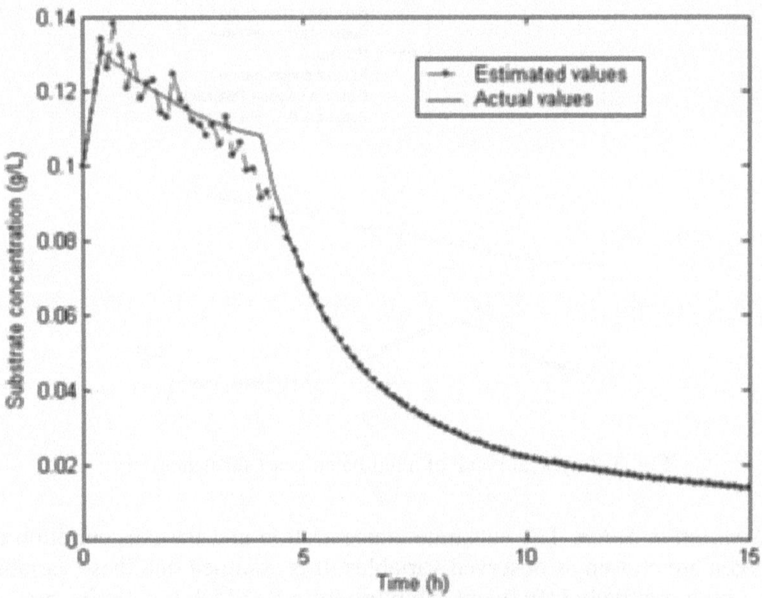

Fig. 5. Substrate concentration estimation

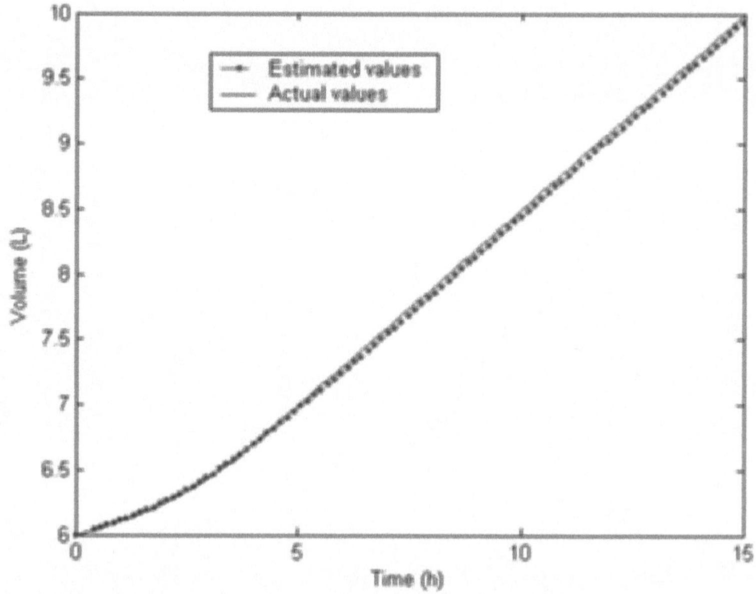

Fig. 6. Volume estimation

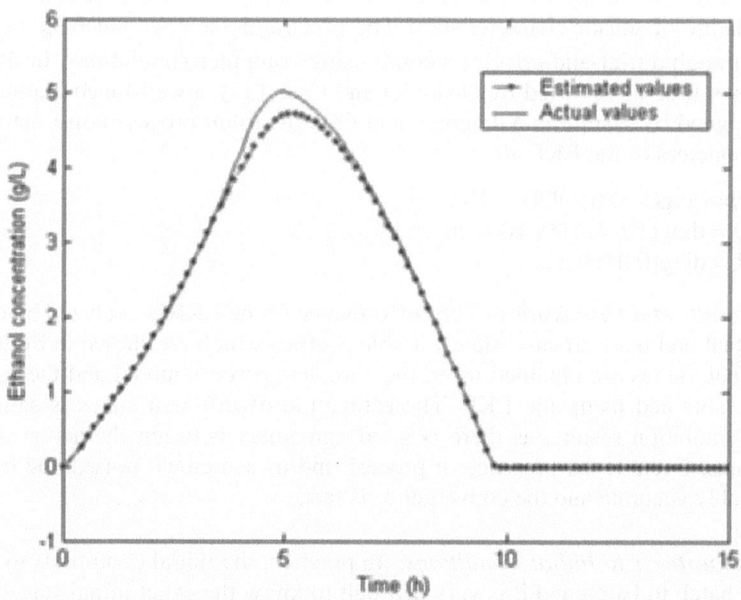

Fig. 7. Ethanol concentration estimation

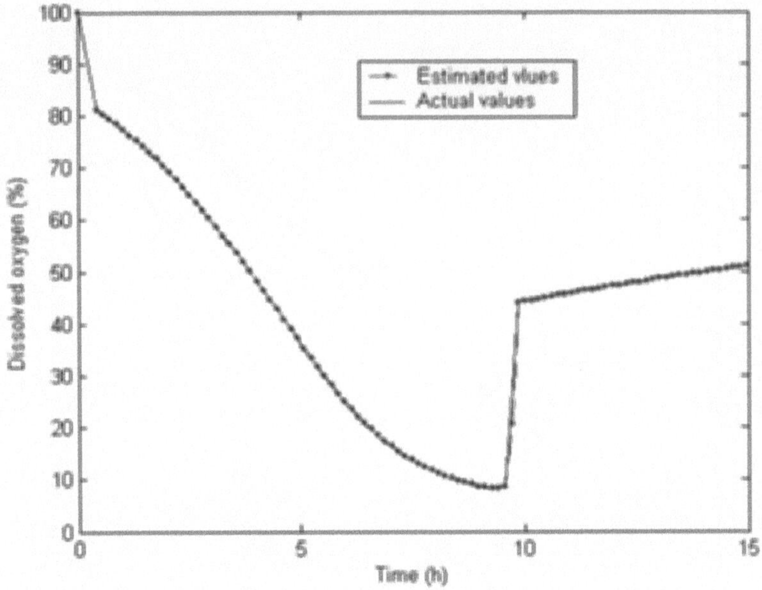

Fig. 8. Dissolved oxygen estimation

The measurement noise covariance **R** can be obtained from the measurement data and knowledge of sensor characteristics. The process noise covariance **Q** is usually selected through a trial-and-error procedure using computer simulations. In this work it was shown that, as reported by Oisiovici and Cruz [34], a well-tuned Kalman filter can be designed by assuming a diagonal and time-invariant process noise matrix. The tuning parameters of the EKF are:

$\mathbf{P}_0$= diag(1, 0.01, 4, 0.1, 10),
Q = diag(10, 1, 100, 100, 1);
R = diag(0.01, 0.1).

3.2.2.2 Results and Discussions. The performance of the EKF is analyzed by comparing the actual and the estimated state variable profiles which are shown in Figures 4 to 8. The actual values are obtained using the complete process model and the estimated values are obtained using the EKF. The estimation of different states is satisfactory from the simulation results, as there is good agreement between the actual and estimated values. Even in the presence of process and measurement noises, the estimates are reasonably accurate and the convergence is fast.

3.2.2.3 Robustness to Initial Conditions. In practice, the initial conditions may often vary from batch to batch and it is very difficult to know the exact initial states. Therefore, the robustness of the estimator to initial conditions is of great importance. To test this, the initial conditions in the following simulations have been changed deliberately as shown in Table 1.

Table 1. Change of initial conditions

Variable	Initial value	Test 1	Test 2 (20% change)	Test 3 (50% change)
X- Biomass (g/L)	5	4.5	4	2.5
S - Substrate (g/L)	0.1	0.15	0.08	0.05
V- Volume (L)	6	5.8	4.8	3
DOT (%)	100	95	80	50
E - Ethanol (g/L)	0	0	0.2	0.5

The initial conditions are changed a little in Test run 1 with 20% change in Test 2 and 50% in Test 3. Since much of the work in this chapter focuses on the development of a software sensor for biomass concentration then only this variable will be compared here between different test runs. Figure 9 shows the estimations of biomass concentration during the 3 test runs. The EKF is found to be robust because it reproduced the behavior of the process despite the change of initial conditions.

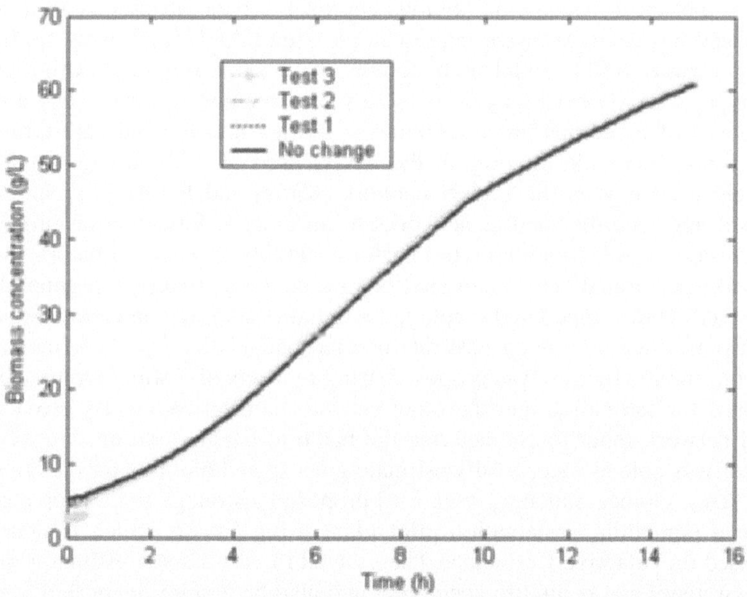

Fig. 9. Biomass concentration estimations in different initial conditions

3.2.3 Summary

In this section, a recursive estimator of state variables (biomass, substrate, ethanol and dissolved oxygen concentrations, and the fermenter volume) using two observed variables (the substrate and dissolved oxygen concentration) implemented in a fed-batch yeast fermentation process based on the use of the extended Kalman filter is developed and tested. The state estimates are accurate, and the EKF has confirmed its ability to incorporate the effects of measurement and modeling noises. It is further found that the EKF is robust to the changes in initial conditions. However, as it is true to all EKF type of estimators, good filter performance relies heavily on the accurate description of mathematical models. The design of a well-tuned EKF is an iterative process of system modeling and trading off between the performance capabilities and the computational constraints and requirements. For practical applications, it is very important to validate the model with experimental data.

Although EKF is a generic and elegant approach to cope with the problem of recursive estimation, it should be noted that successful applications of EKF depend largely on the accuracy of the process model. It requires a large design effort and a priori estimates of measurement noise and model uncertainty characteristics. The EKF can also suffer from some numerical problems and convergence difficulties due to approximations associated with the model linearization [6][30]. As a result of these limitations alternative nonlinear estimation, such as ANNs, have been investigated.

3.3 ANN-Based Software Sensors

As reviewed earlier, ANNs have been proven to be successful in modeling bioprocesses, thus they have become an obvious approach for the development of software sensors. There has been extensive interest in the use of ANNs and many applications, employing various ANN model architectures, have been reported. Although many differing types of ANN exist they do possess some common features. They are generally composed of numerous process elements, termed nodes, which are arranged together to form a network. Among all the ANNs used for inferential estimation, the most widely used type is the FANN network. Karim and Rivera [35] applied both feedforward and recurrent neural networks to an ethanol fermentation process. The results showed that although both types perform suitably as state estimators, the recurrent networks performed better than feedforward networks in terms of generalization. Dacosta *et al.* [4] developed and employed a radial basis function neural networks to estimate biomass and unutilized substrate in a fed-batch baker's yeast fermentation. A new network input structure was proposed using two sets of distinct inputs where one set was used for estimation and the other set for characterization. By providing this additional network input to characterize the batch-to-batch variation, the ANN software sensor was able to successfully estimate glucose and biomass for different initial inoculum sizes. Gomez Sanchez, *et al.* [36] proposed a strategy to develop a software sensor for a penicillin fermentation pilot plant using FasArt which is neuro-fuzzy system based on Adaptive Resonance Theory (ART). The FasArt estimator predicted biomass, viscosity and penicillin accurately and also be employed for fault detection, and as the basis for a model predictive controller.

Since there are a large number of textbooks published on the topic of artificial neural networks available, to fully understand and learn how to use artificial neural networks the readers are referred to these textbooks [37][38]. This section only aims to give a short overview of the implementation of neural network as software sensors for bioprocess.

3.3.1 Network Architecture

The architecture of a neural network defines how the network is constructed from layers of interconnected neurons as processing elements. The simplest and most widely used feedforward network comprises an input layer where input information is presented to the network, one or more hidden layers where neuron processing takes place and an output layer from which the network outputs are obtained. Since the outputs from one layer are fed forward as inputs to the subsequent layer thus it is termed as feedforward network. Each connection between two neurons has a weight coefficient attached to it. In such a network the processing element is one which weights the input signals and then sums them together with a bias term [39]. The neuron output is then obtained by passing the summed, weighted inputs through a non-linear activation function, such as a sigmoid function. Figure 10 shows the architecture of a typical FANN.

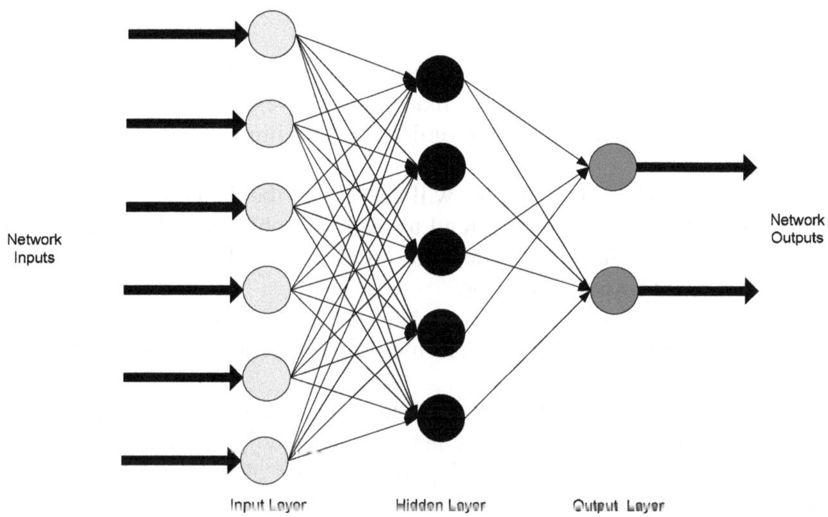

Fig. 10. Feedforward artificial neural network

3.3.2 Backpropagation Learning Algorithm

After the topology of the neural network is decided, the weights for the neuron connections must be determined. Backpropagation algorithm is one of the most well-known and suitable techniques to perform weight selection and adjustment. Backpropagation is based on the delta rule, a gradient descent search technique to minimize a cost function equal to the mean square of the output error. It was well-documented that the backpropagation algorithm can suffer from slow convergence times and can easily get trapped in local minima within the weight space. However, with an addition of a 'momentum' term, the convergence can be speeded up and the local minima can be avoided.

3.3.3 Underfitting and Overfitting of ANN

Besides the training data set, a smaller validation data set is needed. This set is not used for training but to check the network during training. Neural networks, like other flexible nonlinear estimation methods such as kernel regression and smoothing splines, can suffer from either 'underfitting' or 'overfitting' [40]. A network that is not sufficiently complex can fail to detect fully the signal in a complicated data set, leading to 'underfitting'. A network that is too complex may fit the noise and errors, not just the signal, leading to 'overfitting'. To avoid 'overfitting', the validation data set is used to check whether the network is overtrained or needs more training cycles.

Validation can be done by following the root-mean-square (RMS) error on the predictions. Thus, the trained network tries to predict the quality variable from process data given in the validation data set, and the output value is then compared to the real value in the validation data set. Once the sum of RMS errors on the validation data set starts to increase during training, the network can be regarded as a sufficiently trained network if not the result of a local minimum in the process data space.

3.3.4 Implementation Issues

ANNs are capable of learning from process data to predict the value of the key process variables and parameters in real-time. However, implementing them to particular process is not so straightforward and a few implementation issues need to be discussed.

Firstly, selecting ANN architectures will depend on the data representation and the application. Overly complex models tend to lead to problems with learning. The optimal number of neurons/layers also depends on the type and complexity of the task. It is usually determined by trail and error.

Sccondly, there are numerous tradeoffs between learning algorithms. Selecting and tuning an algorithm for training on unseen data requires a significant amount of experimentation.

Thirdly, to successfully implement the neural network software sensor, data has to be collected followed by training of the network. Generally, at least 5 times the number of weights in the net (Number of inputs * number of hidden units + number of hidden units * number of outputs) of data needs to be collected to train a neural network for use in construction of a software sensor [41].

3.3.5 Case Study: FANN-Based Software Sensor for a Penicillin Fermentation Process

Due to its academic and industrial importance, the production of secondary metabolites has been the subject of many studies. For the production of secondary metabolites such as antibiotics, the formation of the target product is usually not associated with cell growth. For this reason, it is common practice to grow the microorganisms in a batch culture followed by a fed-batch operation to promote the synthesis of the antibiotic. There are extensive publications in the literature on penicillin production exhibiting models with varying degrees of complexities.

A fed-batch penicillin fermentation process is used as a case study to show the implementation of FANN-based software sensors. The secondary variables in the process include aeration rate, agitation power, feed flow rates of substrate and oxygen, carbon dioxide concentration, feed coolant and bioreactor temperatures, generated heat and the medium pH. In a typical penicillin production process, the bioreactor is switched to the fed-batch mode of operation after about 40 hours of batch growth phase when the cells enter their stationary phase.

To generate the data necessary for the development of the ANN model, data from 30 batches was collected. Pseudo-random signal (PRS) with the size of up to ±25% of its set-point value is applied to substrate federate during each of the batches to increase variations in the data. 20 of these batches were used to train the FANN model (training batches) with the remainder used to validate the model (validating batches).

A two layered fully connected backpropagation network has been chosen. Besides input units, one hidden layer and an output layer are forming the sensor network. A sigmoid activation function is used and a threshold term is added to the neuron output function preventing the network to be trapped in a local minimum. The following measurements are used as input variables for the software sensor: substrate feed rate, aeration rate, agitator power, substrate feed temperature, dissolved oxygen concentration, pH, fermenter temperature and generated heat. The two outputs from the net are giving the estimated biomass and penicillin concentration.

Figure 11 and 12 show the estimations of biomass and penicillin concentration using the FANN respectively. It can be seen that, for both variables, good agreements between on-line prediction and actual concentration are achieved. This information can be provided to operators for information purposes or used in closed-loop control and condition monitoring.

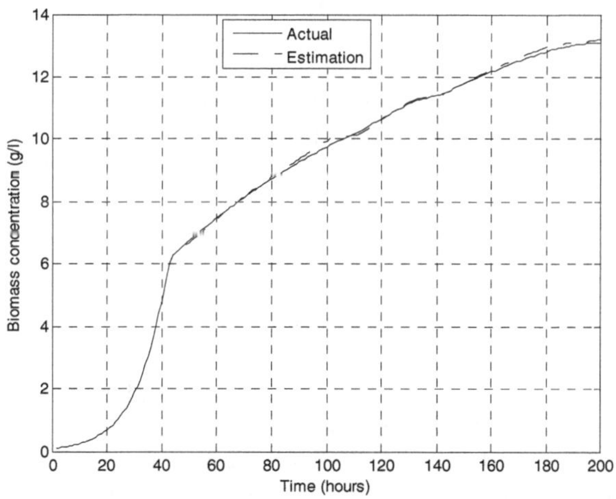

Fig. 11. Estimation of biomass concentration in a Penicillin fermentation process using a FANN-based software sensor

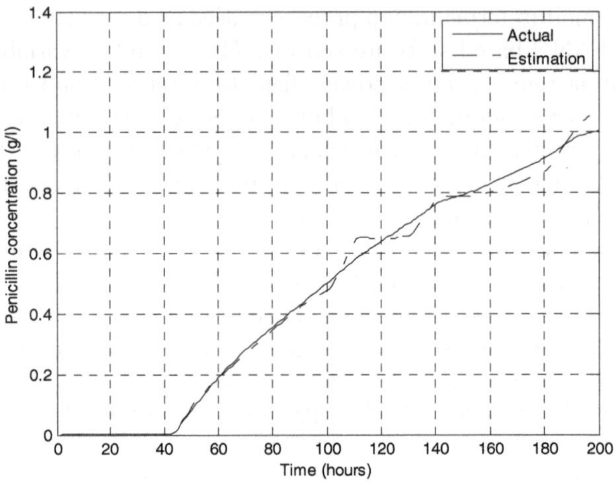

Fig. 12. Estimation of penicillin concentration in a Penicillin fermentation process using a FANN-based software sensor

Apart from the FANN, another popular choice of neural network used as software sensors for bioprocesses is the Radial Basis Function (RBF) neural network [39][42].

The main advantage of ANN-based techniques for software sensors is that they do not need any prior knowledge about the kinetic growth rate, but the disadvantage is the large amount of training and test data that is required for this development.

3.4 MSPC-Based Software Sensors

Multivariate statistical methods based on linear projection, such as PCA and PLS, have attracted considerable interest as a method for producing robust empirical models, particularly when there are high dimensionality and collinearities in the data. PCA and PLS, in particular, and their variations, such as neural network partial least squares (NNPLS) [43] and nonlinear principal components analysis (NLPCA) [44][45], have been applied to many practical regression problems to estimate quality related variables in chemical engineering processes such as distillation columns, combustion processes, the paper and pulp making process and polymerization processes. This section explores and demonstrates MPLS's ability to develop software sensors for fed-batch fermentation processes.

3.4.1 Partial Least Squares

PLS is a system identification tool that is capable of identifying the relationships between cause (X) and effect (Y) variables. The advantage that this approach offers over more traditional identification techniques, such as ordinary least squares, is that it is able to extract robust models even in applications involving large numbers of highly correlated and noisy process variable measurements.

The approach works by selecting factors of cause variables in a sequence that successively maximizes the explained covariance between the cause and effect variables.

Given a matrix of cause data, **X**, and effect data, **Y**, a factor of the cause data, $\mathbf{t}_k$, and effect data, $\mathbf{u}_k$, is evaluated, such that:

$$\mathbf{X} = \sum_{k=1}^{np<nx} \mathbf{t}_k \mathbf{p}_k^T + \mathbf{E} \tag{21}$$

and

$$\mathbf{Y} = \sum_{k=1}^{np<nx} \mathbf{u}_k \mathbf{q}_k^T + \mathbf{F} \tag{22}$$

where **E** and **F** are residual matrices, *np* is the number of inner components that are used in the model and *nx* is the number of causal variables. $\mathbf{p}_k$ and $\mathbf{q}_k$ are referred to as loading vectors.

These equations are referred to as the *outer relationships*. The vectors $\mathbf{t}_k$ are mutually orthogonal. These vectors and $\mathbf{u}_k$ are selected so as to maximize the covariance between each pair, ($\mathbf{t}_k$, $\mathbf{u}_k$). Linear regression is performed between the $\mathbf{t}_k$ and the $\mathbf{u}_k$ vectors to produce the *inner relationship*, such that:

$$\mathbf{u}_k = b_k \mathbf{t}_k + \varepsilon_k \tag{23}$$

where b_k is a regression coefficient, and ε_k refers to the prediction error. The PLS method provides the potential for a regularised model through selecting an appropriate number of latent variables, $\mathbf{u}_k$ in the model (*np*). The number of latent variables is typically made through the use of cross validation.

For further details of the PLS algorithm, the reader is referred to Geladi and Kowalski [16].

3.4.2 Multiway Partial Least Squares

PLS is a linear tool, which unfortunately limits its effectiveness when applied to non-linear fed-batch processes. Two options exist for improving the capabilities of PLS when applied to fed-batch systems. The first is to develop non-linear counterparts to PLS and the second is to transform the fed-batch data in such a way as to remove the non-linear characteristics [18]. Although non-linear PLS techniques exist [43], the transformation of batch data has proved to be a more effective option and has been adopted in this investigation. The most common form of data transformation, termed multiway PLS, was initially proposed by Nomikos and MacGregor [18]. Since then other researchers have adopted the approach and applied it to a variety of processes. For example, Gallagher *et al.* [46] applied the technique to monitor nuclear waste storage vessels and Lennox *et al.* [42] and Lakshminarayanan *et al.* [47] investigated the detection of faults in fed-batch fermentation processes.

In a fed-batch process, the cause and effect data can be thought of as being in two 3-dimensional arrays of size $nb \times nx \times mx$ and $nb \times ny \times my$, where *nb* is the number of batches for which data is available, *nx* and *ny* are the number of cause and effect variables respectively and *mx* and *my* are the number of observations of the cause and effect variables respectively that are made during a batch. Unfortunately, PLS requires that the cause and effect arrays be two-dimensional. To address this problem the three-dimensional arrays are recast into two-dimensional arrays in a process referred to as *unfolding*. The concept of unfolding is illustrated in Figure 13. The original data arrays are unfolded into a cause variable array, **X**, of size $nb \times (nx \times mx)$ and an effect variable array of size $nb \times (ny \times my)$. It should be noted that the number of observations made of the cause variables need not be equal to the number of observations made for the effect variables. In fact, it is relatively common for there to be a single effect measurement made during a batch. This measurement being the final product quality taken at the end of the batch. Following the unfolding of the data, it is then possible to apply PLS to the data in the conventional manner.

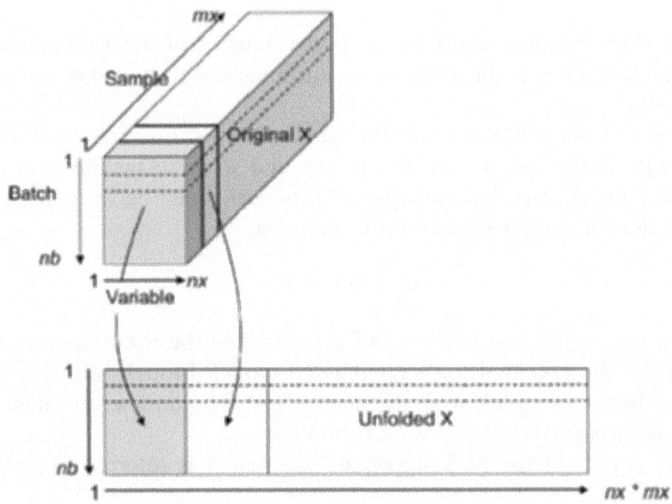

Fig. 13. Unfolding

The subsequent use of this model on-line poses the problem that it is necessary to know the values of all process measurements through to the end of the batch, since the unfolded array contains the measurements of each of the variables throughout the duration of the batch. This means that with the exception of the end point of the batch, it is necessary to estimate the future values of all the measured variables. The prediction of future process values is referred to as *filling up* the array. Of the three methods that were suggested by Nomikos and MacGregor [18] for filling up the array, Lennox *et al.* [42] found that the most appropriate method for an industrial fed-batch process was to assume that the values of all process measurements remain at their current offset from the mean trajectory through to the end of the batch. Whilst the most suitable filling up method is likely to be process dependent, this method was also found to be the most appropriate in the case study below.

3.4.3 Application of MPLS to an Industrial Fed-Batch Fermentation Process

In this section the application of MPLS to an industrial fed-batch fermentation process is demonstrated. In this particular process, biomass measurements are taken at infrequent times during the batch. The quality of these measurements is considered to be poor and it is not uncommon for operations staff to dismiss the results they are provided with from the laboratory. In addition there can be a significant delay between the sample being taken and the laboratory returning the biomass concentration measurement.

Figure 14 illustrates the quality of data that was received from the laboratory during one particular fermentation batch. This figure shows the actual biomass measurements that were made during a typical batch on the industrial process under investigation. These measurements are indicated by the stars. The first observation regarding these measurements is that there is a high level of noise present which makes interpretation of the data difficult (it is believed that the growth of biomass is relatively uniform, which isn't evident from the laboratory assays). For example, in the particular batch that the measurements displayed in Figure 14 were taken, the measurements labeled 1 and 2, and circled, would appear to indicate a significant abnormality. At the sample labeled 1 it can be seen that there appears to be a sharp increase in biomass. However, the following sample, labeled 2 shows a significant drop in biomass concentration. At sample time 2 these results could be interpreted in several ways, two of which are provided below.

- The conditions within the fermenter are unsuitable and the concentration of biomass has begun to reduce, consequently the biomass concentration labeled 2 is significantly lower than that labeled 1.
- One or both of the laboratory measurements are inaccurate.

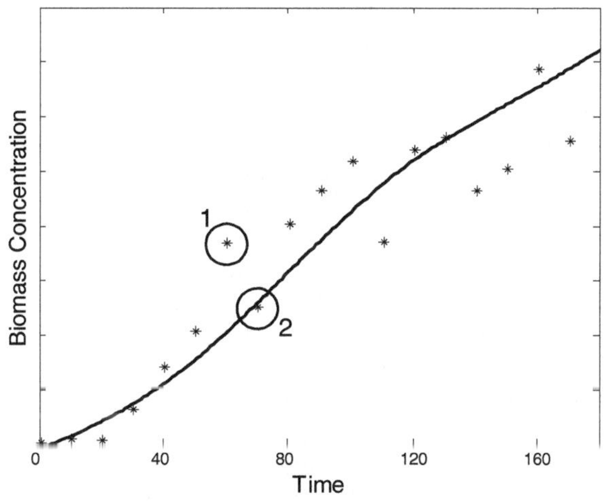

Fig. 14. PLS prediction of biomass

Following the results from the two samples, the response from operations staff would typically be to wait until the next laboratory measurement is made before taking any corrective action. If this measurement also indicated reduced biomass concentrations then it is likely that action would be taken to recover the batch. Unfortunately, the delay in taking any corrective action may result in the batch being irrevocably damaged. The ability to provide a software sensor, capable of inferring the biomass concentration from the frequently measured on-line variables would offer significant advantages to operations staff. MPLS provides one method of inferring the biomass concentration measurement. The application of this approach to this industrial fermentation system is described in the following section.

3.4.4 MPLS Model Development

The first stage in developing a PLS model of the fermentation process is to gather data. For this study, data was made available from 15 historical batches. For a process such as this where there is significant batch to batch variation, it would be desirable to have many more historical batches to ensure that the PLS model could identify the key characteristics of the process. However, it was anticipated that even with this limited data the investigation would produce meaningful results.

For each of the 15 batches, final production levels varied with some of the batches being high yield and others relatively low yield. 10 of the batches were selected to train the model with the remainder being used for validation purposes and to identify a suitable number of latent variables. The sampling frequency of the on-line measurements such as pH and temperature is significantly higher than that for the biomass concentration. To enable the PLS model to provide frequent inference of the biomass concentration it is necessary to provide this measurement at the same sampling frequency as the on-line measurements. To achieve this, the off-line biomass measurements can be interpolated so that a measurement is available at the same sampling frequency as the on-line measurements. In this study the interpolation method proposed by Montague [3], which involves the use of cubic splines to interpolate inaccurate biomass levels was used.

The solid line in Figure 14 shows the inferred biomass measurement that is made by the PLS model, with 3 latent variables (the number of latent variables was selected using cross validation). By analysis of the prediction made by this model there is an indication that the biomass measurement, labeled 2 in Figure 14 was accurate. However, the measurement labeled 1 was probably a little higher than the actual biomass concentration. The soft-sensing capabilities of the PLS model provide important information to operations staff, who will be able to make better informed decisions when trying to maintain maximum productivity from the batch.

3.5 Comparison of Software Sensors Using MPLS and EKF

The previous section has demonstrated the soft-sensing capability of the MPLS in application to an industrial fed-batch fermentation process. In section 3.2, it was similarly shown that an EKF could be implemented in a similar way to a simulation of a fed-batch yeast fermentation. A comparison of these two methods would now seem appropriate to identify their strengths and weaknesses. To this end, MPLS has been applied to the simulated fed-batch yeast fermentation process, described in section 3.2, to develop a soft-sensor for the process. The performance of the MPLS software sensors are then compared with those obtained from using EKF and the results are discussed.

To develop a MPLS model based on the simulation, 15 batches of data have been collected. The first 10 of these batches were used to train the model and the remaining 5 were used for validation purposes. When generating the data, the initial conditions and the culture parameters were selected to be the same as those used in section 3.2. The reason for this is to ensure a fair comparison. PRS sequences were applied to the feed rate in order to excite the process so that the data collected had sufficient variation to identify accurate process models.

A PLS model, containing 2 latent variables, was identified from the training batches. In this model the following measurements were used as input, or cause, variables: substrate feed rate, dissolved oxygen concentration and culture volume. Based on this model, three software sensors have been constructed to estimate the biomass concentration, substrate concentration and ethanol concentration. The accuracy of the estimates provided by these software sensors are illustrated in figures 15 to 17 which compare the actual concentrations with those predicted by the PLS model for a typical batch.

When comparing the previous figures with the estimations provided by EKF, although both methods provide stable and satisfactory estimations, the results using MPLS are favorable to the EKF's. This is also demonstrated by the mean square errors (MSE) in Figure 18. The MSE is calculated using equation (24).

$$\mathrm{MSE} = \frac{\sum_{k=1}^{k_{tend}} [x_k - \hat{x}_k]^2}{k_{tend}} \tag{24}$$

Where x is the actual value, $\hat{x}$ is the estimated value predicted by either MPLS or EKF and k_{tend} is the total number of samples.

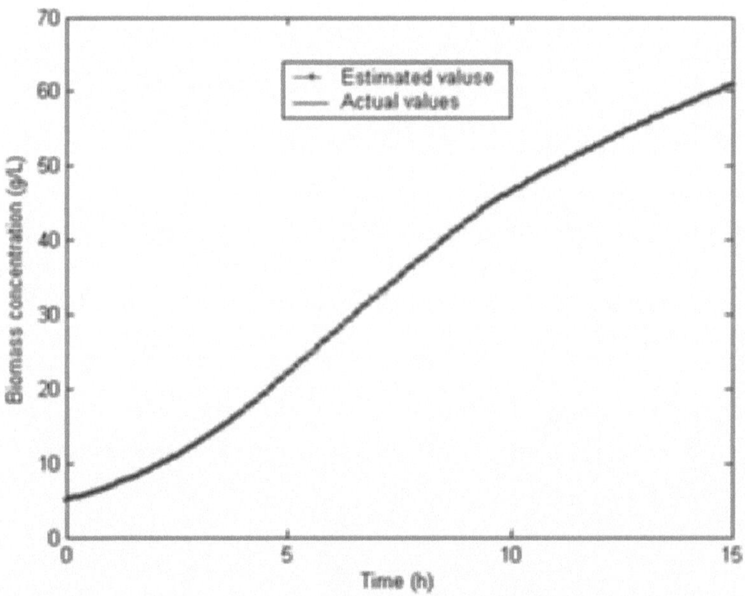

Fig. 15. Biomass concentration estimations using MPLS

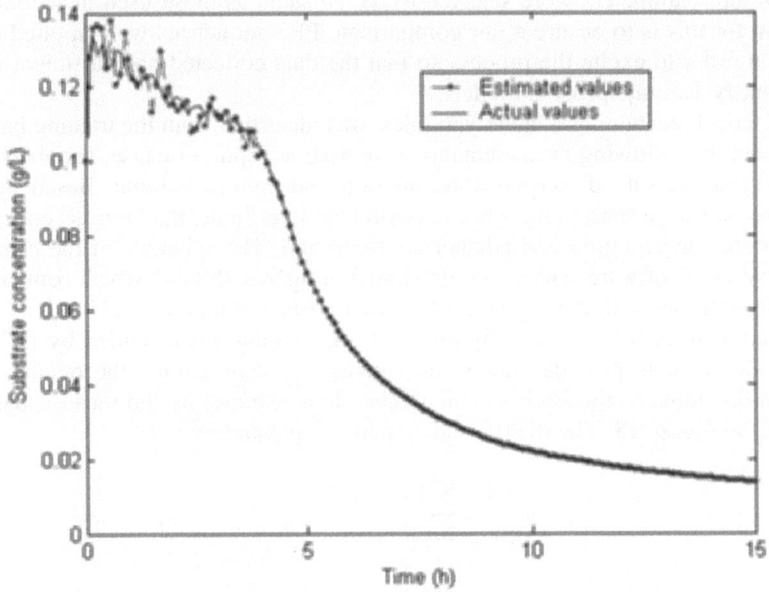

Fig. 16. Substrate concentration estimations using MPLS

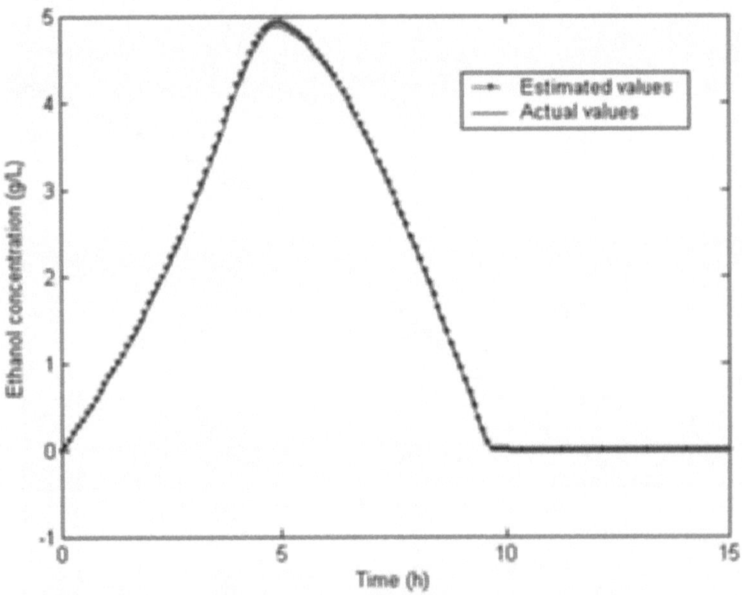

Fig. 17. Ethanol concentration estimations using MPLS

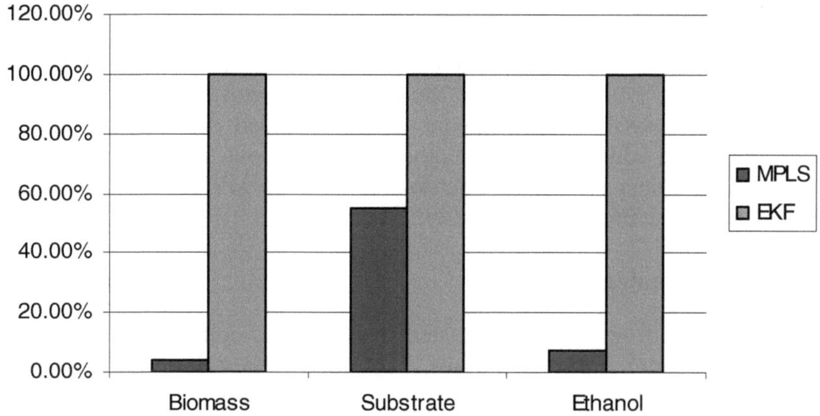

Fig. 18. MSE comparison (scaled in percentage)

The results show that MPLS models can be developed for the on-line prediction of low frequency biomass measurements, as well as other variables, using direct secondary measurements. The advantage of the MPLS technique is that it does not need any prior knowledge regarding the process mechanism or the kinetic growth rate, which was required for the EKF model. The disadvantage is that the accuracy of the MPLS model is related to the quality and quantity of experimental data that is available from the process. In many situations this data might be difficult to obtain. In contrast the EKF approach depends largely on the accuracy of the process model. It requires a large design effort and *a priori* estimates of measurement noise and model uncertainty characteristics. The EKF can also suffer from numerical problems and convergence difficulties due to approximations associated with model linearization. However, it is a generic and elegant approach to cope with the problem of recursive estimation.

3.6 Other Methods for Software Sensors

Apart from the four inferential estimation methods introduced above, other promising methods have been reported include genetic programming [48] and least square support vector machine (LSSVM) [49]. Sensor fusion is another useful tool to improve software sensor's stability and reliability.

4 Practical Considerations

Despite that the concept of software sensor is straightforward and there are plenty of inferential estimation methods available to choose from, when it comes to implement the software into practice, there are still a number of issues to consider.

4.1 Software Sensor Input Selection

To develop a data-driven software sensor, input selection, i.e. selection of secondary variables and process input as inputs to the software sensor, is very important. The number of variables and the choices of variables have a big impact on the reliability and accuracy of the software sensor. To save training and online computation time, it is normal to use the least of number of variables to develop a model of sufficient accuracy. And it is crucial to choose those variables with the strongest relationship with the primary variable and to avoid using correlated variables as inputs to the software sensor. In addition, as Montague [3] pointed out, it is advisable not to use off-line lab assay data as software sensor inputs as it will reduce the frequency at which the sensor can provide information.

4.2 Data Conditioning

Inevitably, there will be noises associated with the data collected as well as outliers caused by, for example, sensor failures. Therefore, before developing a data-driven software sensor, a preliminary data conditioning step has to be carried out to get all the corrupt data out. This step includes filtering out noises using low pass filters and eliminating outliers through the identification of trends in the data. Data cleaning can be carried out by an expert or by automatic validation procedures based on standard signal processing methods (such as filtering, sequential hypothesis testing, etc). On this aspect, Tham and Parr [50] have provided a thorough review of data validation and reconstruction procedures.

4.3 Sensor Validation

When preparing the data set, besides the training data, a smaller validation data set is needed. This set must be independent to the training set and is not used for training but to check the model to see if the software sensor accurately replicates the process. The validation process is an iterative exercise and must be done rigorously prior to online implementation. In addition, end-users need offline verifications, such as analyzer data or laboratory results, to ensure that the software sensors accurately track the real process.

4.4 Industrial Applications

Following extensive research conducted on software sensor technology in the last 20 years also, it becomes more and more attractive for the industry to use software sensors in real applications. Honeywell, a leading advanced control provider, has developed a comprehensive collection of advanced process control and optimization products called Profit Suite™ in which Profit Sensor Pro enables the development of reliable process models for online prediction and soft sensor applications. Profit Sensor Pro provides empirically based software sensors facility and it has been successfully applied to a broad range of industries, including pharmaceutical, refining, oil and gas, chemicals, pulp and paper, power generation and mining and minerals. AspenTech, another leading supplier of process optimization software in the world, has a similar product called Aspen Online Deployment which enables the use of models in process operations for a wide range of real-time process monitoring, software sensor, and operator advisory applications. A UK based company, Perceptive Engineering Limited, has also integrated the capability of soft sensors into their product – MonitorMV which has been deployed in the industries of food and beverage, pharmaceuticals, metals, mining and minerals, pulp & paper, chemicals and petrochemicals.

5 Summary

This chapter addresses some key issues concerning the technique of software sensors and their applications in bioprocesses. Some most popular and well developed techniques have been reviewed and introduced. It has demonstrated that existing technologies, such as EKF, ANN and PLS, provide a suitable tool for the prediction of key primary process variables. This will enable faster return of information in a bioprocess when the quality variables are unmeasured or measured with a long delay time. Furthermore, software sensors can be used for optimization, closed-loop control and condition monitoring purposes in a bioprocess to achieve better process regulation.

Advances in software sensor technology have not only extended measurement capabilities but also made a mark in emerging control approaches. As stated in an industrial overview by Amos [51], the overall trend is: 'Soft sensors will play an increasing role in automated controllers of whatever type'.

Acronyms

ANN	Artificial Neural Network
ART	Adaptive Resonance Theory
CAFNN	Control Affine Feedforward Neural Network
CER	Co2 Evolution Rate
DOT	Dissolved Oxygen Tension
EKF	Extended Kalman Filter
FANN	Feedforward Artificial Neural Network
FHO	Full-Horizon Observer
GA	Genetic Algorithm
GP	Genetic Programming
HPLC	High Performance Liquid Chromatography
JOE	Joint Obscrvcr Estimator
LSSVM	Least Square Support Vector Machine
MPCA	Multiway Principal Components Analysis
MPLS	Multiway Partial Least Squares
MSE	Mean Square Errors
MSHE	Moving Horizon State Estimation
MSPC	Multivariate Statistical Process Control
NLPCA	Nonlinear Principal Components Analysis
NMR	Nuclear Magnetic Resonance
NNPLS	Neural Network Partial Least Squares
PCA	Principal Component Analysis
PLS	Partial Least Squares
PRS	Pseudo-Random Signal
RBF	Radial Basis Function

RMS	Root-Mean-Square
RBPN	Recurrent Backpropagation Neural Network
RTNN	Recurrent Tainable Neural Network

References

1. James, S., Legge, R., Budman, H.: Comparative study of black-box and hybrid estimation methods in fed-batch fermentation. Journal of Process Control 12(1), 113–121 (2002)
2. Golobic, I., Gjerkes, H., Bajsic, I., Malensek, J.: Software sensor for biomass concentration monitoring during industrial fermentation. Instrumentation Science and Technology 28(4), 323–334 (2000)
3. Montague, G.A.: Monitoring and control of fermenters. Institution of Chemical Engineers (1997)
4. Dacosta, P., Kordich, C., Williams, D., Gomm, J.B.: Estimation of inaccessible fermentation states with variable inoculum sizes. Artificial Intelligence in Engineering 11(4), 383–392 (1997)
5. Joseph, B.: Tutorial on inferential control and its applications. In: Proceedings of the American Control Conference, pp. 3106–3118. IEEE, Los Alamitos (1999)
6. De Assis, A.J., Filho, R.M.: Soft sensors development for on-line bioreactor state estimation. In: Computers and Chemical Engineering 7th International Symposium on Process Systems Engineering, July 16-21, 2000, vol. 24(2), pp. 1099–1103 (2000)
7. Bogaerts, P., Hanus, R.: On-line state estimation of bioprocesses with full horizon observers. Mathematics and Computers in Simulation 56(4-5), 425–441 (2001)
8. Saucedo, V.M., Karim, M.N.: Analysis and comparison of input-output models in a recombinant fed-batch fermentation. Journal of Fermentation and Bioengineering 83(1), 70–78 (1997)
9. De Azevedo, S.F., Dahm, B., Oliveira, F.R.: Hybrid modelling of biochemical processes: A comparison with the conventional approach. Computers and Chemical Engineering 21(suppl. 1), 751–756 (1997)
10. Monod, J.: Recherches sur la croissance des cultures bactériennes [Studies on the growth of bacterial cultures]. Actualite's Scientifique et Industrielles 911, 1–215 (1942)
11. Menezes, J.C., Alves, S.S., Lemos, J.M., De Azevedo, S.F.: Mathematical modelling of industrial pilot-plant penicillin-G fed-batch fermentations. Journal of Chemical Technology and Biotechnology 61(2), 123–138 (1994)
12. Zadeh, L.A.: Fuzzy sets. Information and Control 8(3), 338–353 (1965)
13. Postlethwaite, B.E.: Fuzzy state estimator for fed-batch fermentation. Chemical Engineering Research & Design 67(3), 267–272 (1989)
14. Patnaik, P.R.: Artificial intelligence as a tool for automatic state estimation and control of bioreactors. Laboratory Robotics and Automation 9(6), 297–304 (1997)
15. Martin, E.B., Morris, A.J., Papazoglou, M.C.: Confidence Bounds for Multivariate Process Performance Monitoring Charts. In: IFAC Workshop on On-line Fault Detection and Supervision in the Chemical and Process Industries, Newcastle, UK (1996)
16. Geladi, P., Kowalski, B.R.: Partial Least Squares Regression: A Tutorial. Analytica Chimica Acta 185, 1–17 (1986)
17. Lennox, B., Hiden, H.G., Montague, G.A., Kornfeld, G., Goulding, P.R.: Application of multivariate statistical process control to batch operations. Computers and Chemical Engineering 24(2), 291–296 (2000)

18. Nomikos, P., MacGregor, J.F.: Monitoring batch processes using multiway principal component analysis. AIChE Journal 40(8), 1361–1373 (1994)
19. Nomikos, P., MacGregor, J.F.: Multiway partial least squares in monitoring batch processes. Chemometrics and Intelligent Laboratory Systems 30(1), 97–108 (1995)
20. Kulkarni, S.G., Chaudhary, A.K., Nandi, S., Tambe, S.S., Kulkarni, B.D.: Modeling and monitoring of batch processes using principal component analysis (PCA) assisted generalized regression neural networks (GRNN). Biochemical Engineering Journal 18(3), 193–210 (2004)
21. McKay, B., Sanderson, C.S., Willis, M.J., Barford, J.P., Barton, G.W.: Evolving a hybrid model of a fed-batch fermentation process. Transactions of the Institute of Measurement and Control 20(1), 4–10 (1998)
22. Duarte, B.P.M., Saraiva, P.M.: Hybrid models combining mechanistic models with adaptive regression splines and local stepwise regression. Industrial and Engineering Chemistry Research 42(1), 99–107 (2003)
23. Jin, X., Wang, S., Chu, J.: Hybrid modelling and Monitoring of Streptomycin Fermentation Process. In: 42nd IEEE Conference on Decision and Control, Maui, HI, United States, December 9-12, vol. 5, pp. 4765–4769 (2003)
24. Chen, L., Bastin, G., Van Breusegem, V.: Case study of adaptive nonlinear regulation of fed-batch biological reactors. Automatica 31(1), 55–65 (1995)
25. Guilandoust, M.T., Morris, A.J., Tham, M.T.: Adaptive estimation in inferential process control. In: Proceedings of the 1987 American Control Conference, Minneapolis, MN, USA, vol. 3, pp. 1743–1748 (1987)
26. Lant, P.A., Tham, M.T., Montague, G.A.: On the applicability of adaptive bioprocess state estimators. Biotechnology and Bioengineering 42(11), 1311–1321 (1993)
27. Bastin, G., Dochain, D.: Online Estimation and Adaptive Control of Bioreactors. Elsevier, Amsterdam (1990)
28. De Azevedo, S.F., Pimenta, P., Oliveira, F., Ferreira, E.: Studies on on-line state and parameter estimation through real-time process simulator. In: Proceedings of the 2nd IFAC Symposium on Modeling and Control of Biotechnical, Keystone, CO, USA, pp. 453–458 (1992)
29. Nejjari, F., Dahhou, B., Benhammou, A., Roux, G.: Non-linear multivariable adaptive control of an activated sludge wastewater treatment process. International Journal of Adaptive Control and Signal Processing 13(5), 347–365 (1999)
30. Boillereaux, L., Flaus, J.M.: New approach for designing model-based indirect sensors. IEEE Transactions on Control Systems Technology 8(4), 601–608 (2000)
31. Bogaerts, P., Vande Wouwer, A.: Software sensors for bioprocesses. ISA Transactions 42(4), 547–558 (2003)
32. Jazwinski, A.H.: Stochastic Process and Filtering Theory. Academic Press, New York (1970)
33. Bich Pham, H.T., Larsson, G., Enfors, S.-O.: Growth and energy metabolism in aerobic fed-batch cultures of Saccharomyces cerevisiae: Simulation and model verification. Biotechnology and Bioengineering 60(4), 474–482 (1998)
34. Oisiovici, R.M., Cruz, S.L.: State estimation of batch distillation columns using an extended Kalman filter. Chemical Engineering Science 55(20), 4667–4680 (2000)
35. Karim, M.N., Rivera, S.L.: Comparison of feedforward and recurrent neural networks for bioprocess state estimation. In: Computers & Chemical Engineering European Symposium on Computer Aided Process Engineering -ESCAPE-1, May 24-28, vol. 16 (suppl.), pp. 369–377 (1992)

36. Gomez Sanchez, E., Arauzo Bravo, M.J., Cano Izquierdo, J.M., Dimitriadis, Y.A., Lopez Coronado, J., Lopez Nieto, M.J.: Control of the penicillin production using fuzzy neural networks. In: 1999 IEEE International Conference on Systems, Man, and Cybernetics Human Communication and Cybernetics, Tokyo, Japan, October 12-15, vol. 6, pp. 446–450 (1999)
37. Haykin, S.: Neural Networks: A Comprehensive Foundation, 2nd edn. Prentice-Hall, Englewood Cliffs (1998)
38. Bishop, C.M.: Neural Networks for Pattern Recognition. Oxford University Press, Oxford (1996)
39. Warnes, M.R., Glassey, J., Montague, G.A., Kara, B.: Application of radial basis function and feedforward artificial neural network to the Escherichia coli fermentation process. Neurocomputing 20, 67–82 (1998)
40. Artificial Neural Network, Wikipedia, `http://en.wikipedia.org/wiki/Artificial_neural_network` (accessed October 10, 2008)
41. Neural Network Software Sensor, `http://imsb.au.dk/~mha/neural.htm` (accessed October 10, 2008)
42. Lennox, B., Montague, G.A., Hiden, H.G., Kornfeld, G., Goulding, P.R.: Process monitoring of an industrial fed-batch fermentation. Biotechnology and Bioengineering 74(2), 125–135 (2001)
43. Qin, S.J., McAvoy, T.J.: Nonlinear PLS modeling using neural networks. Computers & Chemical Engineering 16(4), 379–391 (1992)
44. Dong, D., McAvoy, T.J.: Nonlinear principal component analysis - based on principal curves and neural networks. In: Proceedings of the 1994 American Control Conference, Baltimore, MD, USA, June 29-July 1, vol. 2, pp. 1284–1288 (1994)
45. Park, S., Han, C.: Nonlinear soft sensor based on multivariate smoothing procedure for quality estimation in distillation columns. In: Computers and Chemical Engineering 7th International Symposium on Process Systems Engineering, July 16-21, vol. 24(2), pp. 871–877 (2000)
46. Gallagher, N.B., Wise, B.M., Stewart, C.W.: Application of multiway principal components analysis to nuclear waste storage tank monitoring. Computers and Chemical Engineering 20S, 739–744 (1996)
47. Lakshminarayanan, S., Gudi, R.D., Shah, S.L., Nandakumar, K.: Monitoring batch proceses using multivariate statistical tools: extensions and practical issues. In: Proceeding of IFAC World Congress, San Francisco, pp. 241–246 (1996)
48. McKay, B., Willis, M., Barton, G.: Steady-state modelling of chemical process systems using genetic programming. Computers & Chemical Engineering 21(9), 981–996 (1997)
49. Zhang, H., Vagapov, Y.: LS-SVM Based Software Sensor for Fed-batch Yeast Fermentation and Comparative Studies. In: IEEE International Conference on Electro and Information Technology, Michigan, USA, pp. 564–568 (May 2006)
50. Tham, M.T., Parr, A.: Succeed at on-line validation and reconstruction of data. Measurement and Control, 46–56 (May 1994)
51. Amos, C.K.: Soft sensors win hard jobs. ChemicalProcessing.com, `http://www.chemicalprocessing.com/articles/2005/606.html` (accessed October 18, 2008)
52. Escalante-Minakata, P., Ibarra-Junquera, V., Rosu, H.C., De Leon-Rodriguez, A., Gonzalez-Garcia, R.: Online monitoring of Mezcal fermentation based on redox potential measurements. Bioprocess and Biosystems Engineering 32(1), 47–52 (2009)

3

Monitoring of Bioprocesses: Mechanistic and Data-Driven Approaches

Laurent Dewasme[1], Philippe Bogaerts[2], and Alain Vande Wouwer[1]

[1] Faculté Polytechnique de Mons, Service d'Automatique, Boulevard Dolez 31, 7000 Mons, Belgium
{Laurent.Dewasme,Alain.Vandewouwer}@fpms.ac.be
http://www.autom.fpms.ac.be/

[2] Université Libre de Bruxelles, Service de Chimie Générale et Biosystèmes, C.P. 165/61
Av F.D. Roosevelt 50,
1050 Brussels, Belgium
Philippe.Bogaerts@ulb.ac.be
http://babylone.ulb.ac.be/

Abstract. Nowadays, bioprocesses play a key role in the production of high-added value products in the pharmaceutical industry and measurements of the main component concentrations are of great importance for monitoring cell cultures. Although some hardware sensors are readily available, they often have several drawbacks, including purchase and maintenance costs, sample destruction, discrete-time measurements (instead of continuous ones), processing delay, calibration, sterilization, disturbances in the hydrodynamic conditions inside the bioreactor, etc. It is therefore of interest to use software sensors which reconstruct on-line some component concentrations in continuous time. Software sensors are based on the theory of state estimation. In this chapter, some state estimation techniques are reviewed, and two important situations are distinguished: (a) some component concentrations can be measured and a dynamic model of the bioprocess can be established and (b) only basic operating signals, such as pH, base addition, stirrer speed, feed rates, can be measured on-line and it is difficult (or even impossible) to build a mechanistic model linking these variables. In the latter case, a neural network approach appears particularly suitable, and is largely illustrated in this chapter by real life experimental applications.

1 Introduction

Monitoring of bioprocesses is a particularly delicate task because on-line hardware sensors measuring the concentrations of the key culture components (biomass, substrates, products) are expensive (in terms of acquisition and maintenance costs), are not always available (for instance, it is difficult to measure the low concentration levels of substrate in yeast or bacteria cultures, or to measure some product concentration such as acetate), and have stringent operational constraints (sterilization, sampling, etc). Hence, it is of significant interest to develop software sensors, which make use of available information sources, e.g., a mathematical process model and available

M. do Carmo Nicoletti, L.C. Jain (Eds.): Comp. Intel. Tech. for Bio. Mod., SCI 218, pp. 57–97.
springerlink.com

on-line measurements. In this chapter, attention is focused on these soft sensor techniques, distinguishing mechanistic and data-driven approaches. On the one hand, the mechanistic approach - which will appear as an "expensive" but "comprehensive" approach - is based on the existence of a physical (biological) process model and some measurements of the main biological variables. Mechanistic models have a biological interpretation, which can be quite useful in understanding the process behaviour, and have a predictive capability which can compensate for the lack of frequent measurements of the biological variables. However this approach suffers from the difficulty and costs associated with the model derivation (which requires process knowledge, collection of experimental data, parameter estimation and model validation), and the costs associated with the sensors measuring the biological variables. On the other hand, the main advantage of data-driven techniques - which will appear as "cheap" but "limited" techniques - is their simplicity of implementation based on inexpensive, basic on-line measurement signals. However, data-driven representations usually lack biological interpretability and may have limited predictive capability (extrapolation in the absence of measurement information).

This chapter is organized as follows. The second section is devoted to bioprocess modelling following either a mechanistic or data-driven approach. In the third section, bioprocess models and sensors are exploited to design software sensors. The classical extended Kalman filter and asymptotic observer are first dealt with, followed by an introduction to partial least squares and neural network techniques. Simulation and real-life experimental data are used to assess the applicability and performance of the several methods.

2 A Brief Overview of Bioprocess Modelling

Any mathematical operation used to extract information from physical measurements can be viewed as a software sensor. State or parameter estimation is then conceivable once one can mathematically model the interactions within a system. The modelling objective is to establish (possibly simple) mathematical relations between explicative variables and explained ones. These relations are not exclusive. Different modelling approaches can be considered, depending on the process characteristics ((non)linearity, structural or parametric time variation, different time scales, stochastic nature of the measured signals), and the operating conditions (batch or continuous, repetitive, open-loop or closed-loop, small or large scale).

In the following, two approaches are distinguished, i.e., the mechanistic approach which makes use of the available a priori knowledge about the process and experimental data to derive a physically (biologically) inspired model, and the data-driven approach which establishes a black-box representation of the process using the observation of input-output data.

2.1 Mechanistic Approaches versus Black-Box Approaches

The mechanistic approach is the most popular bioprocess modelling technique. For optimization, monitoring and control purposes, this approach is usually macroscopic in essence (as opposed to microscopic approaches, which are based on a detailed analysis of the cell metabolic network), i.e., it makes use of the concept of a macroscopic reaction scheme involving a few reactants, products and catalysts considered as macroscopic entities. These reactions are represented as follows [1]:

$$\sum_{i\in R_r} k_{i,r}\xi_i \xrightarrow{\varphi_r} \sum_{j\in P_r} k_{j,r}\xi_j \qquad r\in\{1,\ldots,n_\varphi\}\,. \tag{1}$$

where

- $k_{i,r}$ and $k_{j,r}$ are the pseudo-stoichiometric coefficients or yield coefficients. They are negative when they relate to a reactant and positive when they relate to a product.
- φ_r is the reaction rate of reaction r.
- ξ_i is the ith macroscopic component (N macroscopic components are considered).
- $R_r(P_r)$ is the set of reactants (products) in reaction r.
- n_φ is the number of reactions.

It is important to note that these macro-reactions do not satisfy elementary mass balances, hence the name pseudo-stoichiometric coefficients for $k_{i,r}$ and $k_{j,r}$.

Using macroscopic mass balances, a general differential state-space model can be obtained as follows:

$$\frac{d\xi(t)}{dt} = K\varphi(\xi,t) - D(t)\xi(t) + F(t) - Q(t)\ . \tag{2}$$

where

- ξ is the vector of concentrations of the macroscopic components.
- K is the pseudo-stoichiometric matrix.
- φ is the vector of reaction rates.
- D is the dilution rate ($D(t) = \dfrac{F_{in}(t)}{V(t)}$ where F_{in} is the inlet flow rate and V the medium volume).
- F is the vector of feed rate of selected components ($F_j = D(t)\xi_{in,j}(t)$ if the component is diluted in the feed stream, or $F_j = F_j(t)$ if the component is introduced in the culture in gaseous form) and Q is the vector of outflow rates of the considered components in gaseous form.

Establishing this kind of dynamic model requires the determination of a reaction scheme, the selection of an appropriate kinetic model structure and a parameter identification procedure. The quality of the resulting model will of course depend on the information content of the experimental data at hand, as well as on an appropriate choice of the model structure. For these reasons, bioprocess models are usually uncertain (structural and parametric uncertainties), and while they offer an interesting insight into the system under consideration, they do not always predict the behaviour of the real system in a wide range of operating conditions.

An appealing alternative to bypass the tedious identification steps required by the mechanistic approach is the data-driven, black-box approach which considers the system only through the evolution of its inputs and outputs.

A significant advantage of black-box models are their ability to exploit information contained in every available measurement. While the mechanistic approach requires off-line measurements of the component concentrations for model identification and on-line measurements of some of these components for state estimation, data-driven techniques allow information to be retrieved from signals as diverse as pH, dissolved oxygen, pressure, stirrer speed (signals which would be quite difficult to include in a reasonably complex mechanistic representation of the process).

A classical example of black-box model is the function generated by an *Artificial Neural Network* (*ANN*), which defines a static map, model or function of the inputs. Different *ANN* architectures can be selected, differing in the number of layers and the type of nonlinear activation functions (sigmoids, Gaussian, etc). More details on these alternative structures will be given in the following.

2.2 Observability

Observability is a structural system property that relates to the possibility of estimating the system state on the basis of the available measurement information. If the system under consideration is nonlinear (which is almost always the case in bioprocess applications), observability depends on the system inputs and can be defined as:

Definition. A system is observable if

$$\begin{aligned} &\forall t_0,\ \exists t_1 < \infty \ \mid \ y(t;t_0,\xi(0),u(t)) = y(t;t_0,\xi'(0),u(t)), \\ &\forall u(t), t_0 < t < t_1 \Rightarrow \xi(0) = \xi'(0)\ . \end{aligned} \tag{3}$$

In other words, a system is observable if two identical output trajectories y (function of time t, the initial states $\xi(0)$ and of the input command $u(t)$) over a given finite time horizon implies the equality of the initial states $\xi(0)$ and $\xi'(0)$.

Starting from (2), one defines a general state observer as follows:

$$\frac{d\xi(t)}{dt} = K\varphi(\xi,t) - D(t)\xi(t) + F - Q + \Omega(\hat{\xi})\left[\xi_1 - \hat{\xi}_1\right]. \tag{4a}$$

$$y = L\xi = \xi_1\ . \tag{4b}$$

where $\hat{\xi}$ is the on-line estimate of ξ, $\Omega(\hat{\xi})$ is the gain matrix depending on $\hat{\xi}$, L is an elementary matrix that selects the measured components of ξ (ξ_1 represents the vector of measured components).

The only difference between (2) and (4a) lies in the additional term proportional to the observation error of the measured part of the state $(\xi_1 - \hat{\xi}_1)$. (4a) becomes identical to (2) when the estimation is perfect.

It is now up to the user to find acceptable designs of $\Omega(\hat{\xi})$ such that a fast decay of the observation error is achieved. Omitting the dependence on time t for the sake of clarity, the observation error dynamics corresponds to:

$$\frac{d\left[\xi_1 - \hat{\xi}\right]}{dt} = K\left[\varphi(\hat{\xi} + e) - \varphi(\xi)\right] - D\left[\xi_1 - \hat{\xi}_1\right] - \Omega(\hat{\xi})L\left[\xi_1 - \hat{\xi}_1\right]. \tag{5}$$

where $\left[\xi_1 - \hat{\xi}_1\right] = 0$ appears as an equilibrium point.

A practical way to design $\Omega(\hat{\xi})$ is to consider the linearized approximation of model (5) as follows:

$$\frac{d\left[\xi_1 - \hat{\xi}_1\right]}{dt} = \left[A(\hat{\xi}) - \Omega(\hat{\xi})L\right]\left[\xi_1 - \hat{\xi}_1\right]. \tag{6}$$

with:

$$A(\hat{\xi}) = K\left[\frac{\partial \varphi(\xi)}{\partial \xi}\right]_{\xi = \hat{\xi}} - DI_N . \tag{7}$$

where I_N is the square identity matrix of dimension NxN.

The problem is then reduced to an arbitrary choice of the eigenvalues of the matrix $\left[A(\hat{\xi}) - \Omega(\hat{\xi})L\right]$ fixing the rate of convergence of the observation error to zero. The convergence is ensured if [2]:

- $\left[A(\hat{\xi}) - \Omega(\hat{\xi})L\right]$ is continuously differentiable,
- $\left[A(\hat{\xi}) - \Omega(\hat{\xi})L\right]$ is bounded,
- The real parts of all the eigenvalues of $\left[A(\hat{\xi}) - \Omega(\hat{\xi})L\right]$ are negative.

The solution of (6) takes then an exponential form so that the convergence is said to be exponential. From this, it follows that the system is exponentially observable and (4a) is an exponential observer.

Sometimes, the design of (6) does not allow a free assignment of the eigenvalues (i.e., a free design of the dynamics) but leads to a system converging asymptotically to the equilibrium point $\left[\xi_1 - \hat{\xi}_1\right] = 0$. Actually, this means that the system (6) is still converging but not over a finite time. Such observers are called asymptotic observers.

3 Bioprocess State Estimation

In the next sections, a few popular state estimation techniques are presented (for more details and techniques see [1], [3] and [4]). Their application to a mechanistic model of culture of micro-organisms with an overflow metabolism (e.g. yeasts or bacteria cultures) is investigated in order to discuss the advantages and also the possible limitations of general mechanistic model-based observers.

3.1 Classical State Estimation Method

3.1.1 An Example of Exponential Observer: The Kalman Filter

The Kalman filter [5], which is by far the most popular state estimation technique used for bioprocess monitoring, is an exponential observer that minimizes the variance of the estimation error. It was first developed in a linear framework before being extended to the nonlinear case.

Consider the following nonlinear system derived from (2):

$$\frac{d\xi_1}{dt} = K_1\varphi(\xi) - D\xi_1 + D\xi_1^{in} + \eta_1(t); \qquad \xi_1(0) = \xi_{1,0} \ . \tag{8a}$$

$$\frac{d\xi_2}{dt} = K_2\varphi(\xi) - D\xi_2 + D\xi_2^{in} + \eta_2(t); \qquad \xi_2(0) = \xi_{2,0} \ . \tag{8b}$$

$$y = L\xi + \varepsilon(t) = \xi_1 + \varepsilon(t) \ . \tag{8c}$$

where ξ_1 represents the vector of measured states while ξ_2 represents the unmeasured states that complete the partition of ξ. $\varepsilon(t)$ is the measurement noise, whereas $\eta(t) = [\eta_1(t) \ \ \eta_2(t)]^T$ is the model noise (both assumed white and normally distributed with zero mean).

Assuming that this system is exponentially observable, the *Extended Kalman Filter (EKF)* is based on a first-order linearization of the process model along the estimated trajectory

$$\frac{d\hat{\xi}}{dt} = K\varphi(\hat{\xi}) - D\hat{\xi} + D\xi^{in} + \Omega(\hat{\xi})(y - L\hat{\xi}); \qquad \hat{\xi}(0) = \hat{\xi}_0 \ . \tag{9a}$$

$$\frac{dP}{dt} = A(\hat{\xi})P + PA(\hat{\xi})^T - PL^T R_\varepsilon^{-1} LP + R_\eta; \qquad P(0) = P_0 \ . \tag{9b}$$

$$\Omega(\hat{\xi}) = PL^T R_\varepsilon^{-1} \ . \tag{9c}$$

where Ω is the observer error gain, P is the solution of the *Riccati equation* (9b), R_η is the covariance matrix of the state (or model) noise, R_ε is the covariance matrix of the measurement noise and A is the linear transition matrix computed along the estimated trajectory.

The previous formulation (9) provides time-continuous estimations from time-continuous measurements. In practice however, concentrations of the main species are generally measured at discrete times only and with relatively low sampling frequencies (the measurements are sometimes collected at different rates, i.e., resulting in an asynchronous measurement configuration). The corresponding formulation is the *continuous-discrete EKF* where continuous estimations are provided from discrete measurements

$$y(t_k) = L\xi(t_k) + \varepsilon(t_k) \ . \tag{10}$$

where $\varepsilon(t_k)$, the measurement noise at time t_k is a normally distributed white noise with zero mean and covariance matrix $R_\varepsilon(t_k)$.

The algorithm now proceeds in two steps: a prediction step (corresponding to the time period between two measurement times) and a correction step occurring each time a new measurement is available.

The first step (prediction between t_k and t_{k+1}) corresponds to:

$$\frac{d\xi(t)}{dt} = K\varphi(\xi(t),t) - D\xi(t) + D\xi^{in}; \quad \xi(t_k) = \xi(t_k^+), \ t_k^+ \le t < t_{k+1}^- \ . \tag{11a}$$

$$\frac{dP(t)}{dt} = A(\xi(t))P(t) + P(t)A(\xi(t))^T + R_\eta \quad P(t_k) = P(t_k^+), \ t_k^+ \le t < t_{k+1}^- . \tag{11b}$$

The correction step occurring at time t_{k+1} corresponds to:

$$\Omega(\xi(t_{k+1})) = P(t_{k+1}^-)L^T\left[LP(t_{k+1}^-)L^T + R_\varepsilon(t_{k+1})\right]^{-1} \ . \tag{11c}$$

$$\xi(t_{k+1}^+) = \xi(t_{k+1}^-) + \Omega(\xi(t_{k+1}))(y(t_{k+1}) - L\xi(t_{k+1}^-)) \ . \tag{11d}$$

$$P(t_{k+1}^+) = P(t_{k+1}^-) - \Omega(\xi(t_{k+1}))LP(t_{k+1}^-) \ . \tag{11e}$$

with t_{k+1}^- and t_{k+1}^+ characterizing respectively the values before and after correction.

The Kalman filter is an optimal exponential observer in the sense that it minimizes asymptotically the mean square estimation error. Unfortunately, the extension to nonlinear systems is realized through a linearization along the estimated trajectory. The *EKF* is no longer optimal and can sometimes lead to biased estimates or, in the worst case of bad initial conditions, to convergence problems.

The Kalman filter, as well as most of the exponential observers, relies on the availability of an accurate process model. This prerequisite is far from being satisfied in many bioprocess applications, where the models are usually uncertain. It is therefore of interest to look for more robust alternatives, which justifies the following approach.

3.1.2 Asymptotic Observer

Even if uncertainties are always hanging over all the parameters of a bioprocess model, their impact is more concentrated on the kinetic terms which are generally badly known. In the context of bioprocess applications, the observers that allow to asymptotically reconstruct unknown states when the kinetics are unknown are called *asymptotic observers*. The asymptotic convergence requires that the yield coefficients are known and the number of measured state variables is equal to or larger than the rank p of the yield matrix K [1].

A partition of the yield matrix (K_a, K_b) is first selected so that the (pxM) submatrix K_a is a full row rank submatrix (i.e., so that the rank of K_a is equal to p). Then, there exists a state transformation

$$z = C\xi_a + \xi_b \tag{12}$$

where the (N-p)xp matrix C is the unique solution of

$$CK_a + K_b = 0 \tag{13}$$

so that

$$\frac{dz}{dt} = -Dz + C(F_a - Q_a) + (F_b - Q_b) \ . \tag{14}$$

Under these conditions, the dynamics of z is independant of the kinetics.

If a partition induced by the measured and unmeasured states is now considered, i.e., (K_1, K_2) corresponding to (ξ_1, ξ_2), the auxiliary vector z can be defined accordingly

$$z = A_1\xi_1 + A_2\xi_2 \tag{15}$$

with appropriate definitions of the (N-p)xq matrix A_1 and (N-p)x(N-q) matrix A_2.

In this latter case, the asymptotic observer structure is as follows:

$$\frac{d\hat{z}}{dt} = -D\hat{z} + C(F_a - Q_a) + (F_b - Q_b) \tag{16}$$

$$\hat{\xi}_2 = A_2^+(\hat{z} - A_1\xi_1) \tag{17}$$

where A_2^+ is a left pseudo inverse of A_2.

The convergence of such an observer is unfortunately ensured only if the dilution rate is different from zero. This is confirmed by the dynamics of the estimation error:

$$\frac{d\left(\xi_2 - \hat{\xi}_2\right)}{dt} = -D\left(\xi_2 - \hat{\xi}_2\right) . \tag{18}$$

Hence, in a bioprocess monitoring application, the convergence is ensured only if the culture is operated in fed-batch or continuous conditions while, for batch processes, any initial error will last over the batch duration. The main drawback of this class of observer is therefore that the convergence is completely determined by the operating conditions.

3.1.3 Application of Observers to an Identified Mechanistic Model of *E. coli*

E. coli is a very important bacteria host strain abundantly used in biopharmaceutical industries for different purposes like vaccine production through biomass growth [6].

This strain presents a catabolism that can be macroscopically described by the following three main reactions:

Substrate oxidation: $$k_1 S + k_5 O \xrightarrow{r_1 X} X + k_7 P \tag{19a}$$

Substrate fermentation: $$k_2 S \xrightarrow{r_2 X} X + k_3 A + k_8 P \tag{19b}$$

Acetate oxidation: $$k_4 A + k_6 O \xrightarrow{r_3 X} X + k_9 P \tag{19c}$$

where X, S, A, O and P are, respectively, the concentration in the culture medium of biomass, substrate (typically glucose), acetate, dissolved oxygen and carbon dioxide. k_i are the yield coefficients and r_1, r_2 and r_3 are the nonlinear specific growth rates (which, once multiplied by the biomass, are the reaction rates $r_1 X$, $r_2 X$ and $r_3 X$) given by:

$$r_1 = \min(r_S, \frac{r_o}{k_5}) \tag{20a}$$

$$r_2 = \max(0, r_S - \frac{r_o}{k_5}) \tag{20b}$$

$$r_3 = \max(0, \frac{r_o - k_5 r_S}{k_6} \cdot \frac{A}{A + K_A})) \tag{20c}$$

where the kinetic terms associated with the substrate consumption r_S, the oxidative or respiratory capacity r_o and the acetate oxidative rate r_A are given by $r_S = \mu_S \frac{S}{S + K_S}$, $r_o = \mu_o \frac{O}{O + K_O} \frac{K_{io}}{K_{io} + O}$ and $r_A = \mu_A \frac{A}{A + K_A}$, respectively. These expressions take the classical form of Monod laws where μ_S, μ_o and μ_A are the maximal values of specific growth rates and K_S, K_O and K_A are the saturation constants of the corresponding element.

This kinetic model, which is often encountered in the literature, is based on Sonnleitner's bottleneck assumption [7] (see Fig. 1). During a culture, the cells are likely to change their metabolism because of their limited respiratory capacity. When the substrate is in excess (concentration $S > S_{crit}$), the cells produce a metabolic by-product (acetate for bacteria) through fermentation, and the culture is said in respiro-fermentative (RF) regime. On the other hand, when the substrate becomes limiting (concentration $S < S_{crit}$), the available substrate (typically glucose), and possibly the metabolic by-product (as a substitute carbon source), if present in the culture medium, are oxidized. The culture is then said in respirative (R) regime.

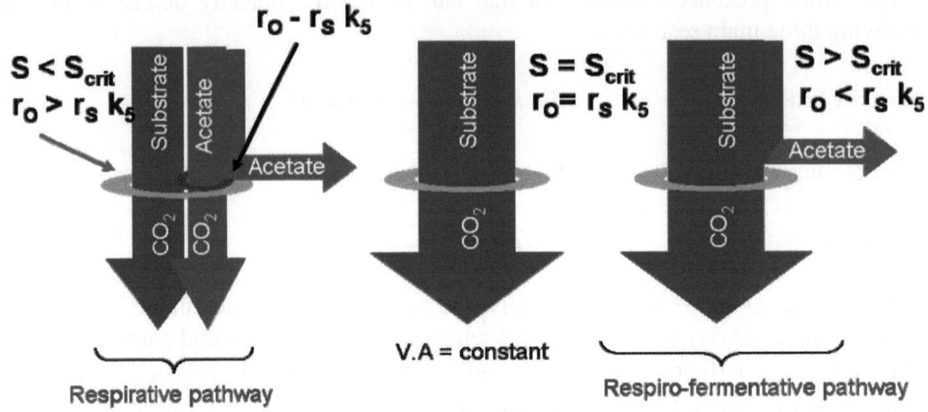

Fig. 1. Illustration of Sonnleitner's bottleneck assumption [7], considering a possible analogy with bacteria limited respiratory capacity

Component-wise mass balances give the following differential equations :

$$\frac{dX}{dt} = (r_1 + r_2 + r_3)X - DX \tag{21a}$$

$$\frac{dS}{dt} = -(k_1 r_1 + k_2 r_2)X - D(S - S_{in}) \tag{21b}$$

$$\frac{dA}{dt} = (k_3 r_2 - k_4 r_3)X - DA \tag{21c}$$

$$\frac{dO}{dt} = -(k_5 r_1 + k_6 r_3)X - DO + OTR \tag{21d}$$

$$\frac{dP}{dt} = (k_7 r_1 + k_8 r_2 + k_9 r_3)X - DP - CTR \tag{21e}$$

$$\frac{dV}{dt} = F_{in} \tag{21f}$$

where S_{in} is the substrate concentration in the feed, D is the dilution rate ($D=F_{in}/V$), F_{in} is the inlet feed rate and V is the culture medium volume. OTR and CTR represent respectively the oxygen transfer rate from the gas phase to the liquid phase and the carbon transfer rate from the liquid phase to the gas phase. Classical models of OTR and CTR are given by:

$$OTR = k_L a_O (O_{sat} - O) \tag{22a}$$

$$CTR = k_L a_P (P - P_{sat}) \tag{22b}$$

where $k_L a_O$ and $k_L a_P$ are respectively the volumetric transfer coefficients of O and P and, O_{sat} and P_{sat} are respectively the dissolved oxygen and carbon dioxide concentrations at saturation.

This model is now used in an illustrative example based on simulated data. First an experimental field (a set of experiments) is defined in order to estimate the model parameters, and assess model accuracy. Rigorously, the choice of this experimental field should rely on optimal experiment design as developed in [8] and [9]. This theory is however not always applicable in practical situations, as it requires input signals that are difficult to achieve in practice (very high or very low concentrations, for instance). Here, operating conditions are chosen based on process knowledge, so as to highlight the influence of the several parameters.

The identification procedure is subdivided into two steps, the first one being devoted to the pseudo-stoichiometry, and the second one to the kinetics, through a decoupling technique based on a state transformation (analog to (12) used in the asymptotic observer; see [1] for more details). For being applicable, this procedure requires that the reaction scheme is C-identifiable (the name of this property comes from matrix C in (12)), which is not the case of the reaction scheme (19). To alleviate this problem, it is required to introduce three additional constraints between the stoichiometric coefficients (one by reaction, which are indeed easy to determine as the ratio of consumed and produced moles of certain components like oxygen-carbon (19a), acetate-carbon (19b) and again oxygen-carbon (19c), are known).

In the first step, the experimental field consists of three simulations. The first one forces the cells to evolve, most of the time, through the respirative pathway, the second one does as well through the respiro-fermentative pathway and the third one reproduces operating conditions in the neighbourhood of the optimum (see Fig. 2, 3 and 4). It is important to notice that noise is added to the samples in order to provide realistic measurement conditions. The sampling period is chosen equal to 1 hour.

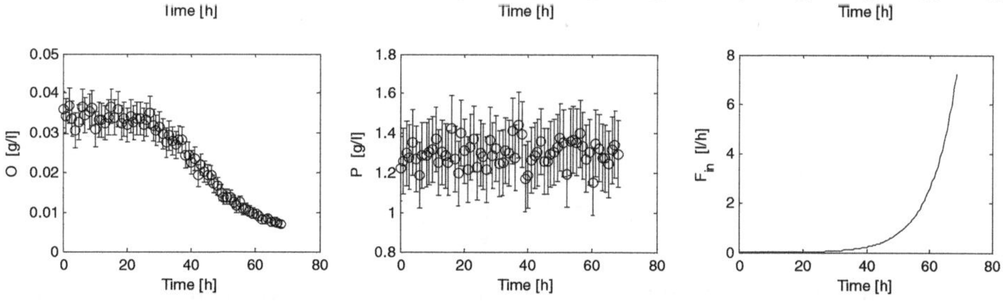

Fig. 2. First simulated experiment – A batch phase is achieved until S, the substrate concentration (initially chosen around 0.1 g/l), decreases to 0.01 g/l. Then, a fed-batch phase starts and the feed flow rate is calculated so that S is maintained at 0.01 g/l. The culture ends when the feed medium (its capacity is estimated to 15kg) is exhausted. The cells evolve most of the time in the respirative regime.

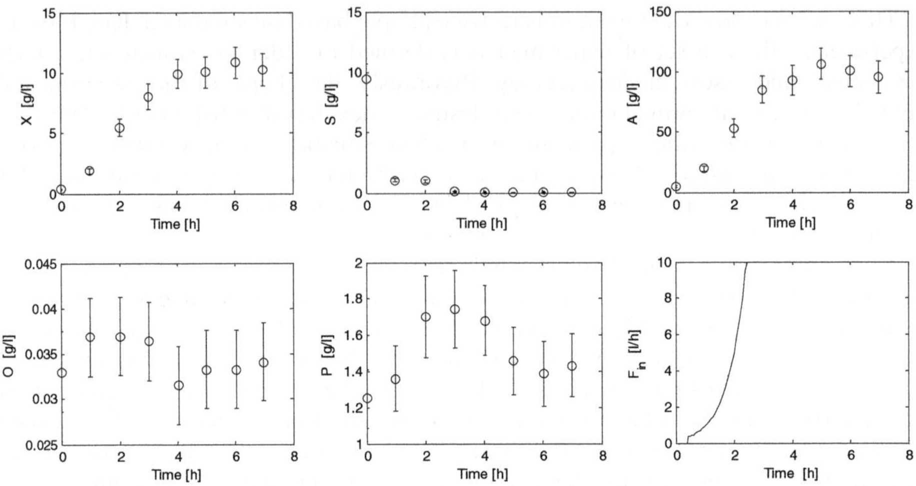

Fig. 3. Second simulated experiment – A batch phase is achieved until S, initialized around 10 g/l, decreases to 1 g/l. Then, a fed-batch phase starts and the feed flow rate is calculated so that S is maintained at 1 g/l. The culture ends when the feed medium (its capacity is estimated to 15kg) is exhausted. Note that the feed flow rate saturates at 10 l/h. This value is considered as the maximum pump speed. After 3 hours, the feed flow rate saturates and the injected quantity of substrate becomes insufficient to maintain S at 1 g/l. However, the accumulation of acetate indicates that the cells evolve through the respiro-fermentative pathway.

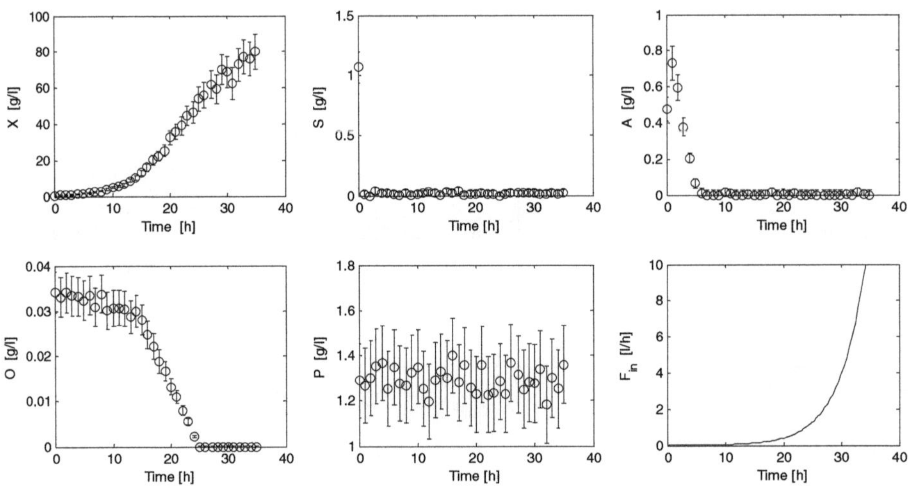

Fig. 4. Third simulated experiment – A batch phase is achieved until S, the substrate concentration (initially chosen around 1 g/l), decreases to 0.02 g/l. Then, a fed-batch phase starts and the feed flow rate is calculated so that S is maintained at 0.02 g/l. The culture ends when the feed medium (its capacity is estimated to 15kg) is exhausted. Once again, the feed flow rate saturates after 34 hours. Nevertheless, the cells evolve in the neighbourhood of the optimal operating conditions. Note how fast the cells grow as compared to the previous experiments (the same concentration is reached within 35 hours instead of more than 60).

Table 1. Yield coefficients values of a mechanistic model of *E. coli*

Yield coefficient	Exact value	Identified value
k_1	3.164	3.211
k_2	25.22	23.911
k_3	10.9	10.247
k_4	6.382	6.35
k_5	1.074	1.088
k_6	6.098	6.207
k_7	1.283	1.3
k_8	19.1	17.872
k_9	6.576	6.694

Table 1 gives values of the yield coefficients found in the literature (considered here as "theoretical" values) and the result of the first identification step based on the previously-defined experimental field. This procedure consists in an identification of the stoichiometric coefficients independently of the kinetics [1].

The maximum deviation from a reference value is *6.9%* and the mean value of the error is *3.2%*. The results of the identification of the yield coefficients can therefore be considered as quite good.

Based on these results, the second identification step is carried out for the kinetic parameters (see Fig. 5). Instead of using the data coming from the previous experiments, a fourth one is achieved in order to put in evidence the influence of some kinetic parameters. For instance, an acetate concentration evolving between 0 and 3 g/l provides a better identification of K_{io} and K_a.

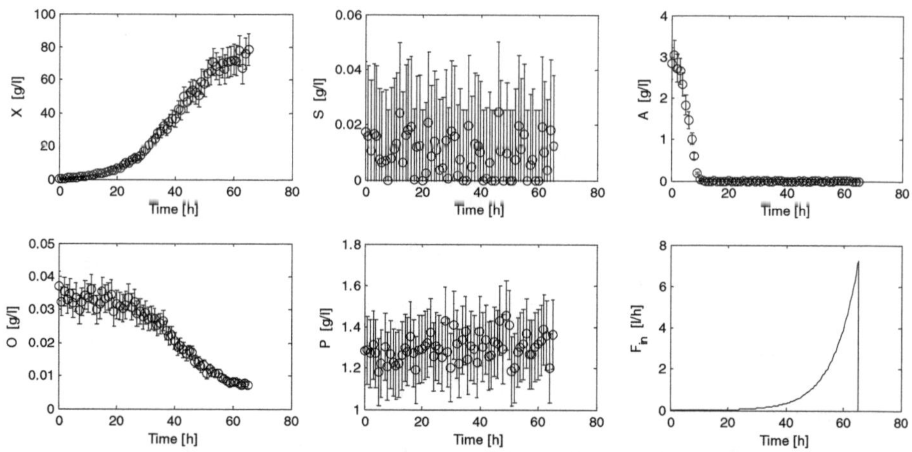

Fig. 5. Fourth simulated experiment – A fed-batch phase is directly started and the feed flow rate is calculated so that S is maintained at 0.01 g/l (i.e., as low as possible). The culture ends when the feed medium (its capacity is estimated to 15kg) is exhausted or if the sum of the reaction rates is close to zero. The cells evolve most of the time in the respirative regime.

Table 2. Kinetic coefficients values of a mechanistic model of *E. coli*

Kinetic coefficient	Exact value	Initial value	Identified value
K_s	0.14	0.1	0.11
μ_s	1.832	1	1.54
μ_o	0.72	1	0.68
K_{io}	7	10	7.66
K_a	0.5	1	0.48

The kinetic values are listed in Table 2. In contrast with the previous step, the minimization of a cost function measuring the deviation between the model prediction and the measured outputs is a nonconvex problem, which usually requires some form of multistart strategy and the selection of an appropriate local (or global) optimizer. Here, for the sake of simplicity, only one starting point is considered, which is chosen far enough from the exact values but allows a satisfactory parameter estimation.

These latter results show that the identification of the kinetic coefficients is more delicate and that experimental data used for identification must be chosen with care, whenever possible. The identified parameters present significant errors from *5* to more than *20%*.

Based on the identified model and a few on-line measurements, an *EKF* can, in principle, be designed. Fig. 6 shows a simulation of the continuous-discrete *EKF* applied to the full model (21). The case where *S, O and P* are measured while *X* and *A*

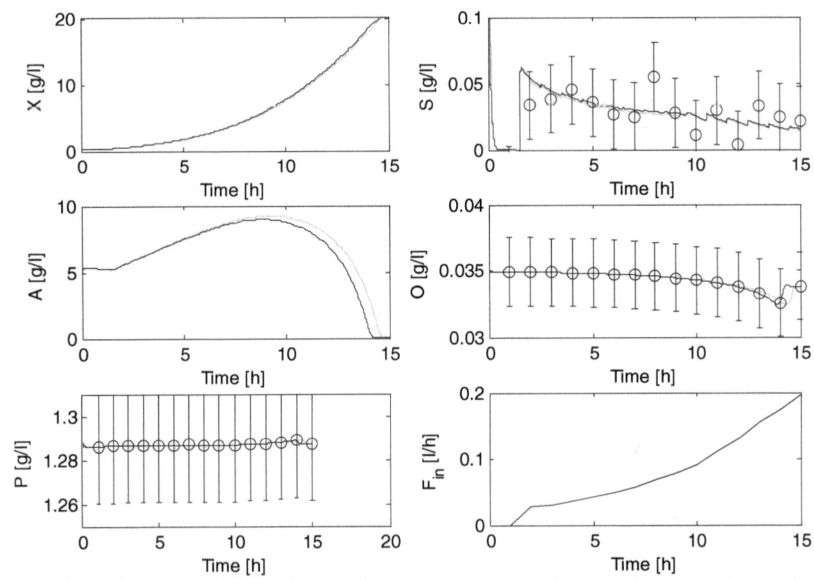

Fig. 6. Application of the EKF to the *E. Coli* model (21) using the parameter identification results – black line: exact simulated model; red-line: EKF estimations; circles: simulated measurements (the sampling period is 1 hour)

are estimated is considered. These results correspond to the ideal case where the initial conditions are perfectly known. However, after 10 hours, the Kalman observer trajectory diverges from the exact simulated model demonstrating a lack of robustness to modelling errors.

In a more realistic situation, errors on the initial states have to be taken into account. Fig. 7 shows the results of the application of the *EKF* to model (21) using an exact predictive model but starting from erroneous initial states. The *EKF* provides very bad estimates and is obviously unable to provide enough correction from the on-line measurements. This latter observation is to be related with observability problems.

A closer look at observability is provided by canonical observability forms as introduced in [10] and [11]:

$$\forall i \in \{1,\ldots,q\},\ x_i \in \Re^{n_i},\ n_1 \geq n_2 \geq \ldots \geq n_q,$$

$$\sum_{1 \leq i \leq q} n_i = n$$

$$\dot{x} = \begin{cases} \dot{x}_1 = f_1(x_1, x_2) \\ \dot{x}_2 = f_2(x_1, x_2, x_3) \\ . \\ . \\ . \\ \dot{x}_{q-1} = f_{q-1}(x_1, \ldots, x_q) \\ \dot{x}_q = f_q(x_1, \ldots, x_q) \end{cases} \qquad (23)$$

$$y = x_1$$

where x is the state vector, y the vector of measured states, f_i a partition of the nonlinear state equations, q the number of partitions.

To assess if the system is observable, the bioprocess model (2) – or (21) in the particular case of bacteria culture – must be put in the form of (23) by defining an appropriate partition, and the following condition must be checked:

$$rank \frac{\partial f_i}{\partial x_{i+1}} = n_{i+1} \qquad \forall i \in \{1,\ldots,q-1\} \qquad (24)$$

In the particular case (21), the partition $x_1 = [S, O, P]$ and $x_2 = [X, A]$, is considered, and therefore:

$$\frac{\partial f_1}{\partial x_2} = \begin{bmatrix} -k_1 r_1 - k_2 r_2 & -X\left(k_1 \frac{\partial r_1}{\partial A} + k_2 \frac{\partial r_2}{\partial A}\right) \\ -k_5 r_1 - k_6 r_3 & -X\left(k_5 \frac{\partial r_1}{\partial A} + k_6 \frac{\partial r_3}{\partial A}\right) \\ k_7 r_1 + k_8 r_2 + k_9 r_3 & X\left(k_7 \frac{\partial r_1}{\partial A} + k_8 \frac{\partial r_2}{\partial A} + k_9 \frac{\partial r_3}{\partial A}\right) \end{bmatrix} . \quad (25)$$

$$rank \frac{\partial f_1}{\partial x_2} = n_2 = 2 . \quad (26)$$

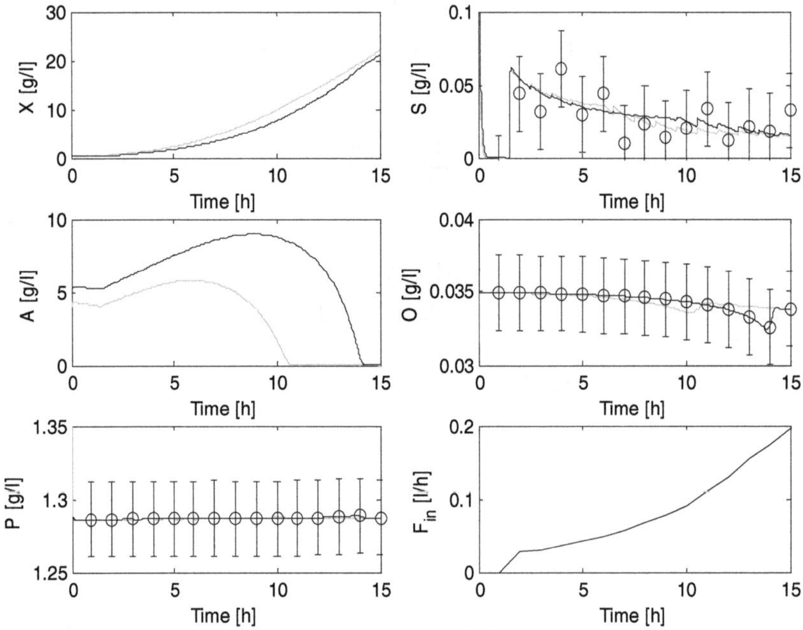

Fig. 7. Application of the EKF to the *E. Coli* model (21) with bad initial estimations – black line: exact simulated model; red-line: EKF estimations; circles: simulated measurements (the sampling period is 1 hour)

Obviously, condition (24) only holds if S, A and X are different from 0. Even if the acetate concentration is vanishing after 15 hours (see Fig. 6 and 7), global observability is thus verified between 0 and 15 hours. It can be shown that even if this test is successful, it only implies *theoretical observability*. To assess practical observability, Bogaerts and Vande Wouwer (2004) [12] have proposed a measure of observability based on square matrices M_i:

$$M_i(\xi, D) = \left(\frac{\partial f_i}{\partial \xi_{i+1}}\right)^T \left(\frac{\partial f_i}{\partial \xi_{i+1}}\right) \tag{27}$$

with $i=1,\ldots,q-1$.

$$F_{obs} = \sum_{k=1}^{N} \sum_{i=1}^{q-1} \sqrt{cond\left(M_i(\xi(t_k), D(t_k))\right)} \tag{28}$$

where "cond" represents the condition number of the matrix, i.e., the ratio of its largest to its smallest eigenvalue and t_k are the discrete measurement times.

Fig. 8 shows the evolution of this criterion in the same situation as Fig. 7. The initial value is very high ($O(10^8)$) and increases gradually with time inducing that the M_i matrices are ill-conditioned and the *EKF* is unable to provide enough correction to the non measured states using information from the measured ones (which is related to the lack of sensitivity quantified by the M_i matrices in (28)).

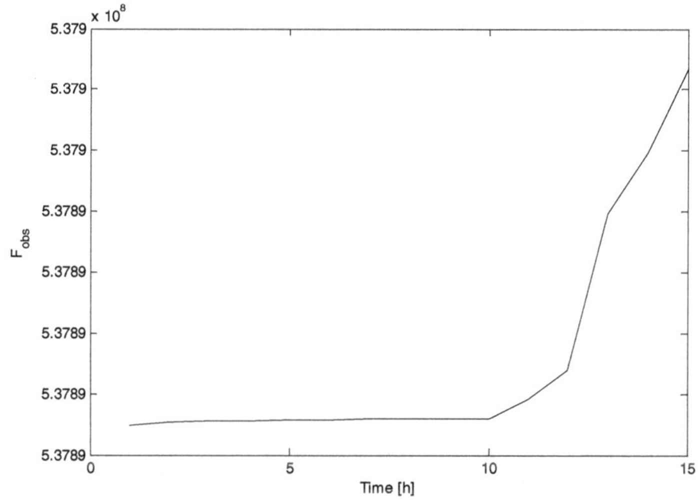

Fig. 8. Application of the EKF to the *E. Coli* model (21) with bad initial estimations – evolution of the "observability measure" (the sampling period is 1 hour).

An asymptotic observer is now designed in order to eliminate the uncertain kinetic model and to estimate in a more robust way the biomass and acetate concentrations. The states and yield matrix partitions are:

$$\xi_1 = \begin{bmatrix} S \\ O \\ P \end{bmatrix} \qquad \text{and} \qquad K_1 = \begin{bmatrix} -k_1 & -k_2 & 0 \\ -k_5 & 0 & -k_6 \\ k_7 & k_8 & k_9 \end{bmatrix} . \tag{29a}$$

$$\xi_2 = \begin{bmatrix} X \\ A \end{bmatrix} \quad \text{and} \quad K_2 = \begin{bmatrix} 1 & 1 & 1 \\ 0 & k_3 & -k_4 \end{bmatrix}. \tag{29b}$$

and the observer equations are given by:

$$\begin{bmatrix} \dot{z}_1 \\ \dot{z}_2 \end{bmatrix} = -D \begin{bmatrix} z_1 \\ z_2 \end{bmatrix} + C \begin{bmatrix} DS_{in} \\ OTR \\ -CTR \end{bmatrix}. \tag{30a}$$

$$\begin{bmatrix} \hat{X} \\ \hat{A} \end{bmatrix} = \begin{bmatrix} z_1 \\ z_2 \end{bmatrix} - C \begin{bmatrix} S \\ O \\ P \end{bmatrix}. \tag{30b}$$

Fig. 9 shows the results of the application of the asymptotic observer, as defined in (30a) and (30b), using the same initial errors on X and A as in Fig. 7. The estimates seem to slowly converge to the right trajectories. Nevertheless, Fig. 10 also shows that the evolution of the estimation error on X is not vanishing.

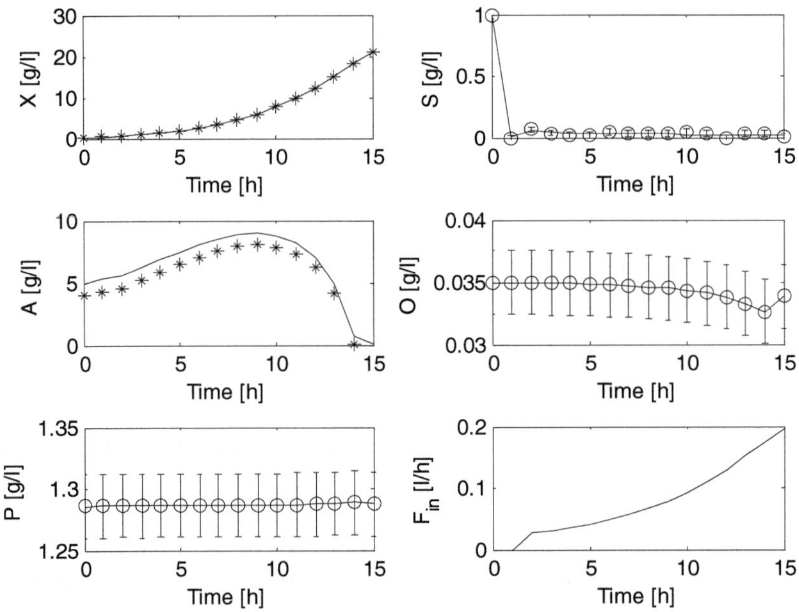

Fig. 9. Application of the asymptotic observer to the *E. Coli* model (21) with bad initial estimations – black line: exact simulated model; red stars: asymptotic observer estimations; circles: simulated measurements (the sampling period is 1 hour)

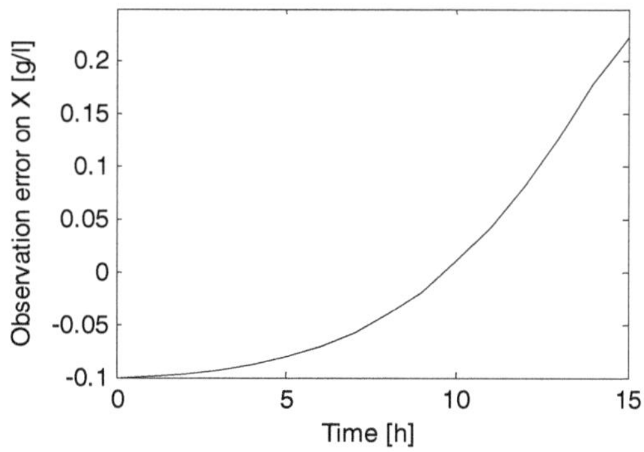

Fig. 10. Application of the asymptotic observer to the *E. Coli* model (21) with bad initial estimations – Evolution of the observation error on X

To stress the fact that the latter results are not an isolated case, we now consider the observation of the substrate and acetate concentrations on the basis of biomass, oxygen and carbon dioxyde measurements, which will lead to even more severe state estimation problems, as will be clear in the following. The states and yield matrix partitions are now:

$$\xi_1 = \begin{bmatrix} X \\ O \\ P \end{bmatrix} \quad \text{and} \quad K_1 = \begin{bmatrix} 1 & 1 & 1 \\ -k_5 & 0 & -k_6 \\ k_7 & k_8 & k_9 \end{bmatrix}. \tag{31a}$$

$$\xi_2 = \begin{bmatrix} S \\ A \end{bmatrix} \quad \text{and} \quad K_2 = \begin{bmatrix} -k_1 & -k_2 & 0 \\ 0 & k_3 & -k_4 \end{bmatrix}. \tag{31b}$$

and the observer equations are given by:

$$\begin{bmatrix} \dot{z}_1 \\ \dot{z}_2 \end{bmatrix} = -D \begin{bmatrix} z_1 \\ z_2 \end{bmatrix} + C \begin{bmatrix} 0 \\ OTR \\ -CTR \end{bmatrix}. \tag{32a}$$

$$\begin{bmatrix} \hat{S} \\ \hat{A} \end{bmatrix} = \begin{bmatrix} z_1 \\ z_2 \end{bmatrix} - C \begin{bmatrix} X \\ O \\ P \end{bmatrix}. \tag{32b}$$

Fig. 11 shows the estimation results where the substrate estimate is not bounded to zero in order to illustrate the dramatically-increasing divergence of the observer (note that the real trajectory is very close but not equal to zero and is, of course, positive).

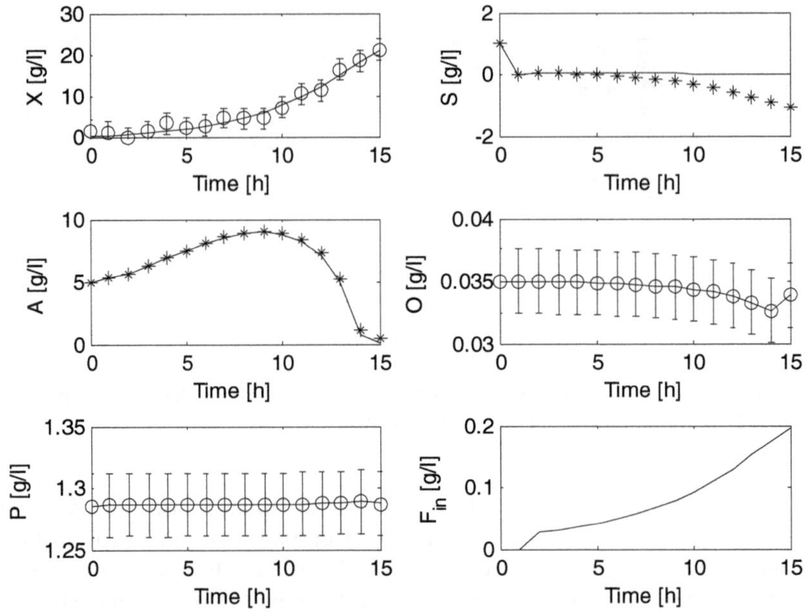

Fig. 11. Application of the asymptotic observer to the *E. Coli* model (21) in order to estimate the substrate and the acetate concentrations – black line: exact simulated model; red stars: asymptotic observer estimations; circles: simulated measurements (the sampling period is 1 hour).

The explanation of this phenomenon is that even though the asymptotic observer is insensitive to the kinetic parameters, it can be quite sensitive to small errors in the pseudo-stoichiometric coefficients as shown by the following developments.

From (32b), the expressions of the estimated states are obtained:

$$\hat{S} = z_1 + \frac{[k_1 k_8 k_6 + k_2(k_8 - k_9)]X + [k_1(k_5 k_9 - k_6 k_7) + k_2(k_9 - k_7)]O + [k_2(k_7 - k_8) - k_1 k_5 k_8]P}{k_5(k_9 - k_8) + k_6(k_8 - k_7)}. \tag{33a}$$

$$\hat{A} = z_2 + \frac{[k_3(k_8 - k_9) + k_4 k_6]X + [k_3(k_9 - k_7) - k_4(k_6 - k_5)]O + [k_3(k_7 - k_8) - k_4 k_5]P}{k_5(k_9 - k_8) + k_6(k_8 - k_7)}. \tag{33b}$$

Some combinations of the yield coefficients can have a significant impact on the accuracy of the estimated states. Consider the sensitivity matrix:

$$\left(\frac{\partial \hat{S}}{\partial k_i}\right) = \begin{pmatrix} \frac{\partial \hat{S}}{\partial k_1} & \frac{\partial \hat{S}}{\partial k_2} & \frac{\partial \hat{S}}{\partial k_3} \\ \frac{\partial \hat{S}}{\partial k_4} & \frac{\partial \hat{S}}{\partial k_5} & \frac{\partial \hat{S}}{\partial k_6} \\ \frac{\partial \hat{S}}{\partial k_7} & \frac{\partial \hat{S}}{\partial k_8} & \frac{\partial \hat{S}}{\partial k_9} \end{pmatrix}$$

$$= \begin{pmatrix} 1.22\ X - 0.008\ O - 0.22\ P & -0.13\ X + 0.06\ O - 0.19\ P & 0 \\ 0 & 0.95\ X - 0.4\ O - 1.35\ P & -0.71\ X - 0.3\ O + 1.1\ P \\ 0.46\ X - 0.38\ O - 0.08\ P & -0.08\ X - 0.07\ O - 0.02\ P & -0.35\ X + 0.29\ O + 0.06\ P \end{pmatrix}. \tag{34}$$

and in particular, the expression of the sensitivity with respect to k_1. When the culture starts, the biomass concentration X is very low so that its impact on the estimated state is negligible. On the other hand, when the culture has been running for hours, the biomass concentration is of the order of *10 to 100* g/l. This means that a small error on k_1 will lead to a variation of the estimate proportional to *1.22x10* or approximately *12 times* the initial error. If the error on k_1 is around *1%* of its theoretical value, the observer is likely to produce an estimation error of *12%* from the correct value, which cannot always be tolerated (for instance, substrate concentrations are generally low in *E. coli* applications, in order to avoid an accumulation of acetate which is an inhibitory product for the cells growth). Consider now that the culture reaches biomass concentrations around *100* g/l, the consequence of a small error on a yield coefficient has a larger effect on the estimated states as the estimation error reaches *100%*. This observation does not hold for the oxygen concentration as its values cannot exceed *0.035* g/l, the concentration at saturation, but it holds for the carbon dioxide in the sensitivities with respect to k_5 and k_6, as the concentration at saturation is *1.293* g/l.

An intuitive way to alleviate this high sensitivity is to separate the model in two sub-models depending on the metabolic pathway [1]. Indeed, this allows the estimation of 3 states on the basis of only 2 measured ones and the biomass terms in (33) disappear so that we can expect less sensitivity problems. In this case, the partitions of the yield matrix of the two submodels are:

For the respirative pathway:

$$K_1^R = \begin{bmatrix} -k_5 & -k_6 \\ k_7 & k_9 \end{bmatrix} \text{ and } K_2^R = \begin{bmatrix} 1 & 1 \\ -k_1 & 0 \\ 0 & -k_4 \end{bmatrix}. \tag{35a}$$

For the respiro-fermentative pathway:

$$K_1^{RF} = \begin{bmatrix} -k_5 & 0 \\ k_7 & k_8 \end{bmatrix} \text{ and } K_2^{RF} = \begin{bmatrix} 1 & 1 \\ -k_1 & -k_2 \\ 0 & k_3 \end{bmatrix}. \tag{35b}$$

For instance, the sensitivities of the estimated substrate concentration with respect to k_5 become for each pathway:

$$\left(\frac{\partial \hat{S}}{\partial k_5}\right)_R = 237\ O + 46.1\ P\ . \tag{36a}$$

$$\left(\frac{\partial \hat{S}}{\partial k_5}\right)_{RF} = 2.74\ O - 0.19\ P\ . \tag{36b}$$

Unfortunately, the terms involving the oxygen and carbon dioxide concentrations in (36a) and (36b) are relatively large so that no real improvement is provided.

Even if the respiro-fermentative pathway does not seem to be sensitive to small errors on k_5, the other pathway does. In fact, the bypass of the biomass concentration measurement transfers the sensitivity to the oxygen and carbon dioxide terms.

As a conclusion, bioprocess modelling and identification are delicate tasks, and uncertainties on the pseudo-stoichiometric and kinetic parameters are almost unavoidable. Standard observers such as the extended Kalman filter and the asymptotic observer are affected to some extent by these uncertainties and of course also by practical observability. More robust state estimation techniques have been proposed, which can alleviate these problems. However, these techniques are beyond the scope of this chapter, and in the continuation of this text, attention is focused on data-driven techniques.

3.2 Data-Driven Approach

The previous chapter presents the potential of classical mechanistic model-based observers. Even though a wide range of state estimation algorithms are readily available, their practical use remains relatively limited. The main reasons previously exposed are: (a) many state estimation algorithms require a dynamic model of the bioprocess, involving a macroscopic reaction scheme and kinetics, which are difficult to accurately establish from prior process knowledge and available measurement data; (b) state estimation algorithms usually rely on some hardware sensors, which are expensive and not always fully reliable; (c) manual operation has a long history in the bioprocess industry and advanced monitoring and control are currently emerging techniques.

It appears legitimate to propose to investigate the use of simple estimation algorithms based on black-box models and standard industrial measurement signals, such as those related to pH, temperature, pressure and dissolved oxygen concentration, as suggested in previous studies (see, e.g., [13], [6] and [14]).

3.2.1 Input/Output Selection and Measurements Data Analysis

It is difficult to establish a mechanistic model linking the abovementioned basic measurements to the main macroscopic species concentrations (biomass, substrate and metabolic products). This, of course, motivates a black-box strategy but also raises the question of which of the available signals could be informative for establishing a black-box model. Indeed, some signals could carry very little or redundant information and play a detrimental role in a learning procedure. That is why the information content of the available measurement signals is first assessed before being used as inputs to a black-box structure reproducing some selected outputs (i.e., some of the macroscopic species concentrations). A way of assessing the information content of the inputs is through *Principal Component Analysis* [15].

For illustration purpose, we consider the cultures of genetically manipulated strains of *S. cerevisiae* and *E. coli* growing in different-scale bioreactors. In standard operations, several regulation loops are active, including temperature, dissolved oxygen, pH (usually by base addition), air flow, pressure and stirring (in order to avoid oxygen limitation). Typical available measurements are represented in Fig. 12. All of them are normalized for the sake of confidentiality.

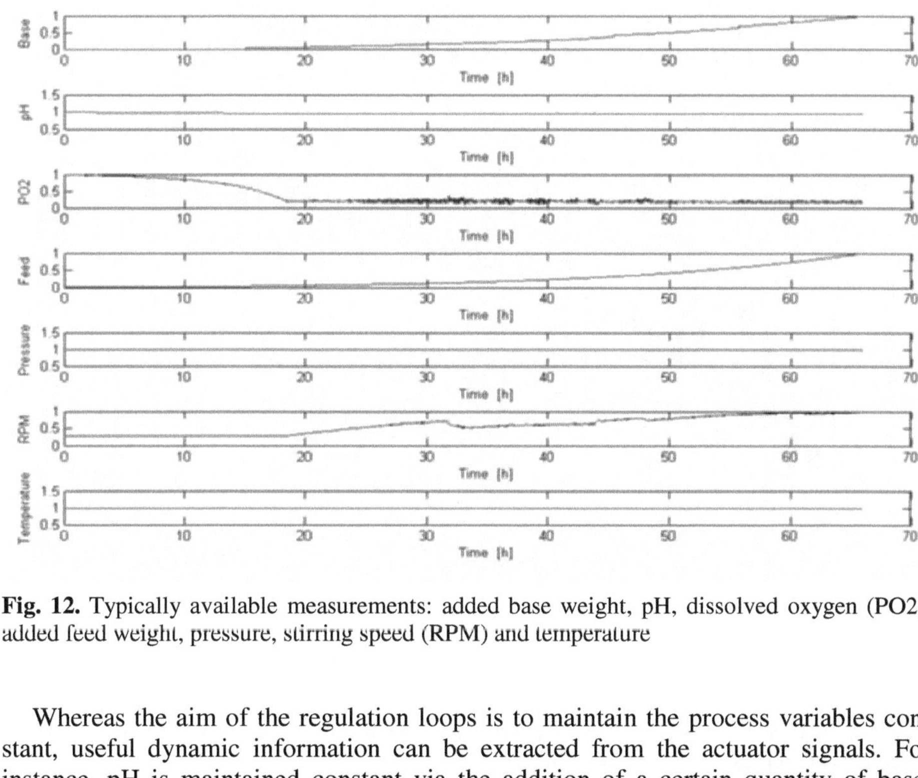

Fig. 12. Typically available measurements: added base weight, pH, dissolved oxygen (PO2), added feed weight, pressure, stirring speed (RPM) and temperature

Whereas the aim of the regulation loops is to maintain the process variables constant, useful dynamic information can be extracted from the actuator signals. For instance, pH is maintained constant via the addition of a certain quantity of base, whose evolution is informative of the culture evolution. In order to assess the correlation and the redundancy existing between the different signals, principal component analysis (*PCA*) is used in order to construct new signals, which are linear combinations of the normalized original ones. This analysis aims at describing a maximum of the data dispersion with a minimum of components. A new representation of the data matrix is obtained through a so-called "score matrix" T:

$$T_{nxa} = X_{nxm} P_{mxa} \tag{37}$$

where X is the initial data matrix of dimension nxm with n and m being respectively the number of measurements and the number of signals. P is the orthogonal loading matrix ($PP'=I$) of dimension mxa where a is the number of considered scores or components.

Geometrically, it consists in finding a space representation spreading as much as possible the cloud of data points in each new axis direction. Fig. 13 gives an idea of what can be these new directions for a specific example where two normalized measurements are considered. The score matrix provided by the principal component analysis is $P = \begin{pmatrix} 0.7071 & -0.7071 \\ 0.7071 & 0.7071 \end{pmatrix}$. The cloud of points can then be represented in a new space defined by the red lines (which, as a matter of fact, correspond to the bisectors of the previous axes).

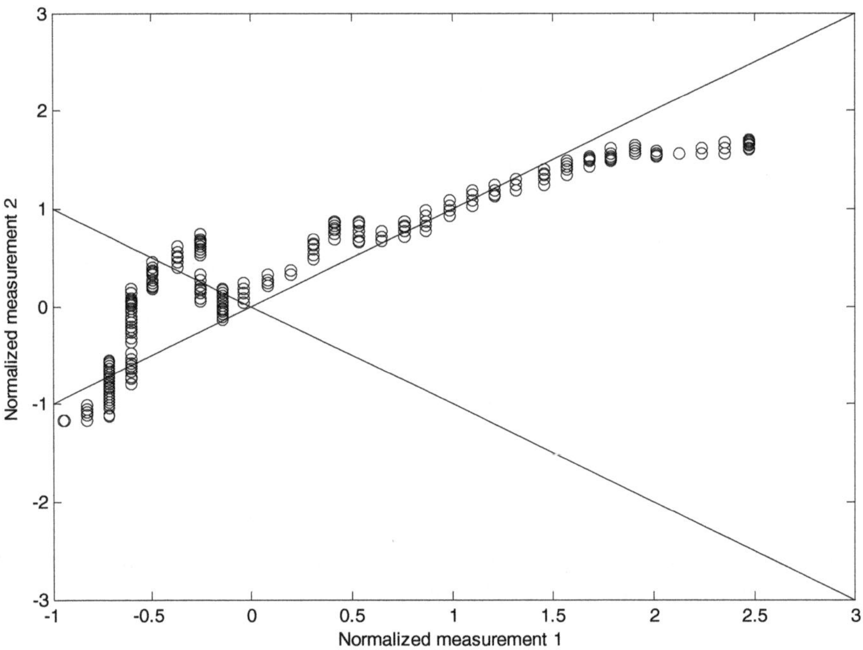

Fig. 13. Example of a PCA performed on two normalized measured signals. The principal directions are represented by the red lines.

A nonlinear iterative partial least squares algorithm (*NIPALS*) is used in order to build the scores in the sense that it calculates the first score t_1 (the first column of T) and the transposed corresponding loadings vector p'_1 from X before subtracting their product to obtain the residual $E_1 = X - t_1 p'_1$. The next scores and loadings are obtained iteratively following the same operation on each residual. As an illustration, Table 3 gives the variances of the scores t_i (i.e., the eigenvalues of the covariance matrix of the measurement data) and also the percentage of the total variance explained by each score (for the specific example illustrated in Fig. 12). The loadings of

Table 3. Score variances and explanations for the specific example illustrated in Fig. 12

	Variances	Explanation (%)
t_1	3.544	50.627
t_2	1.37	19.564
t_3	0.989	14.132
t_4	0.887	12.675
t_5	0.160	2.298
t_6	0.048	0.679
t_7	0.002	0.026

Table 4. Loadings of the measurement signals (illustrated in Fig. 12) in each principal direction

	t_1	t_2	t_3	t_4	t_5	t_6	t_7
Base	0.495	0.263	-0.152	0.055	-0.256	-0.169	0.752
pH	-0.196	0.742	-0.206	-0.1265	0.562	-0.190	-0.015
PO2	-0.429	0.429	-0.120	-0.098	-0.625	0.466	0.007
Feed	0.487	0.279	-0.153	0.060	-0.367	-0.319	-0.650
Pressure	-0.009	-0.320	-0.787	-0.526	0.007	-0.016	-0.003
RPM	0.514	0.090	-0.081	0.039	0.302	0.785	-0.109
Temperature	0.177	0.081	0.520	-0.83	-0.046	-0.013	0.002

the same normalized measurements signals are listed in Table 4 for each score. The stirrer speed (RPM for rotation per minute), the added base weight (Base), the added feed weight (Feed), and the dissolved oxygen (PO2) have the largest weights in the first principal component. Hence, as this component seems to be the most informative (see Table 3), these signals apparently convey more information than the others.

In this example, the notion of data compression is well illustrated as the first 4 scores contain more than 95 percents of the measurement information. Therefore, these 4 signals could be used as inputs to a software sensor if this analysis carries over all data sets.

3.2.2 Linear versus Nonlinear Black-Box Model

Even though a mechanistic bioprocess model is usually nonlinear (the nonlinearity stems from the kinetics), the relation between selected measurement inputs (elementary signals such as RPM, base weight, feed rate and PO2) and outputs (some component concentrations), could incidentally be linear. Before embarking in the construction of a nonlinear black-box model (which is more complex to identify than a linear one), it is therefore of interest to evaluate the linear/nonlinear character of this relationship and to select the best model.

In this section, two classical black-box modelling strategies are applied to bioprocesses and compared. Both of them are built using input and/or output variables represented by their scores matrix as mentioned in the previous section. The first one leads to a linear representation whereas the second is a nonlinear approach.

3.2.2.1 Linear Strategy – Partial Least Squares Regression (PLS). This first strategy [15] is an extension of the multiple linear regression (*MLR*) using the properties of the *NIPALS* algorithm through 2 *PCA* on the input and output data. This strategy is decomposed in 3 relations (2 outer (*PCA*) and 1 inner linear regression (*LR*) relations, see Fig. 14):

One can build the score matrix of the input (*X*) and output (*Y*) data matrix:

$$T_{nxa} = X_{nxn_x} P_{n_x xa} + E_{nxa} \tag{38a}$$

$$U_{nxc} = Y_{nxn_y} Q_{n_y xc} + F_{nxc} \tag{38b}$$

where *T* and *U* are respectively the nxn_x input and nxn_y output score matrix (considering *n* different measurements for n_x input and n_y output signals), *P* and *Q* the loadings matrix (considering again limited numbers $a \le n_x$ and $c \le n_y$ of components) and *E* and *F* the residuals matrix depending on the number of selected components or scores columns (remind that if the number of components is equal to the number of input signals, $E=F=0$).

The third relation is a simple linear regression between the input and output scores following:

$$U_{nxc} = T_{nxa} B_{axc} + G_{nxc} \tag{39}$$

where *B* is the regression coefficient matrix and *G* a new residual matrix.

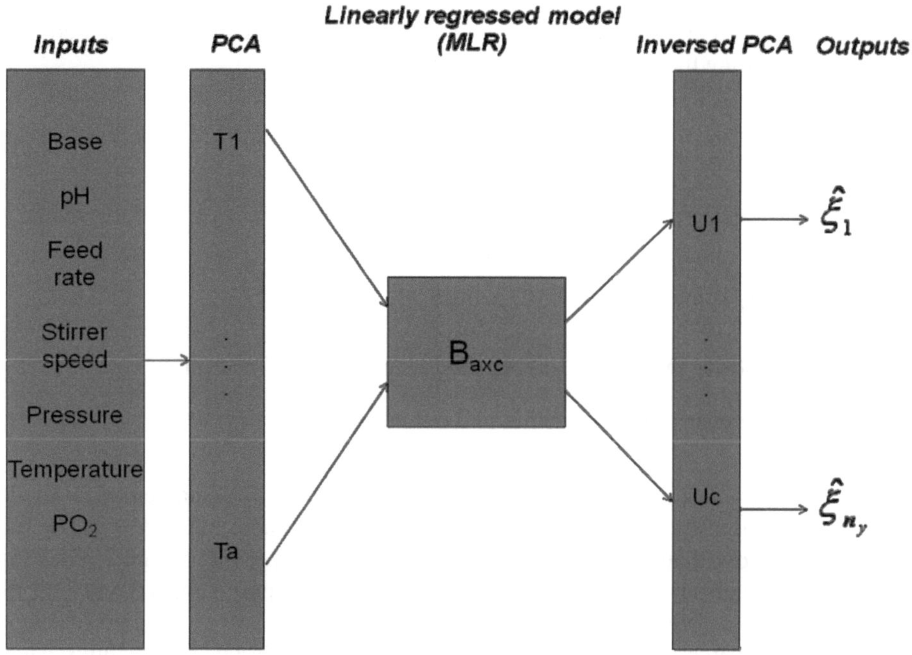

Fig. 14. Linear software sensor structure: PLS

Note that an intermediate algorithm exists and is called *Principal Components Regression (PCR)*. The main difference between *PCR* and *PLS* lies in the way the factor scores are extracted. Indeed, the *PCR* only uses the covariance structure of the input variables while *PLS* uses the covariance structure of the input and output variables. As it is known that bioprocesses present different element concentrations correlated as it appears in (21), using the covariance structure of the input and output variables should be preferable.

3.2.2.2 A Nonlinear Alternative: PCA and Artificial Neural Network (ANN). The previous input-output representation is by essence linear. To describe the behaviour of a bioprocess, it might be necessary to include some nonlinearity in the model representation. Attention is now focused on feed-forward multilayer perceptrons or radial basis function networks, which are universal approximators.

Standard feedforward NNs define a static map between a selected number of inputs (denoted $y(t)$, as they represent the measurement information coming from the bioprocess) and outputs (denoted $\hat{\xi}(t)$, as they correspond to estimates of component concentrations)

$$\hat{\xi}(t) = f(y(t)) \ . \tag{40}$$

One of the most common NN architectures in system modelling is the perceptron (Rosenblatt 1958, [16]). A perceptron consists in an on/off static function (called activation function or decision function) delivering a binary output. The sum of a linear combination of the inputs weighted by synaptic weights is compared to a threshold separating the activation and inactivation zones:

$$\hat{\xi}_i = f(x_i) = \begin{cases} 0 : x_i < 0 \\ 1 : x_i > 0 \end{cases} \tag{41}$$

where

- $x_i = \sum_j w_{ij} y_j$ is the input of the ith neuron.
- w_{ij} corresponds to the weight linking the jth input to the ith neuron.
- y_j and $\hat{\xi}_i$ are respectively the input and output vectors.

This is the simplest neural network structure recalling the first mathematical model of neuron proposed by Mc Culloch & Pitts (1943) [17]. There exists a learning rule developed by Rosenblatt and ensured to converge (i.e., to provide weight values that enable the recognition of each class corresponding to the binary values 0 and 1) if the related data are linearly separable [18].

Networks of perceptrons may then be built using multiple layers structures where all the neurons outputs depend only on the inputs from the previous layer and do not interact with the same-layer neurons. These structures are called *Multilayer perceptrons (MLP)*. This time, the nonlinearity used in the activation function is continuous (for instance, a sigmoid function) and we distinguish the first and the last layers, respectively called input and output layers, from the intermediate layers also called the hidden layers. Actually, only the first and the last layers input and output are available from an external point of view.

The multilayer perceptron learning phase is performed by calculating a least square error criterion first applied to the output layer:

$$E = \frac{1}{2}\sum_{i=1}^{K}(\hat{\xi}_i(X) - \xi_i)^2 \tag{42}$$

where

- K is the number of neurons of the output layer.
- $\hat{\xi}_i$ are the outputs.
- ξ_i are the desired outputs.
- X is the input vector of the output layer neurons.

and by propagating this error back to the previous layers in order to update the weights as follows:

$$w_{l,ij}(k+1) = w_{l,ij}(k) - \alpha \left(\frac{\partial E}{\partial w_{l,ij}} \right)_{l-1} \tag{43}$$

where

- $w_{l,ij}$ corresponds to the weight linking the jth input to the ith neuron of the lth layer.
- α is a strictly positive tuning parameter called the learning rate.
- $\left(\frac{\partial E}{\partial w_{l,ij}} \right)_{l-1}$ is the first derivative of the error calculated for the *l-1* layer with respect to $w_{l,ij}$.

For obvious reasons linked to its principle, this learning algorithm is called *the backpropagation algorithm* [19].

This kind of network is abundantly used in biomedical applications and particularly in speech processing and handwritten pattern recognition ([20], [21], [22] and [23]).

Another classical neural network structure is the Radial Basis Function network, which has proved quite useful in modelling bioprocesses [24].

This structure differs from the last one by the activation principle and the learning rules. This network (Fig. 15) delivers outputs calculated by a continuous mathematical expression of the form:

$$\hat{\xi}_i(t) = \sum_{j=1}^{k} w_{ij} e^{\frac{-\|y(t)-c_j\|^2}{r_j^2}} + b_i \qquad i = 1 \ldots n_y \tag{44}$$

where

- y is the input of length n_y;
- $\hat{\xi}_i$ is the ith component of the output vector;
- k is the number of neurons in the hidden layer;
- w_{ij} is the weights linking the neurons to the outputs;
- c_j are the centers;
- r_j are the radii (i.e., the average spherical distance from the center);
- b_i are the biases.

The learning procedure of this network is divided in several learning steps (see Fig. 16). The number of neurons *k* is taken large enough so as to be able to reproduce the desired input-output behaviour, but small enough so as to avoid overparametrization and lack of generalization (to data sets not considered in the identification phase).

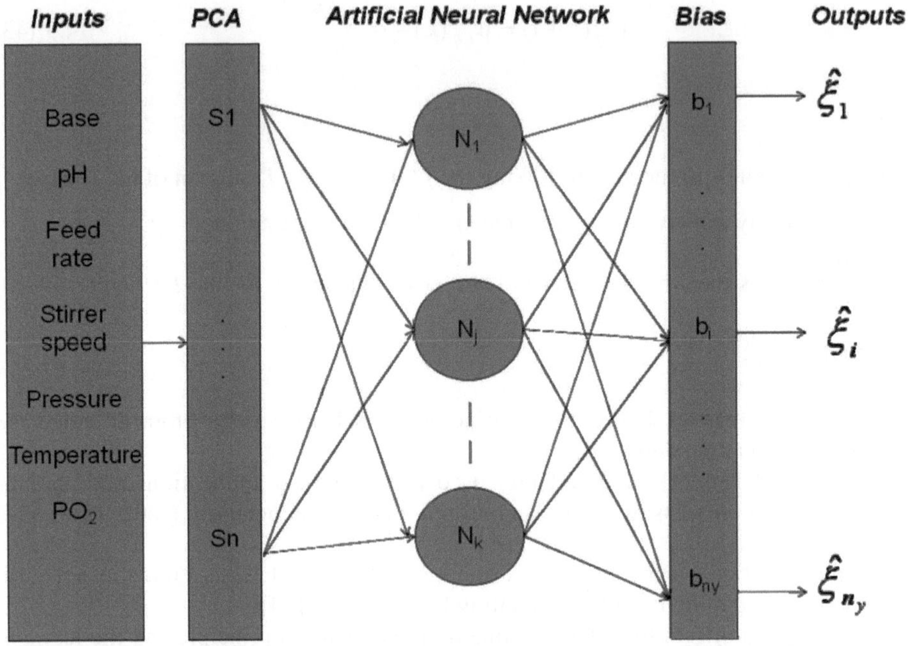

Fig. 15. Nonlinear software sensor structure: PCA and RBF-ANN

A 3-steps learning procedure is used (as in [24], see Fig. 16):

The first step is an unsupervised learning phase in which the first estimates of the centers and the radii are obtained by a *k-means clustering* algorithm. These first estimates allow the *NN* outputs to cover the experimental field.

The second step is a supervised learning phase consisting in an initial linear identification of the weights and biases by minimization of a quadratic cost function.

Finally, a last nonlinear identification of all the parameters is achieved starting from the values obtained in the previous steps and minimizing a *Gauss-Markov criterion* taking the measurement errors into account:

$$\theta^T = \left[w^T b^T r^T c^T\right] = \min_{\theta} J_{ml}(\theta) = \min_{\theta} \frac{1}{2} \sum_{i=1}^{N} (\xi_i - \hat{\xi}_i(\theta))^T Q_i^{-1} (\xi_i - \hat{\xi}_i(\theta)) \qquad (45)$$

where

- ξ_i are the measured output values;
- Q_i is the error covariance matrix on the measured outputs;
- $\hat{\xi}_i$ are the estimated outputs based on θ.

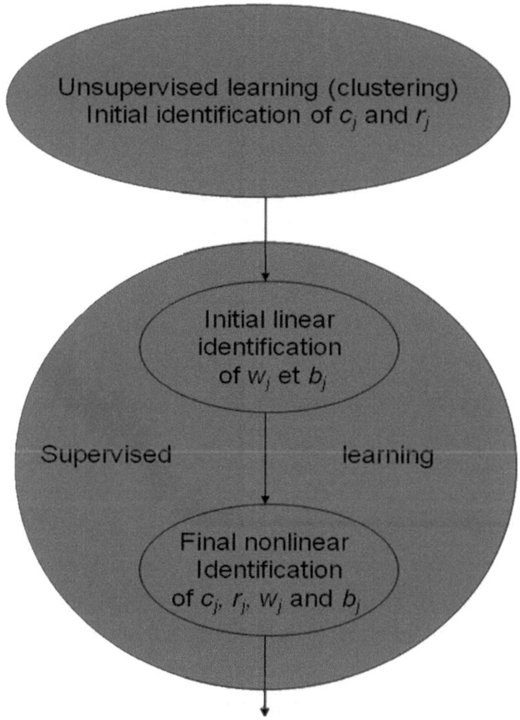

Fig. 16. Parameter identification scheme of the RBF-ANN structure

3.2.2.3 Comparison of the Two Approaches: Linear versus Nonlinear Modelling. To compare the performance of the two proposed methods, a large panel of experimental data collected from different bioreactors operated with different strains at different scales is available. This large data set will ensure the pertinence and the generality of the results.

As a start, four data sets corresponding to fed-batch cultures of *S. cerevisiae* (yeast strain) operated in a laboratory-scale bioreactor are considered. Among the 4 data sets, two are used to identify the black-box model while the two others are used to perform cross-validation tests. Our goal is to estimate the biomass concentration all along the cultures based on four available measurements (base addition, feed flow rate, stirrer speed and dissolved oxygen concentration).

As only one variable need to be estimated, a *PCR* is used instead of a *PLS* for the linear approach while a *RBF-ANN* is used for the nonlinear approach. Fig. 17 shows the evolution of the estimated biomass concentration during one of the two latter experiments (cross-validation). Recall that the standard input signals, on which a *PCA* is achieved, are provided at a high sampling rate and thus provide a frequent biomass estimation which can be very useful for process monitoring.

For the sake of confidentiality, all the experimental results are normalized. Unfortunately, only the nonlinear model seems to reproduce satisfactorily the biomass concentration evolution. This can be explained by the probable existence of nonlinearities

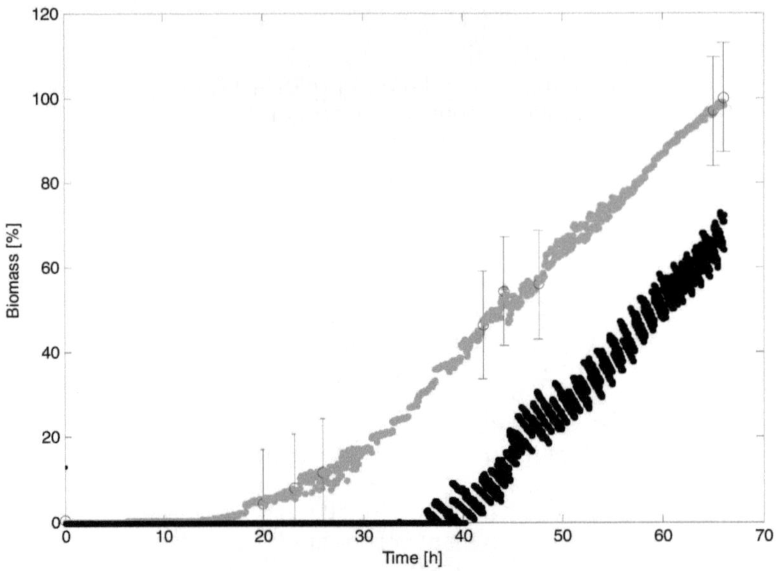

Fig. 17. Biomass estimation in a culture of *S. cerevisiae*. Red line: output of the RBF-ANN nonlinear model; blue line: output of the PCR linear model; circles: real off-line measurements.

in the relation between the chosen input variables and the biomass concentration. 3 neurons are sufficient to deal with these nonlinearities. Note that 3 neurons already represent 12 parameters (4 radii, 4 centers, 3 weights and 1 bias) to identify if 4 inputs and only 1 output are considered. One must be conscious that the number of parameters increases to 16 for 2 outputs instead of 1, to 20 for 3 outputs and so on.

Only three of the four available scores have been used for producing the previous results as it has been observed that the addition of the fourth one is useless (reducing the number of parameters to 10). The best way to understand this observation and justify our choice is to analyse the table of explanation for each score (Table 5).

Table 5. Score explanations of 4 yeast fed-batch cultures data sets

Explanation (%)	Set 1	Set 2	Set 3	Set 4
S1	84.0197	83.2870	83.8876	84.4641
S2	14.4769	14.9386	14.3560	13.7753
S3	1.4563	1.7236	1.6982	1.7162
S4	0.0471	0.0508	0.0582	0.0443

It appears that more than *95%* of the variance information is contained in the first two scores. Anyway, before drawing too early conclusions from this table, we also assess the quality of the NN prediction using different numbers of scores through a criterion calculating an average error:

$$E_m = \left(\frac{1}{N} \sum_{i=1}^{N} \left(\xi_i - \hat{\xi}_i \right)^2 \right)^{\frac{1}{2}} \tag{46}$$

where

- N is the number of experimental biomass concentration measurements;
- ξ is the biomass measurement vector;
- $\hat{\xi}$ is the biomass estimate vector.

Table 6 shows the value of this criterion for different numbers of scores. As expected from Table 5, the projection in only one principal direction of the data space is sufficient to ensure good results. Nevertheless, the minimum is observed for three scores. As this number of inputs corresponds to a reasonable number of parameters (3 radii, 3 centers, 3 weights and 1 bias if only one output is considered), this input structure is selected for the next experimental evaluations.

Table 6. Average error values for the different possible numbers of scores (inputs to the ANN)

Number of scores	E_m (%)
1	6.65
2	4.52
3	2
4	3.5

3.2.3 Monitoring of Fed-Batch Cultures of *E. coli*

Genetically modified strains of *E. coli*, which are commonly used in biopharmaceutical industries to express recombinant proteins and to produce vaccines, are now considered. For these bacteria strains, 16 data sets are available, which correspond to the expression of two different recombinant proteins *RP1* (for 7 of them) and *RP2* (for 9 of them), as well as two different bioreactor scales (20l and 150l). See Table 7 for a description of the experimental field. Fig. 18 shows the time-evolution of the three input signals (scores) and the estimation of the biomass concentration by a RBF-ANN

Table 7. Classification of the data sets in function of the recombinant protein, the scale and the type of validation

Scale\Recombinant protein	RP1		RP2	
	Simple validation	Cross-validation	Simple validation	Cross-validation
20 l	4	2	0	8
150 l	0	1	0	1

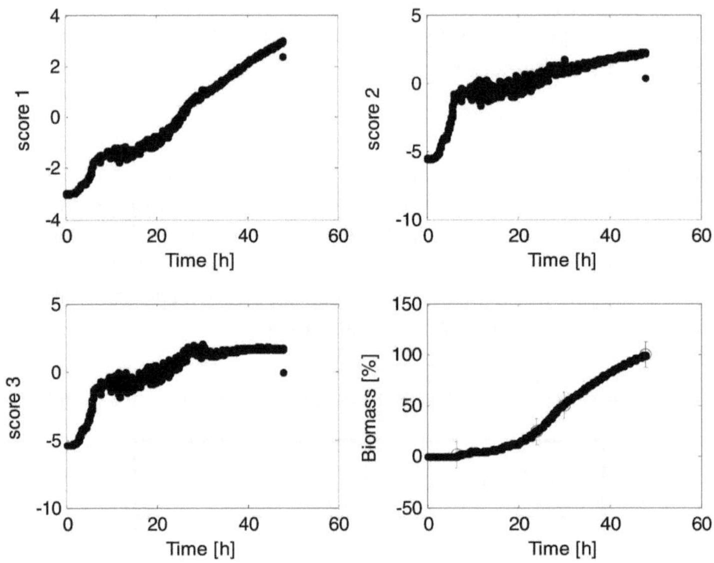

Fig. 18. Biomass estimation in 20-l scale fed-batch culture of *E. coli* (expression of RP1)

trained with 4 data sets corresponding to fed-batch cultures of bacteria in 20-l bioreactor where *RP1* is expressed. Actually, 3 data sets are used in cross-validation tests and Fig. 18 shows one of them. For the represented test, the average error is *1.32%*. Fig. 19 shows the results of the cross-validation test on the 7th data set corresponding to a 150-l scale bioreactor. Again, note that the *ANN* parameters are not modified in between these tests, which constitutes challenging cross-validation tests. The overall average error amounts to only *1.95%*, testifying of the good quality of the results.

The ANN software sensor is then tested with the 9 remaining experiments, where *RP2* is expressed. Fig. 20 shows that the ANN software sensor reproduces again very satisfactorily the evolution of the biomass concentration, the type of protein expressed having little influence on the performance. Moreover, as depicted in Fig. 21, the performance of the ANN is also independent of the bioreactor scale for the strain expressing *RP2*. Note that genetic modifications have significant influence on the cell metabolism. Therefore, the use of a mechanistic model requires a new identification

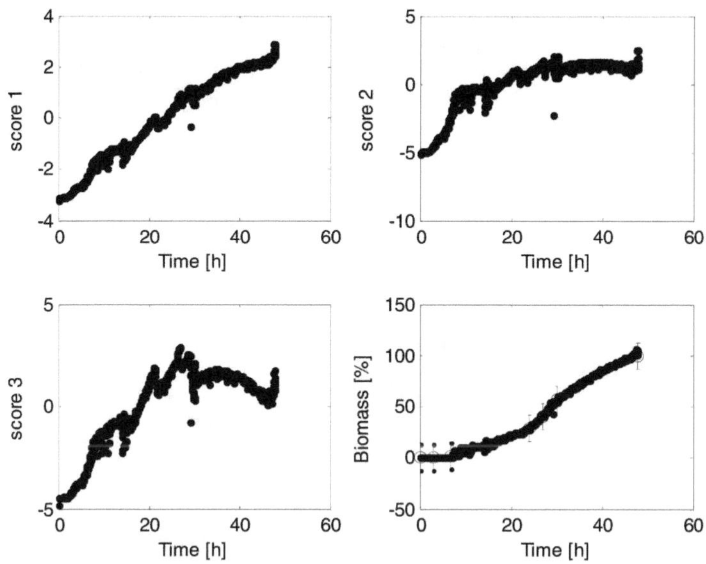

Fig. 19. Biomass estimation in 150-l scale fed-batch culture of *E. coli* (expression of RP1)

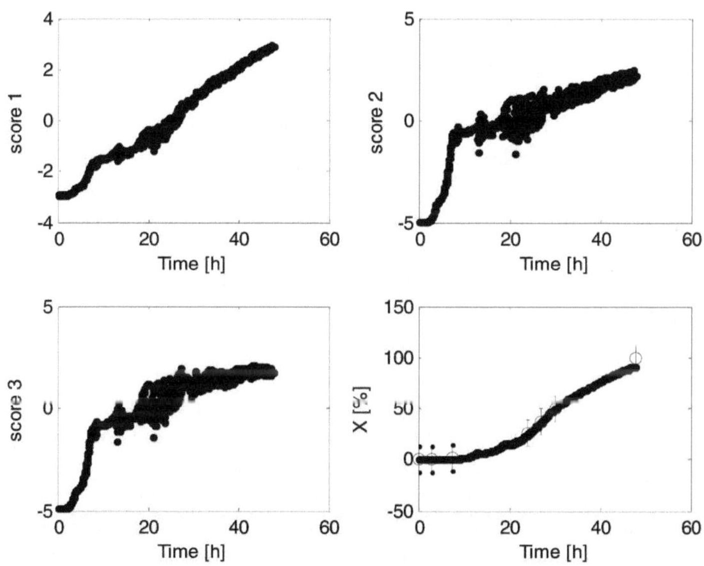

Fig. 20. Biomass estimation in 20-l scale fed-batch culture of *E. coli* (expression of RP2 when trained with RP1)

of the parameters whereas the *ANN* software sensor is still performing well without any parameter reidentification. The *ANN* software sensor appears then as a powerful tool in terms of flexibility and robustness.

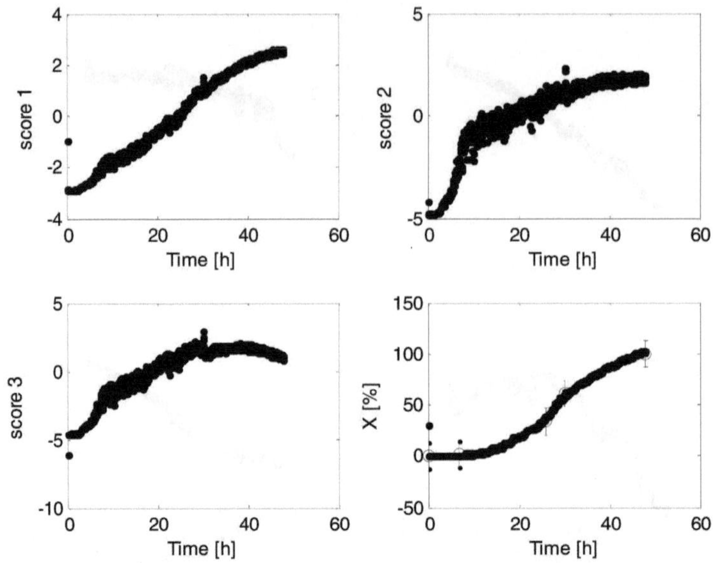

Fig. 21. Biomass estimation in 150-l scale fed-batch culture of *E. coli* (expression of RP2 when trained with RP1).

For the data sets of the strain expressing *RP2* when the ANN sensor is trained with the four data sets corresponding to *RP1*, the overall average error only amounts to *4.4%*.

This method has also been applied to other strains of *E. coli* evolving in different fed-batch processes after reidentification of the parameters. In all the cases, the ANN software sensor performs well, and provides a useful tool for on-line biomass monitoring. The proposed approach is therefore quite versatile and applicable to a broad range of bioprocesses (in some situations, other measurements signals – or additional measurement signals – have to be selected, this selection being achieved efficiently using *PCA*).

3.2.4 Monitoring of Fed-Batch Cultures of *S. cerevisiae*

The ANN software sensor is also able to estimate other key-component concentrations. First, the ethanol concentration, which is another key-component of yeast cultures, is considered. An overfeeding in glucose leads yeasts to switch from the respirative metabolism to the respiro-fermentative one, implying a production of ethanol by fermentation of the excess of glucose. It is common in industrial applications to avoid this accumulation of ethanol because it has an inhibiting influence on cell growth. As the industrial objective is generally to optimize the productivity, i.e. to maximize the production and to minimize the culture time, the ethanol concentration is always under supervision, requiring multiple off-line measurements and the time that it costs. Some hardware sensors are readily available but, unfortunately, they are still very expensive and all the laboratories cannot always afford such equipments. Moreover, on-line ethanol measurements are useful for optimizing control such as the

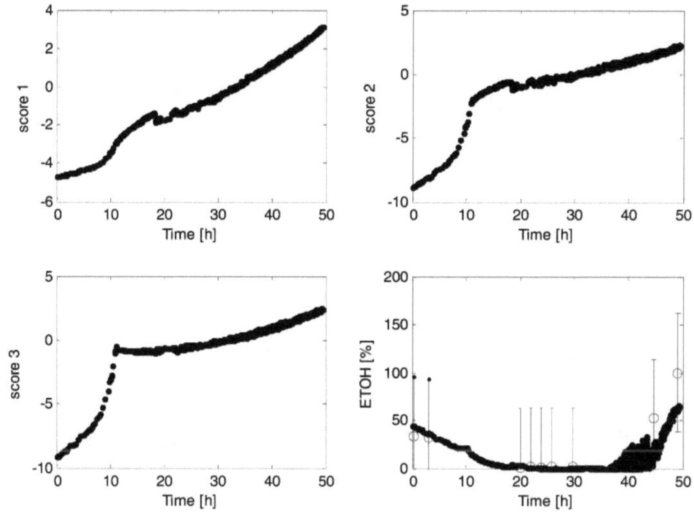

Fig. 22. Ethanol estimation in 15-l scale fed-batch culture of *S. cerevisiae*

robust RST controller presented in [25], [26] and [27]. This motivates the following assessment of the potentialities of the ANN software sensor in estimating other key-components like the overflow product.

Again, the results presented in Fig. 22 are normalized for confidentiality reasons. The 5 data sets that will be used here come from a 15-l scale bioreactor. Two data sets were used for the learning phases and three in cross-validation.

The impact of the noise corrupting the inputs appears clearly on the output between 40 and 50 hours of culture. Indeed, ethanol stays at a low level from the beginning to the end of the culture while some inputs are increasing and the noise that they imply grows as well. On the whole, the output mean value follows the off-line measurements.

The average error confirms these observations as it amounts to *15%* (recall that we deal with very low values), which is very satisfactory.

In many applications, product inhibition is not the main concern as long as the substrate concentration is limited (i.e. for instance in yeast cultures: no overflow, no fermentation). To ensure this last condition, glucose measurements are needed. As for the ethanol, off-line measurements of glucose means a consequent waste of time and monitoring of on-line glucose measurements can be an alternative when no ethanol measurement is available. The following results demonstrate the efficiency of the nonlinear software sensor applied to glucose estimation. The experiment is performed at glucose concentrations of an order of magnitude lower than ethanol concentrations. This explains the more significant impact of noise from 20 hours on to the end of the culture. The average error on the 5 data sets is around *28%*. Remember however that these cultures are operated at a very low glucose concentration level, necessarily leading to a poor estimation accuracy as it is typically the case with available hardware sensors. Therefore, the ANN software sensor needs a learning phase targeting the scale covered by the experimental field. See, for instance, the 7th off-line measurement in

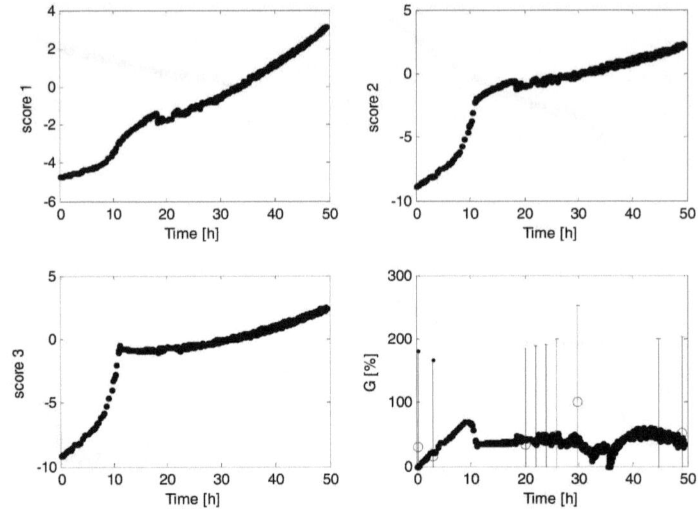

Fig. 23. Glucose estimation in 15-l scale fed-batch culture of *S. cerevisiae*

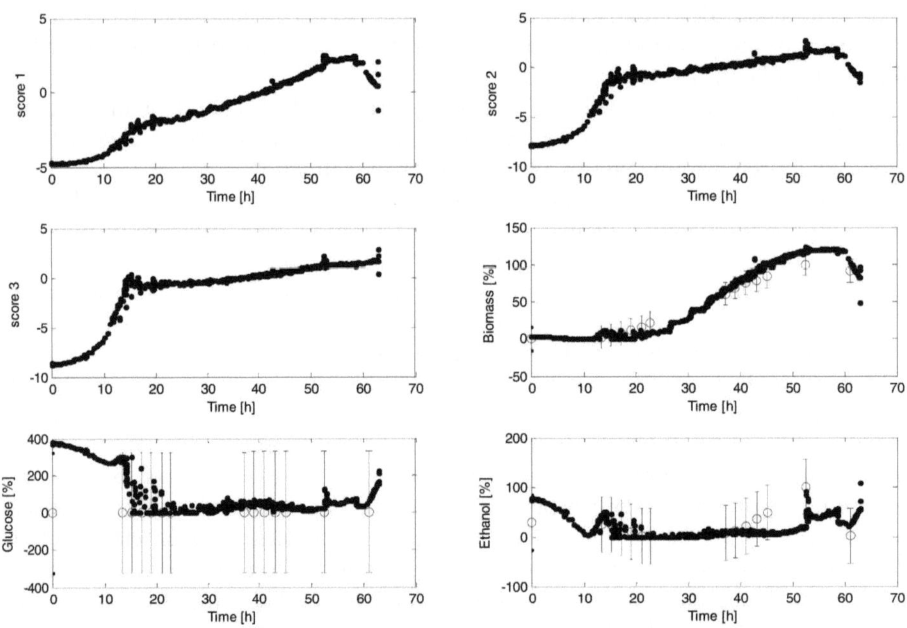

Fig. 24. Simultaneous estimations of biomass, glucose and ethanol concentrations in 15-l scale fed-batch culture of *S. cerevisiae*

Fig. 23, e.g., this point is actually beyond the experimental field so that the ANN calculates a trajectory which is more in accordance with the training data. Note that this deviation remains limited when working at low concentrations.

3.2.5 RBF-ANN Software Sensor Potentiality in Simultaneous Estimations of Different Key-Components

Now that the efficiency of the ANN in estimating key-components of interest is demonstrated, the possibility to estimate multiple outputs with only one RBF-ANN structure is investigated as depicted in Fig. 24. To this end, the 5 same data sets are used and the new structure is trained with 3 scores using only 3 neurons to limit the number of parameters (which amounts now to 18) and the computational expense. As illustrated in Fig. 24: the level of the average error goes from "excellent" for the biomass concentration to "satisfactory" for the glucose and ethanol concentrations.

A *PLS* algorithm is also considered but, again, a linear model seems inadequate for representing the link between the basic signals and the key-components, as observed in Fig. 25.

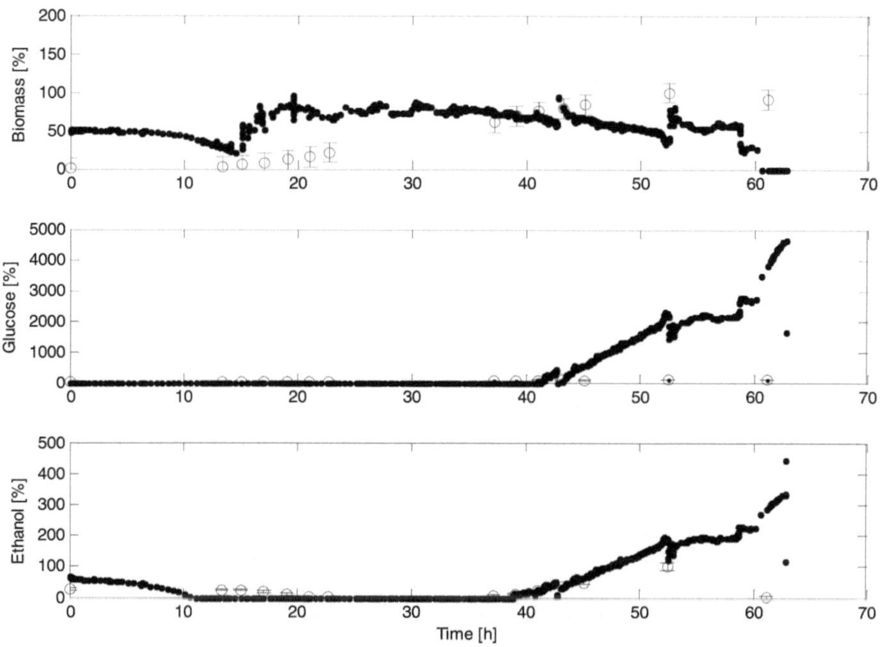

Fig. 25. Simultaneous estimations of key-components in 15-l scale fed-batch culture of *S. cerevisiae* with PLS

4 Conclusion

Soft sensor techniques are particularly significant for monitoring bioprocesses, for which hardware sensors are expensive and delicate to use. Two main design approaches can be distinguished, based on the underlying model that is exploited. On the one hand, mechanistic models provide an interesting insight in the biological system, but require significant efforts for their derivation (measurements of component concentrations, careful design of experiments, parameter identification, model validation). On the basis of the dynamic model and a few on-line measurements, a wide range of observers can be designed, which provides time-continuous estimation of the component concentrations. Parameter uncertainties have to be dealt with adequately. On the other hand, data-driven techniques offer the possibility to exploit basic signals such as base addition, stirrer speed, etc, and to establish a map to some of the component concentrations of interest. In particular, principal component analysis can be used to select informative signals and artificial neural networks can be used to build the desired maps.

These data-driven techniques have proved quite useful and effective in many real life applications, and some of them are discussed in this chapter.

Acknowledgments

This work presents research results of the Belgian Network DYSCO (Dynamical Systems, Control, and Optimization), funded by the Interuniversity Attraction Poles Programme, initiated by the Belgian State, Science Policy Office.

References

1. Bastin, G., Dochain, D.: On-line Estimation and Adaptive Control of Bioreactors. Elsevier Science Publishers B.V, Amsterdam (1990)
2. Vidyasagar, M.: Nonlinear Systems Analysis. Prentice Hall, Englewood Cliffs (1978)
3. Bogaerts, P., Vande Wouwer, A.: Software Sensors for Bioprocesses. ISA Transactions 42, 547–558 (2003)
4. Goffaux, G., VandeWouwer, A.: Bioprocess State Estimation: Some Classical and Less Classical Approaches. In: Symposium on Nonlinear Control and Observer Design, Stuttgart, pp. 111–128 (2005)
5. Gelb, A.: The Analytic Sciences Corporation Technical Staff: Applied Optimal Estimation. MIT Press, Cambridge (1974)
6. Hulhoven, X., Renard, F., Dessoy, S.: Dehottay, P., Bogaerts, P., VandeWouwer, A.: Monitoring and Control of a Bioprocess for Malaria Vaccine Production. In: 5th IFAC Symposium on Robust Control Design, Toulouse, France (2006)
7. Sonnleitner, B., Käppeli, O.: Growth, S. cerevisiae is Controlled by its Limited Respiratory Capacity: Formulation and Verification of a Hypothesis. Biotechnology & Bioengineering 28, 927–937 (1986)
8. Bernaerts, K., Gysemans, K.P.M., Nhan Minh, T., Van Impe, J.F.: Optimal Experiment Design for Cardinal Values Estimation: Guidelines for Data Collection. International Journal of Food Microbiology 100, 153–165 (2005)
9. Versyck, K.J., Bernaerts, K., Geeraerd, A.H., Van Impe, J.F.: Introducing Optimal Experimental Design in Predictive Modelling: A Motivating Example. International Journal of Food Microbiology 51, 39–51 (1999)
10. Gauthier, J.-P., Kupka, I.: Observability and Observers for Nonlinear Systems. SIAM Journal on Control and Optimization 32(4), 975–994 (1994)
11. Zeitz, M.: Observability Canonical (Phase-Variable) Form for Nonlinear Time-Variable Systems. International Journal of System Science 15(9), 949–958 (1984)

12. Bogaerts, P., Vande Wouwer, A.: Parameter Identification for State Estimation – Application to Bioprocess Software Sensors. Chemical Engineering Science 59, 2465–2476 (2004)
13. Karim, M.N., Yoshida, T., Rivera, S.L., Saucedo, V.M., Eikens, B., Oh, G.-S.: Global and Local Neural Network Models in Biotechnology: Application to Different Cultivation Processes. Journal of Fermentation and Bioengineering 83, 1–11 (1997)
14. Dewasme, L., Vande Wouwer, A., Dessoy, S., Dehottay, P., Hulhoven, X., Bogaerts, P.: Experimental Study of Neural Network Software Sensors in Yeast and Bacteria Fed-Batch Processes. In: Preprints of the 10th International IFAC Symposium on Computer Applications in Biotechnology, Cancun, Mexico, vol. 1, pp. 45–50 (2007)
15. Geladi, P., Kowalski, B.: Partial Least Squares Regression. A Tutorial. Analytica Chimica Acta 185, 1–17 (1986)
16. Rosenblatt, F.: The Perceptron: A Probabilistic Model for Information Storage and Organization in the Brain. Cornell Aeronautical Laboratory, Psychological Review 65(6), 386–408 (1958)
17. McCulloch, W.S., Pitts, W.: A Logical Calculus of the Ideas Immanent in Nervous Activity. Bulletin of Mathematical Biophysics 5, 115–133 (1943)
18. Minsky, M., Papert, S.: Perceptrons: A Introduction to Computational Geometry. The MIT Press, Cambridge (1969)
19. Werbos, P.: Beyond regression: New Tools for Prediction and Analysis in the Behavioural Sciences. PhD thesis, Harvard University, Cambridge, MA (1974)
20. Gosselin, B.: Application de Réseaux de Neurones Artificiels à la Reconnaissance Automatique de Caractères Manuscrits. PhD thesis, Faculté Polytechnique de Mons, Belgium (1996)
21. Gosselin, B.: Multilayer Perceptrons Combination Applied to Handwritten Character Recognition. Neural Processing Letters 3(1), 3–10 (1996)
22. Morgan, N., Bourlard, H.: Neural Networks for Statistical Recognition of Continuous Speech. Proceedings of the IEEE 83(5), 741–770 (1995) (Invited paper)
23. Bourlard, H.: Continuous Speech Recognition: from Hidden Markov Models to Neural Networks. Invited Keynote Speaker. In: Proc. EUSIPCO 1992, 6th European Signal processing Conference, Brussels, Belgium, pp. 63–70 (1992)
24. Vande Wouwer, A., Renotte, C., Bogaerts, P.: Biological Reaction Modelling Using Radial Basis Function Networks. Computers & Chemical Engineering 28, 2157–2164 (2004)
25. Valentinotti, S., Srinivasan, B., Holmberg, U., Bonvin, D., Cannizzaro, C., Rhiel, M., Von Stockar, U.: Optimal Operation of Fed-Batch Fermentations via Adaptive Control of Overflow Metabolite. Control Engineering Practice 11, 665–674 (2003)
26. Renard, F., Vande Wouwer, A., Valentinotti, S., Dumur, D.: A Practical Robust Control Scheme for Yeast Fed-Batch Cultures – an Experimental Validation. Journal of Process Control 16, 855–864 (2006)
27. Dewasme, L., Renard, F., Vande Wouwer, A.: Experimental Investigations of a Robust Control Strategy Applied to Cultures of S. cerevisiae. In: Preprints of the European Control Conference, Kos, Greece, pp. 4785–4791 (2007)

4

Novel Computational Methods for Modeling and Control in Chemical and Biochemical Process Systems

Petia Georgieva[1] and Sebastião Feyo de Azevedo[2]

[1] Department of Electronics Telecommunications and Informatics /IEETA, University of Aveiro, 3810-193 Aveiro, Portugal
petia@ua.pt
[2] Department of Chemical Engineering, Faculty of Engineering, University of Porto, 4200-465 Porto, Portugal
sfeyo@fe.up.pt

Abstract. This chapter is focused on developing more efficient computational schemes for modeling and control of chemical and biochemical process systems. In the first part of the chapter a theoretical framework for estimation of general process kinetic rates based on Artificial Neural Network (ANN) models is introduced. Two scenarios are considered: i) Partly known (measured) process states and completely known kinetic parameters; ii) Partly known process states and kinetic parameters. The two scenarios are successfully tested for two benchmark problems, estimation of the precipitation rate of calcium phosphate and estimation of sugar crystallization growth rate. In the second part of the chapter the developed ANN-based models are integrated into the structure of a nonlinear model predictive control (MPC). The proposed ANN-MPC control scheme is a promising framework when the process is strongly nonlinear and input-output data is the only process information available.

1 Introduction

This work is focused on recent achievements in the field of knowledge based modeling and control of chemical and biochemical processes. The problem of overcoming difficulties related with poorly known kinetic mechanistic descriptors is discussed in great details.

In the first part of the chapter (section 2) we introduce a systematic Artificial Neural Network (ANN) training procedure for estimation of process kinetic rates. The ANNs are computational paradigm whose initial structure (multilayer perception) and recent variations (spiking neural networks) are inspired on the latest achievements in neurobiology. ANNs can learn on-line, based on local information. Their main characteristics are their universal approximation properties and generalization ability with respect to unknown data. Also, due to the parallel, distributed processing structure and high degree of connectivity among the units, they exhibit smooth performance degrading when or if the application moves away from their training regions. All these remarkable properties of the ANNs have been intensively exploited in various fields and recently also in biologically inspired control engineering solutions[1], [2].

M. do Carmo Nicoletti, L.C. Jain (Eds.): Comp. Intel. Tech. for Bio. Mod., SCI 218, pp. 99–125.
springerlink.com

Two scenarios are considered: i) the kinetic parameters of the process are completely known and the process states are partly known (measured); ii) The kinetic parameters and the states of the process are partly known. The concept developed is illustrated with two case studies taken respectively from lab scale and industrial scale applications. Our main contributions are the determination of a hybrid (ANN & phenomenological) model that outperforms the traditional kinetic rate estimation approaches and the procedure for ANN supervised training when target outputs are not available.

In the second part of the chapter (section 3), we discuss the development of intelligent control systems for (fed-) batch processes, where we have combined the mapping and learning ability of ANNs with advanced control methods to build an ANN-based model predictive controller (MPC). Here, we combine techniques from the field of artificial intelligence with those of already established control engineering practice and design an autonomous self-learning system that can act in an intelligent manner (i.e. adapt to changing surroundings and learn new experience). The general Model Predictive Control (MPC) formulation is briefly presented in section 3.1.Then, in section 3.2, our main contribution is discussed, namely the error tolerant ANN-based MPC, which is illustrated in section 3.3 on an industrial sugar crystallization process. Finally, in section 4, our conclusions on the benefits of the proposed computational methods are summarized.

2 Artificial Neural Networks for Modeling in Chemical and Biochemical Process Systems

The dynamics of chemical and biochemical processes are usually described by mass and energy balance differential equations. These equations combine two main type of elements, i) the phenomena of conversion of one component into another and/or the growth or decay of chemical species (i.e. the process kinetics) and ii) the transport dynamics of the components in the process. The identification of such mathematical models from experimental input/output data is still a challenging issue due to the inherent nonlinearity and complexity of this class of processes. The most difficult problem is usually how to model the kinetic rates. Traditionally, they are estimated in the form of analytical expressions [3], [4], [5]. First, the parameterized structure of the process kinetics is determined based on data obtained by specially designed experiments. Then the respective parameters of this structure are estimated. Reliable parameter estimation is only possible if the proposed model structure is correct and theoretically identifiable [6]. Therefore the analytical structure of process kinetics is usually determined after a huge number of expensive laboratory experiments. It is further assumed that the initial values of the identified parameters are close to the real process parameters [7], which is typically satisfied only for well known processes. The above considerations motivated a search for alternative estimation solutions based on computationally more attractive paradigms as are the neural networks (NN). Over the last 20 years, the NNs became a well-established methodology as exemplified by their applications to identification and control of nonlinear systems [8], [9]. Their increasing use not only in modeling and control but also in various fields including pattern recognition, feature extraction, classification, speech and vision, is in great part due to their universal approximation properties and generalization ability with respect to unknown data [2]. Moreover the NNs can learn on-line, based on local information, can treat easily multi input multi output (MIMO) systems and their performance degrades gracefully due to the parallel, distributed processing structure and high degree of connectivity among the units. All these remarkable biologically inspired properties made the NNs successful alternative for dynamic modeling and estimation.

In this work, instead of an exhaustive search for the most appropriate parameterized process kinetics structure, a NN is applied to estimate the kinetic rates. In the next section a hybrid model of a general chemical or biochemical process is introduced, with a NN as a kinetic rate sub-model in the framework of a nonlinear state space analytical process model. In section 2.2 a systematic NN training procedure is formulated assuming that all kinetic parameters are available but not all process states are measured. In section 2.3 a similar procedure is developed assuming that the kinetic parameters and the states of the process are partly known. Alternative NN learning algorithms are discussed in 2.4. The new approach is tested for two benchmark problems, estimation of the precipitation rate of calcium phosphate (section 2.5) and estimation of sugar crystallization growth rate (section 2.6).

2.1 Knowledge Based Hybrid Model (KBHM)

A large number of chemical and biochemical processes can be described by the following model, [1], [10].

$$\frac{dX}{dt} = K\varphi(X,T) - DX + U_X \tag{1.1}$$

$$\frac{dT}{dt} = b\varphi(X,T) - d_0 T + U_T \tag{1.2}$$

$$D = \frac{q_{in}}{V}$$

where, for $n, m \in \mathrm{N}$, the constants and variables denote

$X = (x_1(t),.......x_n(t)) \in R^n$	Concentrations or total amounts of n process components
$K = [k_1,.......k_m] \in R^{n \times m}$	kinetic parameters (yield, stoichiometric, or other)
$\varphi = (\varphi_1,.......\varphi_m)^T \in R^m$	Process kinetic rates
T	Temperature
$b \in R^{1 \times m}$	Energy related parameters
d_0	Heat transfer rate related parameter
q_{in} / V	Feeding flow / Volume
D	Dilution rate

U_X and U_T are the inputs by which the process is controlled to follow a desired dynamic behaviour. The nonlinear state-space model (1) proved to be most suitable form of representing several industrial processes. The vector (φ) defines the rate of mass consumption or production of the components. It is usually time varying and dependent of the stage of the process. In the specific case of reaction process systems φ represents the reaction rate vector typical for chemical or biochemical reactions that take place in several processes, such as polymerization, fermentation, biological waste water treatment, etc. In nonreaction processes as for example crystallization and precipitation, φ represents the growth or decay rates of chemical species. In both cases (reaction or nonreaction systems) φ models the process kinetics and is the key factor for reliable description of the components concentrations. In this work a feedforward neural network (FFNN) is chosen to estimate the vector of kinetic rates. The NN submodel is incorporated in the general dynamical model (1) and the mixed structure is termed knowledge-based hybrid model (KBHM), see Fig.1. A systematic procedure for NN-based estimation of kinetic rates is discussed in the next section.

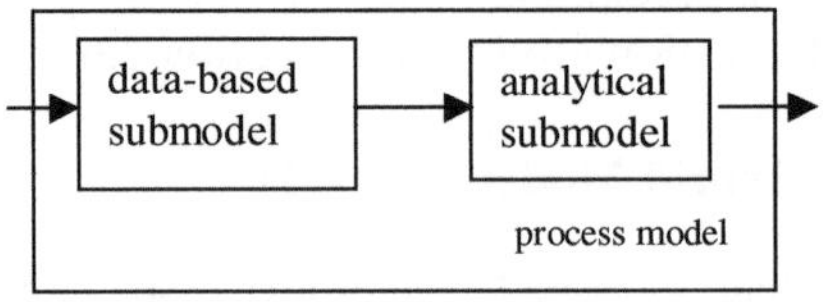

Fig. 1. Knowledge-based hybrid model (KBHM)

2.2 NN Kinetic Rate Estimation with Known Kinetic Parameters

The main challenge for the application of any data-based modeling technique is that the process kinetic rates are not measured variables. The NN supervised learning paradigm is based on error correction principle and in order to update the network weights, an error signal between the network output and the corresponding target (reference output) is required. However, for the kinetic rates, targets (measurements) are not available. Therefore, an alternative NN training procedure is proposed in this work. The NN training structure is augmented with an *Anal. model* module (see Fig.2) with output X for which measurements (target) are available. The proper choice of the partial analytical model and the formulation of the error signal for NN updating are discussed below. The procedure is based on the following assumptions:

(A1) Not all process states of model (1) are measured.
(A2) All kinetic coefficients are known, that is b and all entries of matrix K are available.

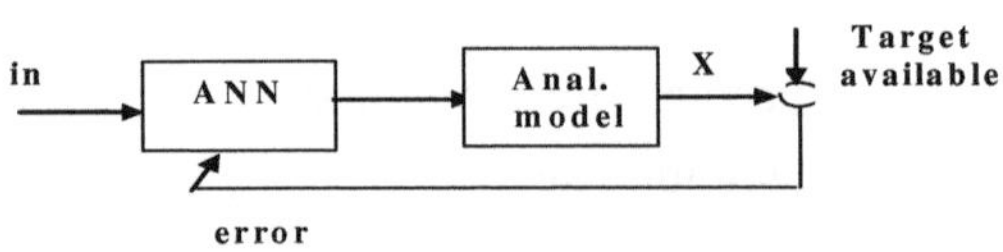

Fig. 2. Hybrid NN training structure

For more convenience, the model (1) is reformulated based on the following augmented vectors

$$X_{aug} = \begin{bmatrix} X \\ T \end{bmatrix}, \qquad X_{aug} \in R^{n+1}, \qquad K_{aug} = \begin{bmatrix} K \\ b \end{bmatrix}, \qquad K_{aug} \in R^{(n+1)\times m}.$$

Then (1) is rewritten as

$$\frac{dX_{aug}}{dt} = K_{aug}\varphi(X_{aug}) - \overline{D}X_{aug} + U, \quad \text{with} \quad \overline{D} = \begin{bmatrix} D & 0 \\ 0 & d_0 \end{bmatrix}, \; U = \begin{bmatrix} U_X \\ U_T \end{bmatrix} \tag{2}$$

Step 1: State vector partition A

The general dynamical model (2) represents a particular class of nonlinear state-space models. The nonlinearity lies in the kinetic rates $\varphi(X_{aug})$ that are nonlinear functions of the state variables X_{aug}. These functions enter the model in the form $K_{aug}\varphi(X_{aug})$, where K_{aug} is a constant matrix, which is a set of linear combinations of the same nonlinear functions $\varphi_1(X_{aug}),.......\varphi_m(X_{aug})$. This particular structure can be exploited to separate the nonlinear part from the linear part of the model by a suitable linear state transformation. More precisely, the following nonsingular partition is chosen [11]

$$LK_{aug} = \begin{bmatrix} K_a \\ K_b \end{bmatrix}, \; rank(K_{aug}) = l, \tag{3.1}$$

where $L \in R^{n\times n}$ is a quadratic permutation matrix, K_a is a lxm full row rank submatrix of K_{aug} and $K_b \in R^{(n-l)\times m}$. The induced partitions of the vectors X_{aug} and U are

$$LX_{aug} = \begin{bmatrix} X_a \\ X_b \end{bmatrix}, \; LU = \begin{bmatrix} U_a \\ U_b \end{bmatrix}, \tag{3.2}$$

with $X_a \in R^l$, $U_a \in R^l$, $X_b \in R^{n-l}$, $U_b \in R^{n-l}$.

According to (3), model (2) is also partitioned into two submodels

$$\frac{dX_a}{dt} = K_a\varphi(X_a, X_b) - \overline{D}X_a + U_a \tag{4}$$

$$\frac{dX_b}{dt} = K_b\varphi(X_a, X_b) - \overline{D}X_b + U_b \tag{5}$$

Based on (3), a new vector $Z \in R^{n+1-l}$ is defined as a linear combination of the state variables

$$Z = A_0X_a + X_b, \tag{6}$$

where matrix $A_0 \in R^{(n+1-l)\times l}$ is the unique solution of

$$A_0 K_a + K_b = 0, \tag{7}$$

that is

$$A_0 = -K_b K_a^{-1}, \tag{8}$$

Note that, a solution for A_0 exist if and only if K_a is not singular. Thus, a necessary and sufficient condition for the existence of a desired partition (3), is that K_a is a *pxm* full rank matrix, which was the initial assumption. Then, the first derivative of vector Z is

$$\begin{aligned}\frac{dZ}{dt} = A_0 \frac{dX_a}{dt} + \frac{dX_b}{dt} &= A_0\left[K_a \varphi(X_a, X_b) - \overline{D}X_a + U_a\right] + K_b \varphi(X_a, X_b) - \overline{D}X_b + U_b \\ &= (A_0 K_a + K_b)\varphi(X_a, X_b) - \overline{D}(A_0 X_a + X_b) + A_0 U_a + U_b\end{aligned} \tag{9}$$

Since matrix A_0 is chosen such that eq. (7) holds, the term in (9) related with φ is cancelled and we get

$$\frac{dZ}{dt} = -\overline{D}Z + A_0 U_a + U_b \tag{10}$$

The state partition A results in a vector Z whose dynamics, given by eq. (9), is independent of the kinetic rate vector φ. In general, (3) is not an unique partition and for any particular case a number of choices are possible.

Step 2: State vector partition B (measured & unmeasured states)

Now a new state partition is defined as sub-vectors of measured and unmeasured states X_1, X_2, respectively. The model (2) is also partitioned into two submodels

$$\frac{dX_1}{dt} = K_1 \varphi(X_1, X_2) - \overline{D}X_1 + U_1 \tag{11.1}$$

$$\frac{dX_2}{dt} = K_2 \varphi(X_1, X_2) - \overline{D}X_2 + U_2 \tag{11.2}$$

From state partitions A and B, vector Z can be represented in the following way

$$Z = A_0 X_a + X_b = A_1 X_1 + A_2 X_2 . \tag{12}$$

The first representation is defined in (6), then applying linear algebra transformations A_1 and A_2 are computed to fit the equality (12). The purpose of state partitions A and B is to estimate the unmeasured states (vector X_2) independently of the kinetic rates (vector φ). The recovery of X_2 is defined as state observer .

Step 3: State observer

Based on (10) and starting with known initial conditions, Z can be estimated as follow (in this work the estimations are denoted by hat)

$$\frac{d\hat{Z}}{dt} = -\overline{D}\hat{Z} + A_0 U_a + U_b \tag{13}$$

Then according to (12) the unmeasured states X_2 are recovered as

$$\hat{X}_2 = A_2^{-1}(\hat{Z} - A_1 X_1) \tag{14}$$

Note that, estimates $\hat{X}_2$ exist if and only if A_2 is not singular [3]. Hence, a necessary and sufficient condition for observability of the unmeasured states is that A_2 is a full rank matrix.

Step 4: Error signal for NN training

The hybrid structure for NN training is shown in Fig. 3, where the adaptive hybrid model (AHM) is formulated as

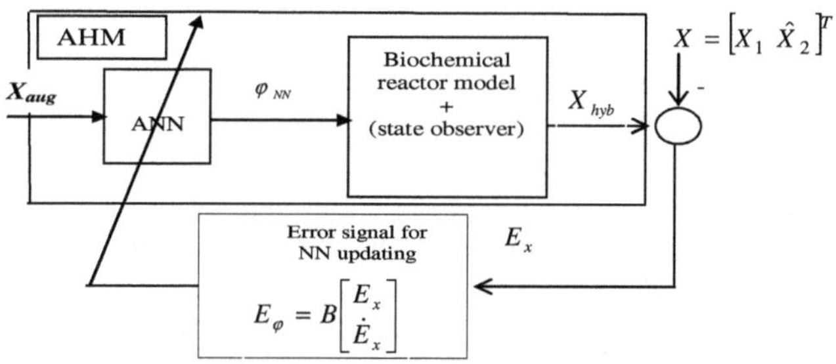

Fig. 3. Hybrid NN-based kinetic rates identification structure

$$\frac{dX_{hyb}}{dt} = K_{aug}\varphi_{NN} - \overline{D}X_{hyb} + U + \Omega(X_{aug} - X_{hyb}) \tag{15}$$

The true (but unknown) process behavior is assumed to be represented by (2). Then the error dynamics is modeled as the difference between (2) and (15)

$$\frac{d(X_{aug} - X_{hyb})}{dt} = K_{aug}(\varphi - \varphi_{NN}) - \overline{D}(X_{aug} - X_{hyb}) + \Omega(X_{aug} - X_{hyb}) \tag{16}$$

The following definitions take place:

$E_x = (X_{aug} - X_{hyb})$ is termed as the observation error,

$E_\varphi = \varphi - \varphi_{NN}$ is the error signal for updating the NN parameters.

X_{aug} consists of the measured (X_1) and the estimated ($\hat{X}_2$) states. Thus, (16) can be rearranged as follows

$$\frac{dE_x}{dt} = K_{aug} E_\varphi - (\overline{D} - \Omega) E_x \tag{17}$$

and from (17) the error signal for NN training is

$$E_\varphi = K_{aug}^{-1} \begin{bmatrix} \overline{D} - \Omega & 1 \end{bmatrix} \begin{bmatrix} E_x \\ \dot{E}_x \end{bmatrix} = B \begin{bmatrix} E_x \\ \dot{E}_x \end{bmatrix}, \quad B = K_{aug}^{-1} \begin{bmatrix} \overline{D} - \Omega & 1 \end{bmatrix} \tag{18}$$

Ω is a design parameter which defines the speed of the observation error convergence. Note that the necessary identifiability condition for the kinetic rate vector is the non singularity of matrix K_{aug}. The cost function to be minimized at each iteration of network training is the sum of squared errors, where N is the time instants over which the optimization is performed (batch mode of training)

$$J = \frac{1}{N} \sum_{k=1}^{N} \left[E_\varphi(k) \right]^2$$

The network parameters (W^{NN}) are updated applying a gradient method such that

$$W_{new}^{NN} = W_{old}^{NN} + f\left(\frac{\partial J}{\partial W_{old}^{NN}} \right) \tag{19}$$

There are a number of algorithms to determine $f\left(\frac{\partial J}{\partial W_{old}^{NN}} \right)$. In our study the Levenberg-Marquardt back-propagation is the preferred algorithms [12], which is a version of Newton method. However alternative methods can be also applied and they are discussed in section 2.4. Note that, the error signal for updating the network parameters is a function of the observation error (E_x) and the speed of the observation error ($\dot{E}_x$). The intuition behind is that the network parameters are changed proportionally to their effect on the prediction of the process states and the prediction of their dynamics.

2.3 NN Kinetic Rate Estimation with Partly Known Kinetic Parameters

In this scenario the following assumptions hold true:

(A1) Not all process states of model (2) are measured.
(A3) Not all kinetic coefficients are known (not all entries of K_{aug} are available).

The states partition A, in the previous section, made possible estimation of the unmeasured states independently and prior to the network training. This enabled the batch mode of NN training providing parameter adaptation only after the complete data set has been processed and the respective errors computed. However, in the present case, when the kinetic parameters (K_{aug}) are partly unknown, the unmeasured states (X_2) has to be estimated jointly with the estimation of K_{aug} and the estimation of the kinetic rates (φ_{NN}). Therefore, the network training is performed simultaneously with the estimation of X_2 and K_{aug}. As a result, the sequential (on-line) mode of NN training is the only possible alternative. The procedure for NN kinetic rate estimation can be summarized as follows:

1) Choose initial guess for the unknown kinetic parameters $\hat{K}_{estim}$ and define the vector of estimated kinetic parameters as
$\hat{K}_{aug} = \lfloor K_{known} \quad \hat{K}_{estim} \rfloor$

2) Choose an appropriate state partition A: $L\hat{K}_{aug} = \begin{bmatrix} \hat{K}_a \\ \hat{K}_b \end{bmatrix}$, and based on (7) compute

$\hat{A}_0 = -\hat{K}_b \hat{K}_a^{-1}$.

3) Define the state partition B:
$Z = \hat{A}_0 X_a + X_b = \hat{A}_1 X_1 + \hat{A}_2 X_2$ and compute matrices $\hat{A}_1$ and $\hat{A}_2$.

4) Estimate the unmeasured states (X_2) as

$$\frac{d\hat{Z}}{dt} = -\overline{D}\hat{Z} + \hat{A}_0 U_a + U_b \tag{20}$$

$$\hat{X}_2 = \hat{A}_2^{-1}(\hat{Z} - A_1 X_1) \tag{21}$$

5) Provide the NN with input data (the measured and the estimated states), compute the network output (φ_{NN}), the error signal for NN training (E_φ) and update the network parameters by (19).

6) Update the values of $\hat{K}_{aug}$ by

$$\frac{d\hat{K}_{estim}}{dt} = \Gamma(X_1 - \hat{X}_1) \tag{22}$$

where

$$\frac{d\hat{X}_1}{dt} = \hat{K}_1 \varphi_{NN} - \overline{D}\hat{X}_1 + U_1 - \Omega(X_1 - \hat{X}_1) \tag{23}$$

7) Go back to step 2 or stop if the quality criteria is satisfied.

Eq. (23) controls the convergence of kinetics coefficients estimation. It is a copy of the model (11.1) with an extra term, proportional to the error between the measured (X_1) and the estimated state ($\hat{X}_1$). In case of perfect estimation this term is equal to zero. Ω and Γ are design parameters that can significantly speed up the adaptation procedure.

Error signal for NN training

The adaptive hybrid model (AHM) is formulated as

$$\frac{dX_{hyb}}{dt} = \hat{K}_{aug}\varphi_{NN} - \overline{D}X_{hyb} + U + \Omega(X_{aug} - X_{hyb}) \tag{24}$$

Subtracting (2) from (24) we obtain the error dynamics

$$\frac{dE_x}{dt} = K_{aug}\varphi - \hat{K}_{aug}\varphi_{NN} + (\Omega - \overline{D})E_x \tag{25}$$

Define an extended error signal for NN training ($E_{K\varphi}$) as follows:

$$E_{K\varphi} = K_{aug}\varphi - \hat{K}_{aug}\varphi_{NN}\,, \tag{26}$$

and the final form of the NN training error is

$$E_{K\varphi} = B\begin{bmatrix} E_x \\ \dot{E}_x \end{bmatrix}, \quad B = \begin{bmatrix} \overline{D} - \Omega & 1 \end{bmatrix} \tag{27}$$

The intuition behind (27) is that the network adaptation depends on the reliable estimation of $\hat{K}_{aug}$. If $K_{aug} = \hat{K}_{aug}$ the error reduces to (18)

$$E_{\varphi} = \varphi - \varphi_{NN} = \tilde{B}\begin{bmatrix} E_x \\ \dot{E}_x \end{bmatrix}, \quad \tilde{B} = K_{aug}^{-1}\begin{bmatrix} \overline{D} - \Omega & 1 \end{bmatrix} \tag{28}$$

Remark: The structural identifiably of the kinetic coefficients is comprehensively studied in [11]. It is proved that the identifiability properties of matrix K_{aug} are invariant with respect to the choice of partition A. The authors define necessary condition according to which if the rank of K_{aug} is *l* at most *n-l* coefficients in each reaction can be identified without the knowledge of the kinetic rates.

2.4 Neural Networks Learning Alternatives

Backpropagation (BP) algorithm – steepest gradient descent

Among various NN training algorithms the BP algorithm is the most widely implemented for modeling purposes [13]. In fact, BP is a generalization of least mean square (LMS) algorithm of Widrow and Hoff, [14] for multilayer feedforward ANN. Standard BP is a gradient descent algorithm in which the network weights are moved in the steepest descent direction i.e. along the negative of the gradient of the performance index. This is the direction in which the performance index is decreasing most rapidly. The term backpropagation refers to the manner in which the gradient is computed for nonlinear multilayer networks. Properly trained BP networks tend to give reasonable answers when presented with inputs that they have never seen. This generalization property makes it possible to train a network on a representative set of input/target pairs and get good results without training the network on all possible input/output pairs. One (k) iteration of the gradient steepest descent algorithm can be written as

$$w_{k+1} = w_k - \alpha_k g_k, \quad g_k = \frac{\partial P_k}{\partial w_k}, \tag{29}$$

where w_k is a vector of current weights and biases, g_k is the current gradient of the performance index P_k and α_k is the learning rate. There are two different ways in which this algorithm can be implemented: incremental mode and batch mode. In the incremental mode, the gradient is computed and the weights are updated after each input is applied to the network. In the batch mode all of the inputs are applied to the network before the weights are updated. It turns out that, although the performance function decreases most rapidly along the negative of the gradient, this does not necessarily produce the fastest convergence.

Conjugate gradient algorithm

An alternative to the steepest descent algorithm is the conjugate gradient algorithm (with many variations) [15] where a search is performed along conjugate directions, which produces generally faster convergence than steepest descent directions

$$w_{k+1} = w_k + \alpha_k r_k \tag{30}$$

All of the conjugate gradient algorithms start out by searching in the steepest descent direction (negative of the gradient) on the first iteration

$$r_{01} = -g_0 \tag{31}$$

A line search is then performed to determine the optimal distance to move along the current search direction. Then the next search direction is determined so that it is conjugate to previous search directions. The general procedure for determining the new search direction is to combine the new steepest descent direction with the previous search direction

$$r_k = -g_k + \beta_k r_{k-1} \tag{32}$$

The various versions of conjugate gradient are distinguished by the manner in which the constant β_k is computed. For the Fletcher-Reeves update the procedure is

$$\beta_k = \frac{g_k^T g_k}{g_{k-1}^T g_{k-1}} \tag{33}$$

This is the ratio of the norm squared of the current gradient (g_k) to the norm squared of the previous gradient (g_{k-1}). For the Polak-Ribiére update, the constant β_k is computed as

$$\beta_k = \frac{\Delta g_{k-1}^T g_k}{g_{k-1}^T g_{k-1}} \tag{34}$$

This is the inner product of the previous change in the gradient (Δg_{k-1}^T) with the current gradient divided by the norm squared of the previous gradient.

Newton's method

A third alternative to the steepest descent algorithm for fast optimization is the Newton's method [9]. The basic step of Newton's method is

$$w_{k+1} = w_k - H_k^{-1} g_k , \; H_k = \frac{\partial P_k^2}{\partial w_k \partial w_k} \tag{35}$$

where H_k is the Hessian matrix (second derivatives) of the performance index P_k at the current values (k) of the weights and biases. Newton's method often converges faster than steepest descent and conjugate gradient methods. Unfortunately, it is complex and expensive to compute the Hessian matrix for feedforward ANN.

Quasi-Newton method - Levenberg-Marquardt algorithm

There is a class of algorithms that is based on Newton's method, but which doesn't require calculation of second derivatives. These are called quasi-Newton methods. They update an approximate Hessian matrix at each iteration of the algorithm. The update is computed as a function of the gradient. Levenberg-Marquardt algorithm, implemented in our application, is one of the celebrating quasi-Newton methods. It is designed to approach second-order training speed without having to compute the Hessian matrix. When the performance function has the form of a sum of squares of the errors (as it is typical in training a feedforward NN), then the Hessian matrix can be approximated as

$$H_k = J_k^T J_k \tag{36}$$

where J_k is the Jacobian matrix that contains first derivatives of the network errors (e_k) with respect to the weights and biases

$$J_k = \frac{\partial e_k}{\partial w_k}; \quad e_k = y_m - t . \tag{37}$$

where y_m is the network output vector and t is the target vector. Then the gradient can be computed as

$$g_k = J_k e_k \tag{38}$$

The computation of the Jacobian matrix is less complex than computing the Hessian matrix. The Levenberg-Marquardt algorithm updates the weights in the following way

$$w_{k+1} = w_k - \left[J_k^T J_k + \mu I \right]^{-1} J_k^T e_k \tag{39}$$

When the scalar μ is zero, this is just Newton's method, using the approximate Hessian matrix. When μ is large, this becomes gradient descent with a small step size. Newton's method is faster and more accurate near an error minimum, so the aim is to shift towards Newton's method as quickly as possible. Thus, μ is decreased after each successful step (reduction in performance function) and is increased only when a tentative step would increase the performance function. In this way, the performance function will always be reduced at each iteration of the algorithm.

2.5 Case Study 1 – Precipitation Rate Estimation of Calcium Phosphate

The precipitation of calcium phosphate was studied by many authors under different conditions [16], [17]. Depending on the temperature, the level of supersaturation, pH and initial concentration of reagents, one can obtain different calcium phosphate phases. One of them is the dicalcium phosphate dehydrate (DCPD) known also as brushite. DCPD is recognized as an important product in the application of fertilizers to soil and is studied mainly for its role in the physiological formation of calcium phosphates.

For the present study, the precipitation of DCPD was performed in a batch laboratory crystallizer. The precipitation was carried out by mixing equimolar quantities of calcium hydroxide suspension and orthophosphoric acid solution. Five successive stages were identified during a number of experiments performed with different initial reagent concentrations [21]: i) spontaneous precipitation of hydroxyapatite (HAP); ii) complete dissolution of calcium and HAP growth; iii) appearance of first nuclei of brushite; iv) coexistence of HAP and brushite and v) transformation of HAP into brushite and growth of brushite. The last stage represents the main challenge with respect to the precipitation rate modeling because it has to take simultaneously into account two kinetics phenomena: brushite grows first, due to direct consumption of calcium in the solution and second, due to the transformation of HAP into brushite.

According to the general dynamical model (2), the state space representation of the last stage, namely the transformation of HAP into brushite and growth of brushite, is

$$\frac{dM_c}{dt} = -q_{m1}\varphi(M_c, M_B) \tag{40.1}$$

$$\frac{dM_{HAP}}{dt} = -K_{HAP}(M_{HAP})^2 \tag{40.2}$$

$$\frac{dM_B}{dt} = \varphi(M_c, M_B) + 10q_{m2}K_{HAP}(M_{HAP})^2 \tag{40.3}$$

where M_c is the mass of calcium into solution, M_{HAP} is the mass of HAP and M_B is the mass of brushite. φ is the precipitation rate (the main modeling challenge), K_{HAP}, q_{m1}, q_{m2} are kinetics coefficients assumed as known. Note that M_{HAP} is not a function of φ , therefore only M_c (the measured state) and M_B (the unmeasured state) explicitly depend on the precipitation rate. The objective is to estimate M_B independently of the precipitation rate. In this simple illustrative case the state partitions A and B are equal

$$X_a = X_1 = M_c, \; X_b = X_2 = M_B, \tag{41}$$

with

$$K_a = -q_{m1}, \; K_b = 1 \tag{42}$$

$$A_0 = A_1 = 1/q_{m1} \qquad A_2 = 1. \tag{43}$$

According to (13-14), the state observer of M_B is the following

$$\frac{d\hat{M}_{HAP}}{dt} = -K_{HAP}\left(M_{HAP}\right)^2, \tag{44}$$

$$\frac{d\hat{Z}}{dt} = 10q_{m2}K_{HAP}\left(\hat{M}_{HAP}\right)^2, \tag{45}$$

$$\hat{M}_B = \hat{Z} - \frac{1}{q_{m1}}M_c. \tag{46}$$

The observation error is

$$E_x = \begin{bmatrix} M_c - M_{chyb} \\ \hat{M}_B - M_{Bhyb} \end{bmatrix}, \tag{47}$$

with respective matrices

$$\text{D=0}, \quad \Omega = \begin{bmatrix} -\omega_1 & 0 \\ 0 & -\omega_2 \end{bmatrix}. \tag{48}$$

Finally, the NN error is

$$E_\varphi = \frac{1}{q_{m1}}\begin{bmatrix} \left(\dot{M}_c - \dot{M}_{chyb}\right) - \omega_1\left(M_c - M_{chyb}\right) \\ \left(\dot{\hat{M}}_B - \dot{M}_{Bhyb}\right) - \omega_2\left(\hat{M}_B - M_{Bhyb}\right) \end{bmatrix}. \tag{49}$$

Note, that the error is composed of two terms with ω_1, ω_2 being the weights of each first order error term. The usual choice of the weights is $\omega_1 = \omega_2 = 0.5$ when no extra information for the importance of each term is not available, and these are the values selected.

For the numerical realization, the error at each iteration is determined as follows

$$E_{\varphi(i)} = \frac{1}{q_{m1}}\left[\frac{M_{c(i)} - M_{c(i-1)}}{T} - \frac{M_{c(i)hyb} - M_{c(i-1)hyb}}{T} - \omega_1\left(M_{c(i)} - M_{c(i)hyb}\right)\right] + \left[\frac{\hat{M}_{B(i)} - \hat{M}_{B(i-1)}}{T} - \frac{M_{B(i)hyb} - M_{B(i-1)hyb}}{T} - \omega_2\left(\hat{M}_{B(i)} - M_{B(i)hyb}\right)\right] \tag{50}$$

A feedforward NN with 2 inputs (M_c, M_B) , 1 output (φ_{NN}), one hidden layer with 7 sigmoid nodes was trained to minimize the error (50) and to model the precipitation rate. The Levenberg-Marquardt (LM) algorithm was chosen as the training method due to its advantages in terms of execution time and robustness [12]. Since the LM algorithm requires a lot of memory, a powerful (in terms of memory) computer is the main condition for successful training. In order to solve the problem of several local minima, that is typical for all gradient based optimization algorithms (including the

LM method), we have repeated several time the optimization specifying different starting points.

The hybrid model is evaluated for its ability to predict process behavior for various initial concentrations of reagents (0.05, 0.1, 0.2, 0.3, and 0.4 M). Data for the main system states (M_c - measured, M_{HAP} and M_B - estimated) are denoted by dashed line in all subplots of Fig. 4. Data for average particle size (AM) are denoted by stars in all subplots b). The model time trajectories of M_c (subplots a)) and AM (subplots b)) are direct indicators of the model quality since measurements for them are available. The mass of HAP and brushite are not directly measured variables but they can be inferred by the available measurements. Thus, the plots of M_{HAP} (subplots c)) and M_B (subplots d)) are indirect indicators of the model quality. Data from experiments with initial concentrations of reagents 0.2 and 0.3 M were used to train the NN because more data with these concentrations were available. On Fig.4 are depicted only results where the model is tested on new 'unseen' validation data corresponding to experiments with initial concentrations (0.05, 0.1 and 0.4 M). Data and model trajectories match quite well which confirms the ability of the hybrid NN to estimate the precipitation rate of calcium phosphate. We suppose, different results may be obtained if other concentrations would have been used for NN training, however we do not expect qualitative difference.

For the purpose of comparative analysis, three existing analytical models were also applied on the same task, see Table1. The parameters of the fixed models were tuned with the same data obtained by experiments with (0.2 and 0.3 M) initial concentrations of reagents. Here, we take the nondimensional error index (NDEI), which is defined as the root mean square error (RMSE) divided by the standard deviation of the target series.

Table 1. Estimation results for 3 analytical models and the hybrid model

Calcium phosphate precipitation rate models	NDEI for model validation data
Monod type model [3] $\varphi = \frac{\lambda_{HAP1} M_{HAP}}{\beta_{HAP1} + M_{HAP}} + \frac{\lambda_{c1} M_c}{\beta_{c1} + M_c}$ $\lambda_{HAP1}, \lambda_{c1}, \beta_{HAP1}, \beta_{c1}$ - tuning parameters	0.404
Contois type model [1] $\varphi = \frac{\lambda_{HAP2} M_{HAP}}{\beta_{HAP2} M_B + M_{HAP}} + \frac{\lambda_{c2} M_c}{\beta_{c2} M_B + M_c}$ $\lambda_{HAP2}, \lambda_{c2}, \beta_{HAP2}, \beta_{c2}$ - tuning parameters	0.096
"Logistic" type model [1] $\varphi = \exp(-\lambda_3 M_B)$, λ_3 - tuning parameter	0.062
Hybrid model (this paper) FFNN with 2 inputs (M_c, M_B) , 1 output (φ_{NN}) one hidden layer with 7 sigmoid nodes	0.023

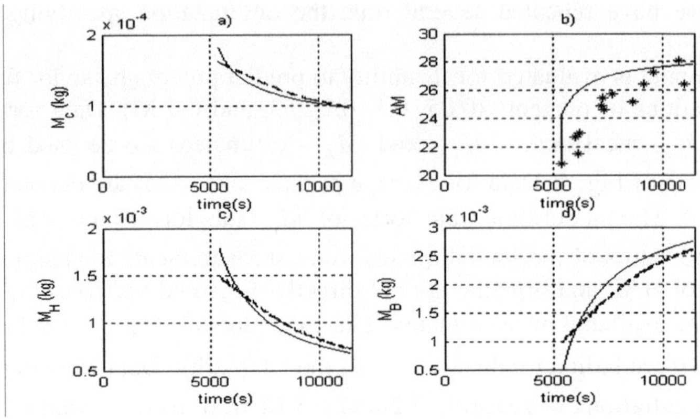

4.1) Initial concentration of reagents 0.05 M

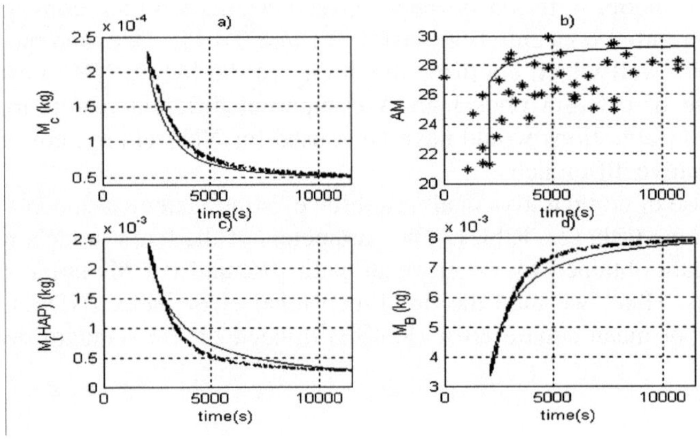

4.2) Initial concentration of reagents 0.1 M

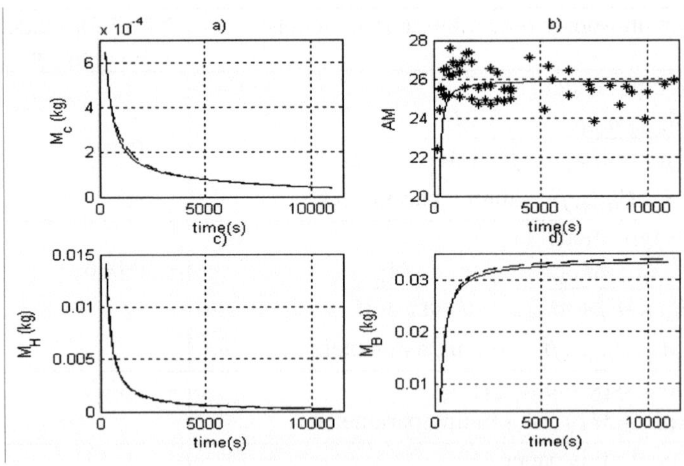

4.3) Initial concentration of reagents 0.4 M

Fig. 4. Model validation. Data points (dashed line or stars), model estimations (solid lines) a) mass of calcium, b) average particle size [μm]; c) mass of HAP; d) mass of brushite.

2.6 Case Study 2 – Estimation of Sugar Crystallization Growth Rate

Sugar crystallization occurs through mechanisms of nucleation, growth and agglomeration that are known to be affected by several not well-understood operating conditions. The search for efficient methods for process description is linked both to the scientific interest of understanding fundamental mechanisms of the crystallization process and to the relevant practical interest of production requirements. The sugar production batch cycle is divided in several phases. During the first phase the pan is partially filled with a juice containing dissolved sucrose. The liquor is concentrated by evaporation, under vacuum, until the supersaturation reaches a predefined value. At this point seed crystals are introduced into the pan to induce the production of crystals (crystallization phase). As evaporation takes place further liquor or water is added to the pan. This maintains the level of supersaturation and increases the volume contents. The third phase consists of tightening which is controlled by the evaporation capacity [18]. Since the objective of this paper is to illustrate the technique introduced in section 4, the following assumptions are adopted:

i) Only the states that explicitly depend on the crystal growth rate are extracted from the comprehensive mass balance process model;
ii) The population balance is expressed only in terms of number of crystals;
iii) The agglomeration phenomenon is neglected.

The simplified process model is then [5]

$$\frac{dM_s}{dt} = -k_1 G + F_f \rho_f B_f Pur_f \tag{51.1}$$

$$\frac{dM_c}{dt} = k_1 G \tag{51.2}$$

$$\frac{dT_m}{dt} = k_2 G + bF_f + cJ_{vap} + d \tag{51.3}$$

$$\frac{dm_0}{dt} = k_3 G \tag{51.4}$$

where G is the crystallization growth rate (the principal modeling challenge), M_s is the mass of dissolved sucrose, M_c is the mass of crystals, T_m is the temperature of the massecuite, m_0 is the number of crystals. Pur_f and ρ_f are the purity (mass fraction of sucrose in the dissolved solids) and the density of the incoming feed. F_f is the feed flowrate, J_{vap} is the evaporation rate and *b, c, d* are parameters incorporating the enthalpy terms and specific heat capacities. They are derived as functions of physical and thermodynamic properties. The full state vector is

$$X_{aug} = [M_s \quad M_c \quad T_m \quad m_0]^T \text{ , with } K_{aug} = [-k_1 \quad k_1 \quad k_2 \quad k_3]^T .$$

Remark: The crystallization is not a reaction process, however, the describing model (Eqns. 51) fits the framework of the formalism developed in sections 2.2 and 2.3.

Scenario 1: Completely known kinetic coefficients

We chose the following state partition A : $X_a = M_c$, $X_b = [M_s \quad T_m \quad m_0]^T$ and the solution of equation (7) is

$$A_0 = \begin{bmatrix} 1 & -\frac{k_2}{k_1} & -\frac{k_3}{k_1} \end{bmatrix}^T \tag{52}$$

M_c and T_m are the measured states, then the unique state partition B is

$$X_1 = [M_c \quad T_m]^T, \; X_2 = [M_s \quad m_0]^T,$$

Taking into account (52), the matrices of the second representation of vector Z in (12) are computed as

$$A_1 = \begin{bmatrix} 1 & -\frac{k_2}{k_1} & -\frac{k_3}{k_1} \\ 0 & 1 & 0 \end{bmatrix}^T, \quad A_2 = \begin{bmatrix} 1 & 0 & 0 \\ 0 & 0 & 1 \end{bmatrix}^T$$

For this case D=0, then the estimation of the individual elements of Z are

$$\hat{Z}_1 = M_c + \hat{M}_s, \quad \hat{Z}_2 = -\frac{k_2}{k_1} M_c + T_m, \quad \hat{Z}_3 = -\frac{k_3}{k_1} M_c + \hat{m}_0 \tag{53}$$

The analytical expression for the estimation of the unmeasured states is then

$$\begin{bmatrix} \hat{M}_s \\ \hat{m}_0 \end{bmatrix} = \begin{bmatrix} 1 & 0 & 0 \\ 0 & 0 & 1 \end{bmatrix} \left(\begin{bmatrix} \hat{Z}_1 \\ \hat{Z}_2 \\ \hat{Z}_3 \end{bmatrix} - \begin{bmatrix} 1 & 0 \\ -\frac{k_2}{k_1} & 1 \\ -\frac{k_3}{k_1} & 0 \end{bmatrix} \begin{bmatrix} M_c \\ T_m \end{bmatrix} \right) \tag{54}$$

The observation error is defined as

$$E_x = \begin{bmatrix} \hat{M}_s - M_{shyb} \\ M_c - M_{chyb} \\ \hat{m}_0 - m_{0hyb} \\ T_m - T_{mhyb} \end{bmatrix} \tag{55}$$

In the numerical implementation the first derivative of the observation error is computed as the difference between the current $E_x(k)$ and the previous value $E_x(k-1)$ of the observation error divided by the integration step (Δt)

$$\dot{E}_x = \frac{E_x(k) - E_x(k-1)}{\Delta t} \tag{56}$$

Scenario 2: Partly known kinetic coefficients

Apart from k_3, all other kinetic coefficients are usually badly known, therefore this is the real scenario in practice. The procedure is similar to *Scenario 1*, however in all expressions k_1, k_2 are substituted by their estimations $\hat{k}_1, \hat{k}_2$. In order to really compare the two scenarios the same state partitions A and B are chosen. An extra step is introduced to estimate the two unknown coefficients (k_1, k_2)

$$\begin{bmatrix} \dfrac{d\hat{k}_1}{dt} \\ \dfrac{d\hat{k}_2}{dt} \end{bmatrix} = \begin{bmatrix} \gamma_1 & 0 \\ 0 & \gamma_2 \end{bmatrix} \begin{bmatrix} M_c - \hat{M}_c \\ T_m - \hat{T}_m \end{bmatrix} \tag{57}$$

$$\begin{bmatrix} \dfrac{d\hat{M}_c}{dt} \\ \dfrac{d\hat{T}_m}{dt} \end{bmatrix} = \begin{bmatrix} \hat{k}_1 \\ \hat{k}_2 \end{bmatrix} G_{NN} + \begin{bmatrix} 0 \\ bF_f + cJ_{vap} + d \end{bmatrix} - \begin{bmatrix} \omega_1 & 0 \\ 0 & \omega_2 \end{bmatrix} \begin{bmatrix} M_c - \hat{M}_c \\ T_m - \hat{T}_m \end{bmatrix} \tag{58}$$

The initial values assigned for the unknown coefficients are $\hat{k}_1 = \hat{k}_2 = 0.01$. As for the design parameters, standard values are chosen, $\omega_1 = \omega_2 = 0.5$ and $\gamma_1 = \gamma_2 = 1$.

The observation error and the extended NN error signal are:

$$E_x = \begin{bmatrix} M_c - M_{chyb} \\ T_m - T_{mhyb} \end{bmatrix}, \qquad E_{K\varphi} = \begin{bmatrix} \left(\dot{M}_c - \dot{M}_{chyb}\right) - \omega_1\left(M_c - M_{chyb}\right) \\ \left(\dot{T}_m - \dot{T}_{mhyb}\right) - \omega_2\left(T_m - T_{mhyb}\right) \end{bmatrix} \tag{59}$$

For each scenario, a feedforward NN with 4 inputs (M_c, T_m, m_0, M_s), 1 output (G_{NN}), one hidden layer with 5 sigmoid nodes was trained to minimize the error (59) and to estimate the growth rate. For two of the inputs (M_c, T_m) industrial data of six batches is provided, the rest of the inputs are estimated.

The two hybrid models obtained here are compared with an analytical model of the sugar crystallization, reported in [5]. The difference between the two models is in the way the growth rate is estimated. In the hybrid models G is estimated by a NN following the techniques discussed above. In the analytical model G is given by the following empirical correlation

$$G = K_g \exp\left[-\frac{57000}{R(T_m + 273)}\right](S-1)\exp\left[-13.863(1 - P_{sol})\right]\left(1 + 2\frac{V_c}{V_m}\right), \tag{60}$$

where S is the supersaturation, P_{sol} is the purity of the solution and V_c / V_m is the volume fraction of crystals. K_g is a constant, optimized following a classical non-linear least-squares regression. The performance of the three models is examined with respect to prediction quality of the crystal size distribution (CSD) at the end of the process which is quantified by two parameters - the final average (in mass) particle size (AM) and the final coefficient of particle variation (CV). In Table 2, the predictions given by the models are compared with the experimental data for the CSD, coming from 8 batches not used for training. The hybrid models outperform the analytical model particularly with respect to predictions of CV. The predictions based on *Scenario 1* and *Scenario 2* are almost equal and it is expected because the CSD are computed only at the batch end and at that time the estimated kinetics parameters and the actual values are practically the same.

Table 2. Final CSD – experimental data versus hybrid and analytical model predictions

batch No.	experimental data		Hybrid model (*Scenario 1*)		Hybrid model (*Scenario 2*)		analytical model	
	AM [mm]	CV [%]	AM [mm]	CV [%]	AM [mm]	CV [%]	AM [mm]	CV [%]
1	0.479	32.6	0.5	30.9	0.49	29.2	0.583	21.26
2	0.559	33.7	0.51	31.19	0.51	30.17	0.542	18.43
3	0.680	43.6	0.51	33.13	0.52	33.81	0.547	18.69
4	0.494	33.7	0.53	31.93	0.51	31.7	0.481	14.16
5	0.537	32.5	0.54	30.02	0.55	29.66	0.623	24.36
6	0.556	35.5	0.51	31.21	0.51	29.13	0.471	13.642
7	0.560	31.6	0.58	34.17	0.59	34.34	0.755	34.9
8	0.530	31.2	0.58	34.96	0.58	35.16	0.681	27.39
av. err			**5.7%**	**12.1**	**5.9%**	**10.2%**	**13.7%**	**36.1%**

3 ANN-Based Model Predictive Control

3.1 General MPC Problem Formulation

Nonlinear model predictive control (NMPC) is an optimisation-based multivariable constrained control technique that uses a nonlinear dynamic model for the prediction of the process outputs [19]. At each sampling time the model is updated on the basis of new measurements and state variables estimates. Then the open-loop optimal manipulated variable moves are computed over a finite (predefined) prediction horizon with respect to some performance index, and the manipulated variables for the subsequent prediction horizon are implemented. Then the prediction horizon is shifted or shrunk by usually one sampling time into the future, and the previous steps are repeated. The optimal control problem in the NMPC framework can be mathematically formulated as:

$$\min_{u_{\min}\le u(t)\le u_{\max}} \quad J=\varphi(x(t),u(t),P), \tag{61}$$

subject to:

$$\dot{x}=f(x(t),u(t),P), \quad 0\le t\le t_f, \quad x(0)=x_0 \tag{62.1}$$

$$y(t)=h(x(t),P)$$

$$g_j(x)=0, \quad j=1,2,......p \tag{62.2}$$

$$v_j(x)\le 0, \quad j=1,2,......l$$

where (61) is the performance index, (62) is the process model, function f is the state-space description, function h is the relationship between the output and the state, P is the vector of possibly uncertain parameters and t_f is the final time. $x(t) \in R^n$, $u(t) \in R^m$ and $y(t) \in R^p$ are the state, the manipulated input and the control output vectors, respectively. The manipulated inputs, the state and the control outputs are subject to the following constraints, $x(t) \in \mathrm{X}$, $u(t) \in \mathrm{Z}$, $y(t) \in \mathrm{Y}$ in which X, Z and Y are convex and closed subsets of R^n, R^m and R^p. g_j and v_j are the equality and inequality constrains with p and l dimensions respectively.

3.2 Error Tolerant ANN-MPC Control Framework

We propose a study framework for the ANN-MPC closed loop control system which is summarized in Fig.5. The block *Process (KBHM)* simulates the process subject to control and corresponds to the model introduced in section 2.1. The block *ANN process model* is the predictive model required by the general MPC formulation, eq. (62). It predicts future process responses to potential control signals over the prediction horizon. The predictions are supplied to the block *Optimization procedure* to determine the values of the control action over a specified control horizon according to the following performance index

$$u(t+k) = \begin{cases} u: \begin{cases} \min\limits_{[u(t+k),u(t+k+1,\ldots.u(t+c))]} J = \\ \lambda_1 \sum\limits_{k=1}^{p} \left(e(t+k)\right)^2 - \lambda_2 \sum\limits_{k=1}^{c} \left(\Delta u\right)^2 \end{cases}, & if \quad |e(t+k)| > \alpha \\ \\ u^* & if \quad |e(t+k)| < \alpha \end{cases} \tag{63}$$

where $e(t+k) = r(t+k) - y_{NN}(t+k), \quad \Delta u = u(t+k-1) - u(t+k-2)$

Subject to the following constrains:

$$\begin{aligned} &u_{\min} \le u(t+k) \le u_{\max}, \quad k = 1,2,\ldots c \\ &u_{\min} \le \Delta u \le u_{\max} \\ &y_{\min} \le y_{NN}(t+k) \le y_{\max}, \quad k = 1,2,\ldots p \end{aligned} \tag{64}$$

Eq. (63) is a particular digital format of the general performance index defined by (61) [20]. We term it as an error tolerant MPC formulation because the optimization is performed only when the error $e(t+k)$ is bigger than a predefined α value. In order to reduce the computational burden when the absolute error $e(t+k)$ is smaller than α the control action is equal to u^* which is the last value computed before the error enters the α strip. The prediction horizon p is the number of time steps over which the prediction errors are minimized and the control horizon c is the number of time steps over which the control increments are minimized, r is the desired response and

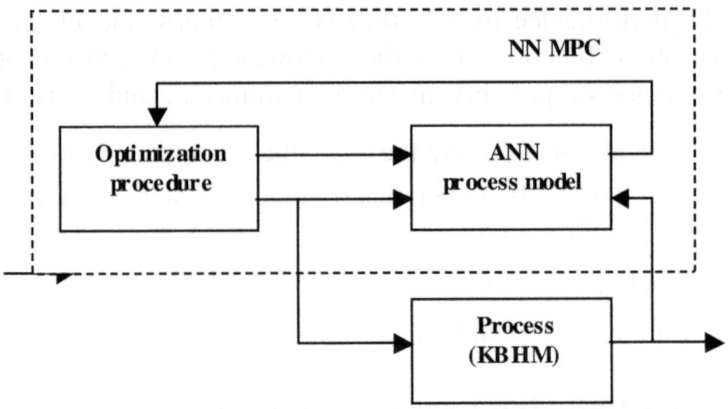

Fig. 5. ANN-based model predictive control (ANN-MPC)

y_{NN} is the network model response. $u(t+k), u(t+k+1),.....u(t+c)$ are tentative values of the future control signal, which are limited by $u_{\min}$ and $u_{\max}$, $y_{\min}$ and $y_{\max}$ are the technological bounds of the controlled output.

Parameters λ_1 and λ_2 determine the contribution of the output error and the contribution of the control increments respectively on the performance index. The length of the prediction horizon is crucial for achieving tracking and stability. For small values of p the tracking deteriorates however for high p values the bang-bang behavior of the process input might be a real problem. The MPC controller requires a significant amount of on-line computation, since the optimization (63) is performed at each sample time to compute the optimal control input. However, the α error tolerance proposed inhere reduces significantly the computational efforts. At each step only the first control action is implemented to the process.

3.3 ANN-MPC Control Applied to a Sugar Crystallization Process

The implementation of ANN-MPC control for the sugar crystallization (Case Study 2) is schematically represented in Fig. 6. The process consists of a number of comparatively independent stages: charging concentration, seeding, setting the grain, crystallization, tightening and discharge. Since the crystallization is the main stage responsible for the final product quality, the control efforts were focused on this stage. The feedback control policy is based on measurements of the supersaturation (by a refractometer), considered as the controlled variable, and measurements of the feed flow, considered as the control action. The closed loop system is simulated in Matlab/Simulink framework as a set of modules. The Matlab NN Toolbox is also required. The controller is designed as an independent block and the crystallization process is simulated as the KBHM model obtained in section 2.6. The KBHM model is coded as an S-function required by Simulink. First, the ANN process model is identified in a specific Plant Identification window (Fig. 7). In this window are specified the network architecture (number of inputs, outputs, layers, type and number of neurons in a layer), training data limitations and training parameters (number of epochs, training algorithm).

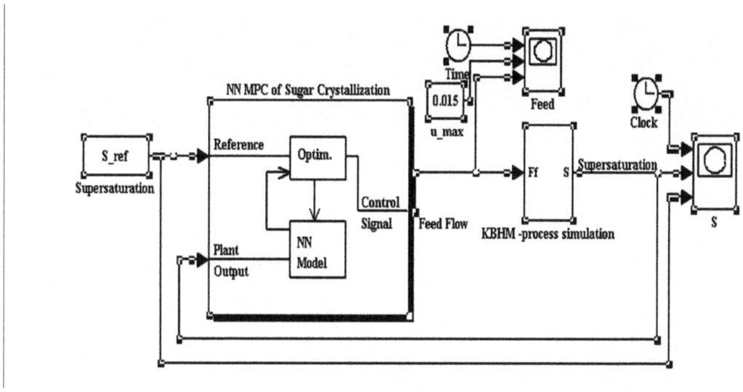

Fig. 6. ANN MPC control of sugar crystallization

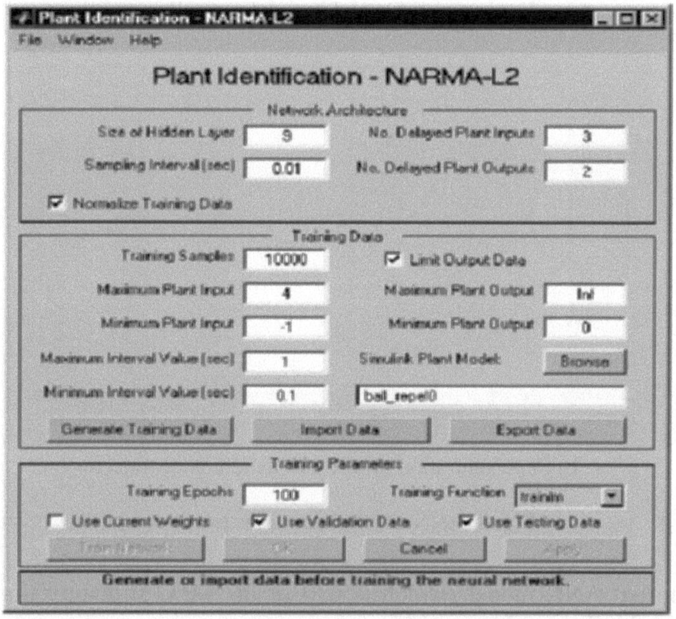

Fig. 7. Plant Identification block

The ANN is trained in a batch mode by Levenberg-Marquardt algorithm, eq.(39). Data are divided into training, validation and simulation portions. Before introducing to the ANN, it is normalized in the range (-1,1) and after processing over the network, the network outputs are denormalized. While the training and the validation errors are significantly different, the generalization properties of the network are not reliable and the training has to continue. The learning is stopped when the training and the validation errors are sufficiently close.

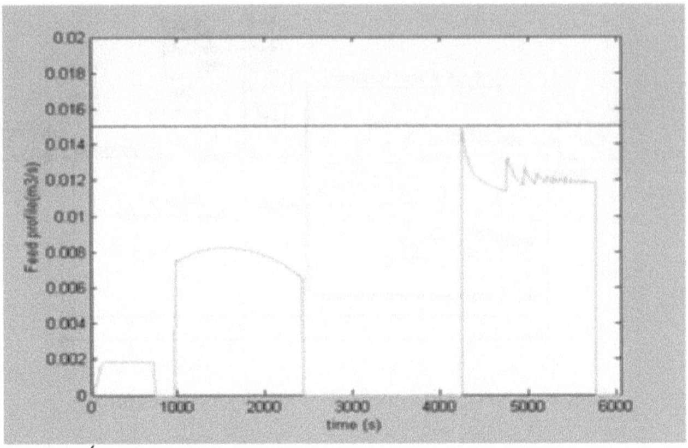

a) Feed rate profile and its maximum allowed value

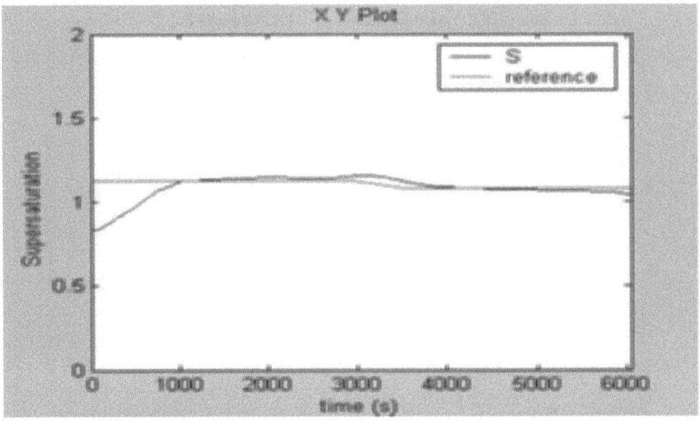

b) Supersaturation profile and its reference trajectory

Fig. 8. ANN-MPC simulations p=10, c=4

The identified ANN model (which is part of the MPC structure) consists of one hidden layer with 7 sigmoid activation functions and one output linear layer. Simulation results are summarized in Fig. 8 and Fig. 9. The process manipulated input (the feed flow rate) is depicted in subplot (a) and the process controlled output (the supersaturation) is depicted in subplot (b) for control horizon c=4 and prediction horizon p=10 (Fig. 8) and p=4 (Fig. 9). During the stages of concentration and feeding with liquor, the reference was set at Sref=1.15 and afterwards was reduced to Sref=1.05. A smooth transition between the two levels was determined to overcome possible overreaction of the tracking controller. The graphics show satisfactory reference tracking with an acceptable smooth behaviour of the control input which stays within the technological constrains defined with u_{max}=0.015 [m3/s]. Higher the prediction horizon better is the tracking but to the expense of more vivid manipulated input. For higher values of p (>10), the control action faced saturation problems.

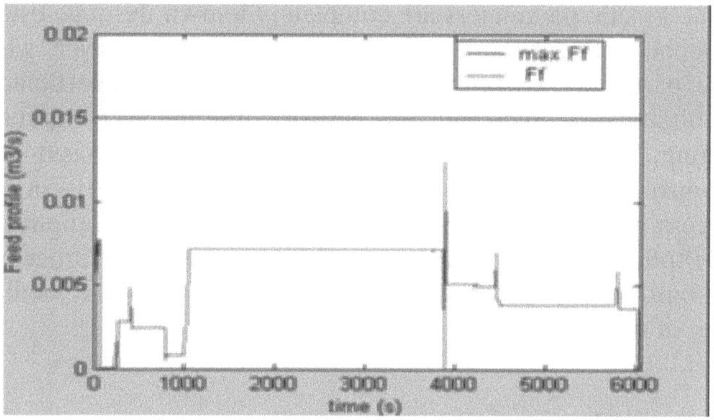

a) Feed rate profile and its max allowed value

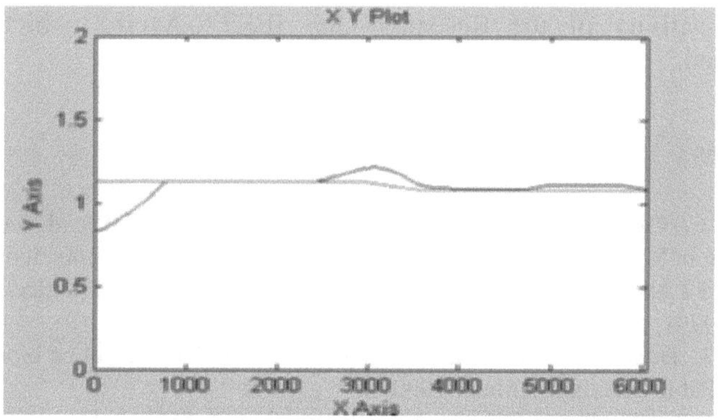

b) Supersaturation profile and its ref. trajectory

Fig. 9. ANN-MPC simulations p=4, c=4

4 Conclusions

This chapter is focused on presenting a more efficient computational scheme for estimation of process kinetic rates based on artificial neural network (ANN) models and an error tolerant computational framework for ANN-based control. Two scenarios are considered: i) the kinetic coefficients of the process are completely known and the process states are partly known (measured). Such a scenario is very common in less complex chemical/biochemical systems; ii) The kinetic coefficients and the states of the process are partly known. This is the most general case in process engineering.

The concepts developed here concern three aspects. From one side we formulate a hybrid (ANN + analytical) model that outperforms the traditional kinetic rate estimation approaches. From other side a novel procedure for ANN supervised training is discussed when target (reference) outputs are not available. The ANN is embedded in the framework of a first principle process model and the error signal for updating the network weights is determined analytically. In both scenarios, the unmeasured states are first estimated and then the ANN is trained with the estimated and the measured data. In case the kinetic parameters are completely known the state estimation can be performed independently of the kinetic rates estimation. However, when the kinetic parameters are partly known both kinetic rates and kinetic coefficients have to be estimated simultaneously, therefore the network training is computationally much more challenging. The third aspect of the chapter is the proposed α error tolerant ANN-MPC control framework which is quite promising alternative in case no reliable analytical process model is available. In this case based on the input output data, an ANN model is trained and imbedded in the MPC optimization procedure. The MPC computational burden is reduced significantly due to the proposed modification (small error tolerance) of the optimization framework.

Acknowledgements

This work was financed by the Portuguese Foundation for Science and Technology within the activity of the Research Unit IEETA-Aveiro, which is gratefully acknowledged.

References

1. Oliveira, C., Georgieva, P., Rocha, F., Feyo de Azevedo, S.: Artificial Neural Networks for Modeling in Reaction Process Systems. Neural Computing & Applications (2008)
2. Haykin, S.: Neural Networks: A Comprehensive Foundation. Prentice-Hall, Englewood Cliffs (1999)
3. Bastin, G., Dochain, D.: On-line estimation and adaptive control of bioreactors. Elsevier Science Publishers, Amsterdam (1990)
4. Lubenova, V., Rocha, I., Ferreira, E.C.: Estimation of multiple biomass growth rates and biomass concentration in a class of bioprocesses. Bioprocess and Biosystems Engineering 25(6), 395–406 (2003)
5. Georgieva, P., Meireles, M.J., Feyo de Azevedo, S.: Knowledge Based Hybrid Modeling of a Batch Crystallization When Accounting for Nucleation. Growth and Agglomeration Phenomena. Chem. Eng. Science 58, 3699–3707 (2003)
6. Walter, E., Pronzato, L.: Identification of parametric models from experimental data. Springer, UK (1997)
7. Noykove, N., Muller, T.G., Gylenberg, M., Timmer, J.: Quantitative analysis of anaerobic wastewater treatment processes: identifiably and parameter estimation. Biotechnology and bioengineering 78(1), 91–103 (2002)
8. Galvanauskas, V., Georgieva, P., Feyo de Azevedo, S.: Dynamic Optimisation of Industrial Sugar Crystallization Process based on a Hybrid (mechanistic+ANN) Model. In: IEEE World Congress on Computational Intelligence, Vancouver, Canada, July 16-21 (2006)
9. Georgieva, P., Oliveira, C., Rocha, F., Feyo de Azevedo, S.: Process modelling strategy combining analytical and data based techniques - I. NN identification of reaction rates. In: International Joint Conference on Neural Networks (IJCNN), Orlando, Florida, USA, August 12-17 (2007)
10. Ilchmann, A., Thuto, M., Townley, S.: Input constrained adaptive tracking with applications to exothermic chemical reaction models. SIAM J. Control Optim. 43(1), 154–173 (2004)

11. Chen, L., Bastin, G.: Structural identifiability of the yeals coefficients in bioprocess models when the reaction rates are unknown. Mathematical Biosciences 132, 35–67 (1996)
12. Hagan, M.T., Mehnaj, M.B.: Training Feed Forward Networks with the Marquardt Algorithm. IEEE Trans. on Neural Networks 5(6), 989–993 (1994)
13. Rumelhart, D.E., McClelland, J.L.: Parallel Distributed Processing. MIT Press, Cambridge (1986)
14. Widrow, B., Hoff, M.E.: Adaptive switching circuits. Convention Record, 96–104 (1966)
15. Hagan, M.T., Demuth, H.B., Beale, M.H.: Neural Network Design. PWS Publishing, Boston (1996)
16. Sorensen, J.S., Lundager Madsen, H.E.: The influence of magnetism on precipitation of calcium phosphate. J. of Crystal Growth 216, 399–408 (2000)
17. Ferreira, A., Oliveira, C., Rocha, F.: The different phases in the precipitation of dicalcium phosphate dehydrate. J. of Crystal Growth 252, 599 (2003)
18. Simoglou, A., Georgieva, P., Martin, E.B., Morris, J., Feyo de Azevedo, S.: On-line Monitoring of a Sugar Crystallization Process. Computers & Chemical Engineering 29(6), 1411–1422 (2005)
19. Mayne, D.Q., Rawlings, J.B., Rao, C.V., Scokaert, P.O.M.: Constrained model predictive control: stability and optimality. Automatica 36, 789–814 (2000)
20. Nagy, Z.K., Braatz, R.D.: Robust nonlinear model predictive control of batch processes. AIChE J. 49, 1776–1786 (2003)
21. Oliveira, C., Georgieva, P., Rocha, F., Ferreira, A., Feyo de Azevedo, S.: Dynamical Model of Brushite Precipitation. Journal of Crystal Growth 305, 201–210 (2007)

5

Computational Intelligence Techniques for Supervision and Diagnosis of Biological Wastewater Treatment Systems

Ana M.A. Dias and Eugénio C. Ferreira

IBB-Institute for Biotechnology and Bioengineering, Centre of Biological Engineering, Universidade do Minho, Campus de Gualtar, 4710-057, Braga, Portugal
{adias,ecferreira}@deb.uminho.pt
http://biopseg.deb.uminho.pt/

Abstract. Wastewater treatment systems (WWTS) are based on complex, dynamic, and highly nonlinear processes. Depending on the design and the specific application, these systems can achieve biological nitrogen and phosphorus removal, besides removal of organic carbon substances. Also, depending on the type/quantity of effluent to be treated, different configurations can be proposed being the most common the aerobic, anoxic and anaerobic schemes working in continuous or sequential modes. In common they have the fact that they deal with communities of different microorganisms which are more or less sensitive to external and/or internal variations of the process conditions. This is a real problem having in mind that usually, influents to be treated are highly inconsistent in flow and concentration being the changes most of the times completely unpredictable. In this way, the knowledge and experience obtained from operational difficulties of one wastewater treatment plant cannot be easily generalized to another.

Recent increased regulation over discharge of nutrients to receiving waterways, associated with operational difficulties of wastewater treatment plants, resulted in an increased need for tools to evaluate the organic matter and nutrient-removal capabilities of wastewater treatment processes. However, the description of its behavior requires complex models that involve a very large number of state variables, parameters and biochemical phenomena that have to be accurately identified and quantified. When deterministic models as the activated sludge model (ASM) and anaerobic digestion model (ADM), fail in predicting the WWT process, alternative modeling methodologies usually known as black-box models, may complement and support the knowledge about the wastewater treatment process and operation. Black-box models are entirely identified based on input–output data without reflecting physical, biological or chemical process knowledge in the model structure. What we purpose in this chapter is to identify and detail the black-box models, also known as artificial intelligent (AI) techniques that are being used for WWT monitoring and control. Particularly focused will be the Multivariate Statistical Methods (MVS), Knowledge Based Systems (KBS), Fuzzy Logic (FL)and Artificial Neural Networks (ANN) as they already proved its potential in different real applications.

M. do Carmo Nicoletti, L.C. Jain (Eds.): Comp. Intel. Tech. for Bio. Mod., SCI 218, pp. 127–162.
springerlink.com

1 Introduction

The last decades have been affected by an increasing rhythm of industrialization, urbanization and population growth that inevitably are changing conditions that are essential to life on Earth. This growth is intimately related with the environmental pollution that is being increasingly affecting the quality of water, air and soil negatively and consequently plant, animal and human life [1]. The solution of this problem is never easy, mostly because of the complexity involving these environmental systems. Uncertainty, variability, multiplicity, as well as political and economical reasons are among the main factors that influence the way these problems are regarded and solved. However, the recurrent climate changes alerts, that are affecting people all over the world, are waking up consciences and leading to social and political changes in the way people and governments should behave in order to avoid major, or even irreversible, damages to the planet. Despite all the controversy, the rules and limits for main polluters are getting more and more restricted and the sanctions to the ones that do not respect them, more severe.

Being essential to human life, the recent news referring the water shortage and general pollution of rivers and oceans make the water consume and treatment a global concern. In this way, scientists are looking for actual and efficient ways to contribute with solutions for water preservation. One of the main aspects of aquatic resources protection is related with wastewater treatment and a great effort is being dedicated in recent years in the improvement of the traditional wastewater treatment, monitoring and control methodologies.

Wastewater treatment is a complex process. Although progress in control engineering and process sensors has enabled significant automatic control improvements of wastewater treatment plants (WWTP), general solutions for plant-wide control are still lacking. Performance of WWTP control systems is especially poor when facing problematic situations of biological origin, such as foaming or sludge bulking. The main reason for poor performance of plant-wide control systems in these situations is the lack of basic knowledge about the complex interactions between the microorganisms' communities in the WWTP and their reaction to disturbances related to the influent composition or plant operating conditions.

A WWTP usually consists of a set of activated sludge tanks, combined with a sedimentation tank, with a range of electron acceptor conditions occurring in the tanks. In the activated sludge process, a bacterial biomass suspension (the activated sludge) is responsible for the removal of pollutants. Depending on the design and the specific application, an activated sludge WWTP can achieve biological nitrogen (N) removal and biological phosphorus (P) removal, besides removal of organic carbon substances. Evidently, many different activated sludge process configurations have evolved during the years [2]. For instance, depending on the concentrations of dissolved oxygen (DO) and nitrate present in the tanks, aerobic (oxygen present), anoxic (nitrate present, no oxygen) or anaerobic (no oxygen, no nitrate) tanks can be used. Figure 1 represents a typical activated sludge WWTP lay-out, not considering the different pre-treatment steps that normally precede the activated sludge tanks.

Wastewater treatment systems have traditionally being modelled using white-box models, also called deterministic models, because they are based on first engineering principles, meaning that the model equations were developed from general balance

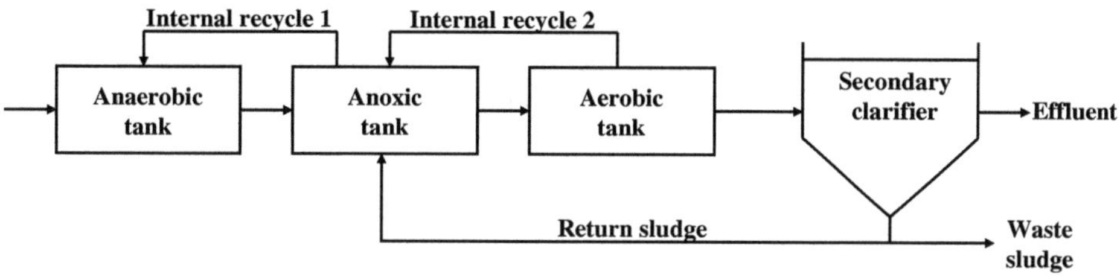

Fig. 1. General scheme of an activated sludge treatment system

equations applied to mass and other conserved quantities, resulting in a set of differential equations.

The Activated Sludge Model No. 1 (ASM1) [3] can be considered as a reference model, since it has found acceptance first in the research community and later also in industry as effective to model WWT systems despite its approximations and limitations. ASM1 has become a reference for many scientific and practical projects, and has been implemented in most of the commercial software available for modelling and simulation of WWT systems. Although the model has been extended over the last years, for example to describe growth and population dynamics of floc forming and filamentous bacteria and to include new processes for describing enhanced biological phosphorus removal [4,5], the original model is probably still the most widely used for describing WWT processes all over the world. The model-based control or the optimization of controllers by simulation using an analytical model made a lot of progress within the last years [6,7]. The advantages of using such models include the possibility of using simulations to increase process understanding, to evaluate several design alternatives for new WWTP installations and process optimisation and control by performing several tests without effective time and cost losses [8]. However, the practical application of these methods requires scientifically educated staff that is availably only at large wastewater treatment plants. Moreover, to make these models usable, some assumptions have to be made and some limitations have to be assumed. For instance, the influence of environmental effects (like temperature, pH, toxic components and wastewater composition) and the biodegradation kinctics arc among the main parameters and/or concerns that the modeller has to deal with [8]. A further drawback is that the use of such models always requires the knowledge of some data before starting.

The available data set can vary considerably, depending on the specific WWTP (size, process lay-out, presence of industrial activity) and on the model purpose. Nevertheless, design data (volumes, aeration capacities, lay-out), operational data (flow rates, recycles, trajectories, pH, temperatures, influent, effluent and sludge compositions, kinetic parameters, yields, etc), hydraulic and sedimentation data are essential information that have to be known before using a given model. In some cases, the modeller and/or the operators experience can propose initial start guesses for some of the parameters but in most of the cases previous data collection has to be performed using for example respirometry and batch experiments as the knowledge obtained for one WWT system is not usually directly transferred to another, even if similar, as was demonstrated by Cinar et al. [9]. Finally, but not least important, is the model calibration and optimization. The experience of the modeller is also a key point in this case as it will be essential to identify the right solution among the several possible ones. A major problem facing in calibration of WWTP models is indeed the lack of identifiability of the model: more than one combination of influent characteristics and model parameters can give a description of the available data with similar quality. It therefore becomes important to obtain informative data that allow constraining the model parameters within realistic boundaries.

Because of these constrains, over the years, alternatives are being purposed and among them, black-box models are gaining acceptability and are being increasingly applied even in real systems. Black-box models, also called artificial intelligence (AI) models, are models entirely identified based on input–output data without reflecting physical, biological or chemical process knowledge of the model structure. These models are particularly suitable when a lot of sensorial information about the system is available and in situations where the white-box plant model assumptions are not valid or insufficient data are available for model calibration. Models based on the historical data of a system have the further advantage of being able to identify and easily react to different conditions. The developments in sensor equipment and automatic control techniques observed in the last years resulted in a significant improvement of the monitoring and control performance of WWTP. But much information is not always a signal of quality information. In fact, the maintenance and the analysis of information provided by a large number of sensors can be an arduous and inefficiently task. In these cases, the characteristics inherent to AI methods represent a great potential as is intended to be demonstrated along this chapter.

The great advantage of these methods is that, in contrast to the model-based approaches, where a priori knowledge (either quantitative or qualitative) about the process is needed, in process history based methods, only the availability of large amount of historical process data is needed. When well designed and calibrated, these models are able to systematically analyze non-structured large amounts of information collected by the several sensors that are usually used to monitor WWTP and extract only the valuable information, finding relationships between potentially high dimensional, highly nonlinear data sets and, most important, predicting outputs from selected input variables. There are different ways in which this data can be transformed and presented as valuable knowledge to a diagnostic system. This is known as feature extraction. This extraction process can be either qualitative or quantitative. Methods that extract quantitative information can be further classified as non-statistical or statistical methods. Neural networks are an important class of non-statistical classifiers. Principal component analysis (PCA), partial least squares (PLS), and statistical pattern classifiers form a major component of statistical feature extraction methods. Most of the times a single method does not present all the desirable features stipulated for a diagnostic system and is the association of these methods that complement one another resulting in better diagnostic systems. The development and application of these hybrid systems will also be addressed in this chapter.

Despite their differences, it is important to retain that both AI and white-box models can play an important role in supervisory control systems: AI methods can maximise the knowledge extracted from data and operator experience and subsequently apply this knowledge to improve WWTP control, whereas white-box models allow evaluating scenarios based on the available process knowledge about the WWTP. The advantage of AI based prediction tools is that qualitative information can also be considered. Even though measuring and control technologies are improving, the problem of incomplete or missing data still exists because many parameters are difficult to determine or cannot be determined at all. Furthermore, in specific cases, the measured data might not be representative for the overall system. Therefore, it often happens that the WWTP operator must control the plant rather with his experience from past events than with sophisticated machines. In other words, these modeling methodologies are available to complement the process knowledge included in white-box models with predictions based on data in areas where the white-box model assumptions are not valid or where white-box WWTP models do not provide accurate predictions. In addition they can generate data, for example as influent flow rate and concentration profile predictions, which can subsequently be applied as input for simulations with the white-box model. The combined use of several methods can result in reasonable prediction models, even in cases where data are missing [10].

In the last decade AI methods have gained increasing attention to solve control problems characterised by ill-defined, nonlinear and time-varying biological wastewater treatment processes. Besides the advantages already mentioned, other advantages achieved by these models include the potential to learn from failure, adapt to different system conditions, predict changes in advance (providing proactive process control actions), optimize operation cycles decreasing process costs and incorporating operator's experiences from past events translating that information to actuator's signals. The models are being used for process monitoring, fault detection and isolation and more recently, several applications are focusing on predictions of quality parameters of the WWTP influent or effluent avoiding off-line measurements and making possible the online survey and rapid actuation on the plant. The main disadvantages usually attributed to these models are the large amount of data that is necessary to 'train' the models and the fact that, as they are treated as black-box models, extrapolations may represent a considerable risk. Despite of that, researchers generally agree that the application of these models can indeed contribute with a significant improvement of WWTS monitoring and control.

In this chapter, the potential of MSV, KBS, FL and ANN, as well as some of their combinations, which are among the most used artificial intelligence models for wastewater monitoring and control, will be enhanced. A brief overview of the concept behind each model will be outlined together with some examples of successfully applications in many cases in real systems.

2 Multivariate Statistical Methods

As a consequence of the increasing instrumentation use in wastewater treatment systems, a lot of variables are being collected on-line from wastewater treatment processes through various sensors. This is obviously advantageous as it permits a more intrinsic knowledge of the system however, at the same time, there is a risk for too complex information, and the need to condense data, for the operator and for the process engineer, will increase. Measured data are usually high dimensional, redundant, and occasionally noisy and missing. In addition, unlike most chemical processes, the characteristics of the WWTP are highly nonlinear, dynamic and even non stationary, which makes WWTP monitoring particularly difficult.

The methods for monitoring and detection used today are normally based on time series charts, where the operator can view the different variables as historical trends. It is hard to keep track of more than a few variables and when the number of monitored variables are increasing it is difficult to draw conclusions and to establish relations among that variables. To be able to monitor the process behaviour effectively, an extraction of important information from the large number of measured variables must be performed. It is also important to present the information in an easy and understandable way.

Multivariate statistical techniques are powerful tools able of compressing data and reducing its dimensionality so that only essential information is retained being easier to analyze than the original huge data set. They are also efficient in handling noise and are among the most robust methods of correlation between variables when they exist. These models are usually called as 'soft-sensors' as they can either replace the hardware sensors or be used as a backup in case of sensor failure. In recent years, a number of attempts have been made to implement MVS modeling methodologies on WWTP, being PCA and PLS the most used, depending on the purpose of the analysis.

2.1 Principal Component Analysis

Principal Components Analysis (PCA) is a favorite tool of chemometricians for data compression and information extraction. The PCA method, initially proposed by Pearson [11] and later developed by Hotelling [12], is a standard multivariate technique and has been included in many textbooks as well as research. Theoretically, PCA is based on an orthogonal decomposition of the covariance matrix of the process variables along directions that explain the maximum variation of the data. The main purpose of using PCA is to find factors that have a much lower dimension than the original data set and which can properly describe the major trends in the original data set. For a given data matrix X with m rows and n columns, with each variable being a column and each sample a row, the covariance matrix of X is defined as

$$\mathrm{cov}(X) = \frac{X^T X}{m-1} . \tag{1}$$

by subtracting the original mean of each column. The data matrix must be scaled prior to analysis. This makes it possible to compare variables with different amplitude and variability. If the columns of X have been "autoscaled," *i.e.* adjusted to zero mean and unit variance by dividing each column by its standard deviation, Equation 1 gives the correlation matrix of X.

PCA decomposes the data matrix X as the sum of the outer product of vectors t_i and p_i plus a residual matrix E:

$$X = t_1 p^T{}_1 + t_2 p^T{}_2 + \ldots + t_k p^T{}_k + E . \tag{2}$$

where k represents the number of PCs retained. The t_i vectors are usually known as scores and contain information on how the samples relate to each other. The p_i vectors are known as loadings and contain information on how the variables relate to each other. In the PCA decomposition, the p_i vectors are eigenvectors of the covariance matrix, meaning that for each p_i

$$\mathrm{cov}(X)p_i = \lambda_i p_i \ . \tag{3}$$

where λ_i is the eigenvalue associated with the eigenvector p_i. The relation between the scores and the loadings is established according to the following relation:

$$Xp_i = t_i \ . \tag{4}$$

being the score vector t_i a linear combination of the original X variables defined by p_i.

The (t_i, p_i) pairs are arranged in descending order according to the associated λ_i. The λ_i are a measure of the amount of variance described by the (t_i, p_i) pair. Being the (t_i, p_i) pairs in descending order of λ_i, the first pairs capture the largest amount of information of any pair in the decomposition [13].

A decision that is crucial for the performance of the PCA model is when to stop including more principal components (PCs) and this is where modelling experience and process knowledge comes in. Too few components imply that there are not enough dimensions to represent the process variability, while too many components implies that measurement and process noise will be modelled. For instance, the eigenvalues themselves can be considered when choosing number of factors provided the data has been auto-scaled. In this case, the eigenvalues will sum to the number of variables, and eigenvalues can be interpreted as the number of original variables each PC is worth. A good rule of thumb is that any PC with an eigenvalue less than one is probably not describing any systematic variance in the system and should likely not be included in the model [13]. It is also possible to choose the number of PCs based on cross-validation. In this procedure, the data set is divided into a number of equal size segments. A PCA model is then built on all but one of the segments. The model is then used to estimate variables that are left out of the data. The error for this "reconstruction" is then plotted as a function of the number of PCs used. Generally this is presented as the root-mean-square error of cross-validation (RMSECV). As PCs which describe large amounts of systematic variance are added to the model the error should decrease. When PCs describing only small noise variance are added, the error should increase [14]. More information about this subject can be found in the work of Qin and Dunia [15].

The most basic way of using PCA for monitoring involves identification of a model from data representing normal or desired operation. New data are then projected onto the model and the scores and/or the model residuals are then monitored as new samples are obtained according to:

$$\hat{t}(k) = x(k)P \ . \tag{5}$$

where $\hat{t}$ is the score correspondent to the new sample $x(k)$ multiplied by the loadings matrix P. It is important to note that the new data must be scaled in the same manner as the data used for identification. By plotting, for instance, the first score vector against the second, process changes can be viewed as movement of a point in the plane as new samples are added. Points that cluster generally represent similar process behaviour whilst points in different regions in the PC space generally represent different operational states. Confidence or control limits are used to discern disturbances. At this point it is important to refer that the method permits to detect changes, by identifying outliers to the regular process, however, it is not possible to differentiate between a change in the process and a sensor failure for example. In this way, the method is not self-sufficient as the intervention of an expert on the analysis of the results is always necessary. In addition to monitoring the scores, the statistical fit of the model can also be monitored [16]. Two commonly used measures of fit are the sum of squared prediction error (SPE) and the Hotelling's statistics T^2. A disturbance that involves a change in the relations between the variables will increase the SPE while a disturbance of the same nature as the identification data, will be detected as an increase in T^2[13].

Many further developments of PCA now exist, including dynamic PCA [17], adaptive PCA [18], multiscale PCA [18], adaptive multiscale PCA [19] and kernel-PCA [20]. In all cases, the implementations done relatively to the original PCA model intend to turn it in a robust model to describe complex non-linear, dynamic and time dependent systems as is the case of wastewater systems. In these papers the authors discuss the advantages and disadvantages of using more complex PCA models when compared to the original model. In some cases, the model becomes more accurate but less intuitive to analyze. This must be accounted for depending on the purpose of the model.

2.2 Partial Least Squares

Partial least squares (PLS) is a tool that can be applied whenever plant variables can be divided into cause (**X**) and effect (**Y**) values and is intended to use the process variables X to predict and/or to detect changes in the product quality variables Y. Originating from the pioneering work of Wold [21] between the mid 1960s and early 1980s, PLS method was further developed by Wold and co-workers [22] in the mid and late 1980s. Similar to PCA conceptually, PLS is useful in reducing the dimensions of both process variables and product quality variables to be analyzed. Its advantage is that unlike certain other identification algorithms it is able to produce accurate and robust models in situations where high levels of correlations exist between the cause variables. It models the relationship between two blocks of data while compressing them simultaneously meaning that it attempts to find factors which simultaneously capture variance, in X and Y and achieve correlation between X and Y.

PLS works by selecting factors of cause variables in a sequence which successively maximises the explained covariance between the cause and effect variables. In PLS regression, the decomposition of X and Y is carried out iteratively. By exchanging information between the two blocks in each step, the principal components (or latent variables, which is a more common term in PLS modelling) of the X-space are rotated so that the predictive power of the X-space with regard to the Y-space is enhanced. There are different algorithms to calculate PLS, but a common algorithm for this is the NIPALS algorithm [23]. Given a matrix of cause data, $\mathcal{X}$ (of size $m \times n_x$, where m is the number of observations and n_x is the number of cause variables), and effect data, $\mathcal{Y}$ (of size $m \times n_y$, where n_y is the number of effect variables), a factor of the cause data, t_k (length m), and effect data, u_k (length m), is evaluated, such that

$$X = \sum_{k=1}^{np<nx} t_k p_k^T + E \ . \tag{6}$$

$$Y = \sum_{k=1}^{np<nx} u_k q_k^T + F \ . \tag{7}$$

These t_k and the u_k vectors are selected so as to maximise the covariance between each pair (t_k, u_k). The matrix E and F are errors and p_k and q_k are referred to as loading vectors. Linear regression is performed between the t_k and the u_k vectors, to produce the inner relationship:

$$u_k = b_k t_k + \varepsilon_k \ . \tag{8}$$

where b_k is a regression coefficient, and .$_k$ refers to the prediction error.

In this method, the latent variables u_i (matrix U) are used for modelling the objects separately in the matrix of Y dependent data whereas the t_i variables (matrix T) are used for modelling the objects separately in the matrix of X independent data. The latent variables U and T are the basis of the regression model and are determined in an iterative process with the centred matrices of X and Y as starting points [13]. The PLS method provides the potential for a regularised model through selecting an appropriate number of scores, or latent variables. Furthermore, it is often found that a relatively small number of the low-index latent variables can explain the greater part of the variation in both the cause and effect variables. Cross validation is usually used to select the necessary number of latent variables. For further details of the PLS algorithm, the reader is referred to the work reported by Geladi and Kowalski [24]. As in the case of the PCA model, PLS is also being implemented in order to better describe complex systems. Multi-block PLS [25], nonlinear PLS [26] and kernel PLS [27] are some of the successful approaches.

The methods outlined above take no explicit account of the ordered nature of the data set, i.e. the fact that the data were collected in a sequential manner and reordering the samples would produce identical results. However, there are cases where order is important as is the case of batch processes where the process variables are changing along time. In these cases, where a response is being measured as a function of two or more parameters, multi-way methods, proposed by Nomikos and MacGregor [28] are used instead of the ones presented above. Multi-way methods are statistically and algorithmically consistent with the original PCA and PLS having the same goals and benefits. Multi-way principal component analysis (MPCA) and multi-way partial least squares (MPLS) have been successfully used to address a variety of batch control and monitoring issues. The way to deal with multi-way methods will not be addressed here but a clear description of the methodology as well as some examples of application can be found in the literature [28,29,30].

Before presenting some successful examples of application of PCA and PLS for wastewater monitoring, it is also important to mention that, a previous screening of the raw data is essential before performing any of the above mentioned methods. Corrupted measurements must be found and dealt with, so that false conclusions based on the measurements are avoided. In fact, particularly in the case of wastewater applications, data are affected by:

- missing values: they can be dealt with in several ways. Extrapolation in online situations or interpolations in off-line situations can be done, if the missing values are few and not succeeding each other. If there is an extended period of missing values, the information is lost and the measurements must be disregarded. More information about dealing with missing values was reported by Walczak and Massart [31,32];
- noise: digital filters can be applied. Any filtering will cause some information loss, but digital filters allow a smart compromise between signal information and noise corruption. More reading on filtering can be found in Åström and Wittenmark [33].

The number and diversity of applications of MVS methods to monitor wastewater treatment systems are increasing during the last years. For instance, Tomita et al. [34] analyzed twelve original physical variables of an activated sludge wastewater treatment system and demonstrated how those cross-correlated variables could be transformed in new ones that were not correlated by the use of PCA. The authors demonstrated that with just three principal components it was possible to explain most of the system total variability what facilitates the analysis, monitoring and diagnostic of the system. Moreover, based on the analysis of the calculated loadings, it was possible to quantify the influence of each original variable on the new variables (principal components) describing the process.

Bonté et al. [35] used two methods based on adaptive principal components analysis (APCA) to extract information related to the changes of wastewater characteristics induced by variable weather conditions and/or the presence of toxic substances. The input data were respiratory data obtained by short-term experiments on a batch respirometer, combined with indirect information on soluble pollution as the wastewater absorbance at 254 nm, pH, turbidity and the wastewater flow rate. The Benchmark Simulation Model 1 (BSM1) [36] was used to obtain large data sets and to test the proposed APCA method, which has then been applied to real wastewater characteristics.

Rosen and Lennox [18] compared the monitoring capacity of the original PCA with an adaptive-PCA model in terms of updating of the scale parameters as well as the covariance structure, to handle with changing process conditions. The superiority of the adaptive model was demonstrated but the authors also mention that the interpretation of the results was not so easy when using the improved model. Lee et al. [37] proposed and applied a sensor fault detection and identification method using dynamic PCA based variable reconstruction. The model was able to detect the occurrence of process abnormalities including sensor faults and to calculate the SVI value of each measured variable for artificial scenarios of sensor faults generated from BSM1. Amaral and Ferreira [38] used image analysis techniques and PLS to establish correlations between microscopic and macroscopic properties of the biomass of a wastewater treatment plant. Morphological properties as filaments versus solids content (TL/TSS) and the filaments versus aggregates content (TL/TA), obtained by image analysis, were successfully related with operating parameters such as the total suspended solids (TSS) and sludge volume index (SVI).

More recently, Gut et al. [39] monitored a semi-industrial pilot plant scale two-step partial nitration/Anammox system for nitrogen removal from wastewater using PCA and PLS. The models enabled to discern relationships between different factors for the start-up period and stable operation of the pilot plant and to identify the key variables in the process control and monitoring. Identification of influential variables was done with the aim of applying the knowledge in planning next experiments and improving the pilot plant operation. For instance, it was demonstrated that the system could be easily monitored by conductivity measurements. Knowledge concerning the effect of changes in the process variables gained by multivariate process modeling enables to avoid disturbances in the nitrogen removal process operation. Fuentes et al. [40] used multivariate statistical methods to discriminate urban *vs* highly or low industrial areas according to the levels of polychlorinated pollutants in sewage sludge from different wastewater treatment plants. Efficient correlations were also established between the technological level of the environment and the concentration of pollutants.

A very interesting and useful application of MSV methods to wastewater treatment systems is the correlation of easily accessible variables of the process, as for example UV-visible or NIR spectra, with traditional monitoring parameters as the chemical oxygen demand (COD), biological oxygen demand (BOD), total oxygen demand (TOC), total suspended solids (TSS), volatile fatty acids (VFA), etc. The traditional methods used to measure these parameters include pre-treatments as filtration as well as the use of chemicals and the production of residues which are harmful to the environment. Moreover, there is always a time delay between sample and analysis that can range from 2-3 hours of digestion in the case of COD analysis to 5 days in the case of the BOD_5 analysis.

Spectroscopy and multivariate data analysis form a powerful combination for obtaining both quantitative and qualitative information in an efficient way avoiding sample pre-treatments and allowing online data acquisition. The first attempts of correlation started to be univariate approaches where a single wavelength in conjunction with a linear regression model was used to determine the water quality parameter of interest. However, the rapid increases in computational power over the last years together with the development of the multivariate statistical techniques lead to the use of the entire spectrum, or at least a higher number of wavelengths, which is indubitably more suitable to represent the complexity and heterogeneous nature of wastewater. Also, the development of portable optical fiber based spectrophotometers permit the online acquisition of the spectra what can be an enormous advantage for control purposes. Some examples of the recent applications of this technique can be referred as is the case of Steyer et al. [41] who described the use of a Fourier Transform Infra-Red (FT-IR) spectrometer as an on-line sensor to measure COD, VFA, TOC and alkalinity in anaerobic digestion processes for the treatment of industrial wastewaters. The authors compared the manual offline analysis with an online industrial TOC analyzer and an online titrimetric sensor and demonstrated the advantages of spectral analysis in the mid infra-red domain for the monitoring of anaerobic digestion processes.

Hansson et al. [42] monitored an anaerobic digester fed with the organic fraction of municipal solid waste for two years with online near-infrared (NIR) spectroscopy. The use of PCA permitted to follow the dynamics that occurred due to changes in substrate composition and changes in operating conditions and to readily detect failures in stirring and increased foaming. The use of PLS permitted to accurately correlate the spectra measured with the propionate concentration with a RMSEP lower than 0.53 g/l. The response on changes in the digester fluid was reproducible and could be detected within 2.5 minutes, which can be considered as real-time monitoring.

Sousa et al. [43] proposed a method based on near-infrared reflectance (NIRR) measurements of filtrated suspended particles collected from wastewater samples for determination of chemical oxygen demand (COD) in domestic wastewater. The analysis did not require any special reagent, catalyst or solvent. The standard method for COD measurement was used to calibrate a PLS model which was then used to predict new COD values from new spectral data. The authors claimed a RMSEP for the PLS model lower then the one expected for the reference method. Dias et al. [44] collected NIR spectra, along a pilot installation of an aerobic biological wastewater treatment process that were used to monitor the status of the biological process along time. The descriptive capacity of the score plot of a PCA model using two PCs applied to the NIR spectra of samples collected at the feed, the reactor and the settler of the pilot installation was evident. It was possible to identify induced perturbations and to follow the evolution of the process until a new equilibrium state was reached. The results support the use of on-line monitoring methods in quality assessment of biological wastewater treatment processes monitoring and control.

Langergraber et al. [45] presented a chemometric analysis based on a spectrometric probe and a fully automated PLS calibration. They used a submersible UV/VIS spectrometer for in-situ real-time measurements allowing the simultaneous measurement of COD, filtered COD, TSS and nitrate with just a single instrument. The authors present a global calibration that was valid for typical municipal wastewater compositions however they refer that running a local calibration would result in improvements concerning trueness, precision and long term stability of the results. Lourenço et al. [46] coupled UV spectrophotometric analyses to PCA to characterize samples taken from a fuel park wastewater treatment plant and to attempt preliminary contaminant identification in the treated wastewater. The score plot resulting from PCA permitted to differentiate spectra from different locations of the WWTP and weekday from weekend samples. Lubricants and detergents were also clearly identified in the score plot as residual contaminants using only two principal components. Rieger et al. [47] used an in-situ UV sensor coupled with PLS regression to predict nitrite and nitrate concentrations in the effluent of a pilot-scale plant. The sensor was used with a calibration that was based on data from another WWTP and was operated over a period of 1.5 years. The results showed constant accuracy although the sensor was operated with minimal maintenance what confirms the robustness of the regression model. The accuracy of the predictions was stated to be of 0.32mgN/l and 1.08mgN/l for nitrite and nitrate, respectively during the relatively long term monitored period. These results encourage the use of the method for alarm systems as well as for control concepts at WWTP.

Finally, it is worth noting that the composition of the incoming wastewater in stabilization ponds can vary according to the habits of the population, as well as the weather and other external factors. Therefore, the application of empirical spectroscopic models to different treatment plants and time periods should be handled with care to avoid possible extrapolations of the calibration limits. Seasonal variations can be taken into account by using calibration data from an entire year including periods of low and high rain index. If the model is to be applied to a different treatment plant, the calculated uncertainty limits for the predicted values could be used as a criterion for validating the prediction and checking for possible extrapolation problems. Also, the reference data used for the calibration models are a very important part of the whole calibration procedure. Therefore it is essential to guarantee the quality of the reference measurements (regarding reference analysis method, measurement range, sampling errors, identity of samples and randomization of sampling) to obtain good calibration results.

3 Knowledge Based Systems

Wastewater treatment plant operators encounter in their daily routine, complex operational problems related to the activated sludge process that they try to manage by integrating different types of information (derived from on-line and off-line data, and heuristic information, etc), and by reasoning over specialised WWTP operation manuals or heuristic knowledge. In addition, WWTP operators often apply their intuition and process experience when dealing with operational problems. Indeed, the problem at hand is usually solved by taking advantage of lessons learnt from successes and failures in plant operation that resulted as the consequence of operational decisions taken in the past when reacting to similar problems. The registration of lessons learnt from past experience is essential from the process management point of view. It enables re-use of knowledge whenever similar problems arise and when new operators, with less experience, are controlling the process. However, past process experience is not easily integrated in numerical control systems (e.g. PI or PID controllers), and thus new tools must be developed.

Knowledge based systems (KBS) are a promising tool for this purpose. KBS are intelligent/knowledge-based systems able to store past experiences efficiently and to retrieve them automatically whenever similar problems arise. A KBS is a simple way to accumulate facts, knowledge and even troubleshooting methods for complex problems, providing learning capacities to WWTP management as schematized in Figure 2. The basic idea behind the functionality is that the second time a problem is solved it is usually easier than the first time because previous solutions or mistakes are remembered or avoided. For example, although there may be a quite well-established plan to chlorinate the mixed liquor volatile suspended solids when the facility is facing severe bulking episodes, every time that this solution is applied there are some process particularities that the operator takes into account to adjust the dose, the dosing point, the frequency, etc. This experience will be very useful in the future whenever the process is affected by a similar upset. The integration of such a tool within a control

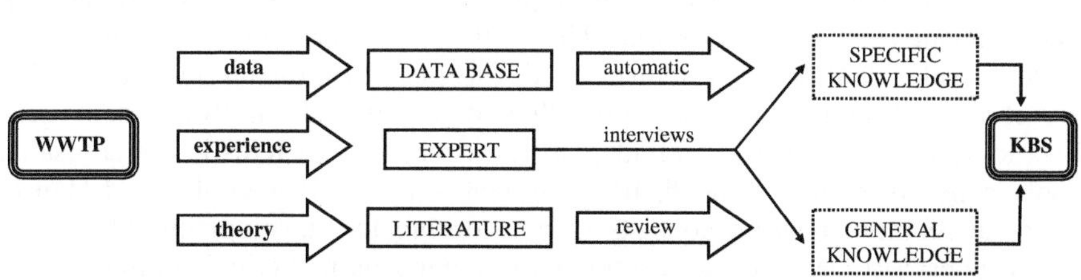

Fig. 2. Methods used to acquire knowledge, according to Poch et al. [1]

system allows the achievement of the management of recurring problems since as it enables combining knowledge and experience re-use with simulation models for similar events [48].

The method starts with the identification of the current disturbance/situation from the influent, effluent and process monitoring data. In the general approach, the situation could represent a rain event, as well as an episode with solids separation problems within the activated sludge process. A case is defined with as much detail as possible including information about the incident, the problem diagnosis, the potential cause(s), the episode duration, a feature vector of some selected process characteristics, the applied control strategy and its evaluation, and the fundamental lesson learnt from the event.

According to Martínez et al. [49] the information to account for in the definition of a new case should include five essential steps, namely: a clear label identifying each new case; the problem description including as much information as possible (diagnosis of the system status, potential cause, episode duration, etc); the control strategy, consisting in a full report of the set of actions carried out to face the problem; daily comments, including any comment that the operator registers as additional information of the episode once the results of the control strategy applied have been evaluated and lesson learnt, where the fundamental lesson drawn from the full episode is registered. It is like a summary of the whole evaluation that should include the correct decisions, but also the side effects and the wrong decisions that were made to operate the process during the problem. After these steps are clearly identified, the case is ready to enter the working cycle.

The traditional working cycle consists of a four-step process: retrieve, re-use, revise and retain [50]. A new problem is (if possible) solved by retrieving one or more previously experienced cases from the case library, re-using those experiences to produce a solution, revising that solution through simulation or test execution, and finally retaining the new experience by incorporating it into the existing case library for future use. The iterative procedure is repeated every time the WWTP is confronted by a new situation. The case library is a plant-specific database that stores the historical cases in an easily retrievable way [51,52].

Despite their potential, KBS still present some shortcomings. Allowing final users to automatically update the KB (e.g., by adding, deleting or modifying rules) proved to be a difficult task. The ambiguity found in the linguistic description of a concept during the decision making process advices the incorporation of a measure of uncertainty into the KBS conclusions to fully exploit the fallible but valuable judgemental knowledge offered by human experts. There are many approaches to cope with this vagueness. Some authors presented optimizing methodologies for KBS results in order to evaluate when similar inputs will actually evolve to similar outputs and consequently to similar solutions. This will not be further developed here but the interested reader can consult for instance the works of Bellmann and Zadeh [53] or Wiese et al. [54]. In addition, the knowledge contained in the KBS is mostly general and static, rather that specific and adaptable. Accordingly, a KBS is not advisable as the only supervisory tool. Indeed, in agreement with Stephanopoulos and Han [55], the general recommendation consists in considering the development of intelligent decision supervisory systems, as a framework integrating numerical computations and KBS to provide reliable tools to solve wastewater problems. In such a framework, the classical management techniques (simulation, control, design, optimization) would be allocated to numerical computations, while delegating the logical analyses and reasoning to KBS [56].

The application of KBS is particularly useful to overcome situations that, although being common in WWTS can not be predicted by simulation. For instance, it is commonly reported that biological nutrient removal processes, in which environmental conditions are continuously alternating between aerobic, anoxic and anaerobic, favor the growth of various filamentous microorganisms which compete with floc-forming organisms [57].

The absence of basic knowledge about the interaction mechanisms between the microorganism communities and operational parameters, which are not described by standard models, is an obvious limitation when evaluating control strategies via simulations. For example, although there are several hypotheses to explain the development of filamentous bacteria, none of them provide a definitive and general solution. Moreover, most of these theories still lack experimental verification [58]. Therefore, heuristic knowledge, qualitative information and operator expertise are still necessary components when dealing with these operational problems in real WWTP. In this context, Comas et al. [59] proposed an extension of the BSM1 [36] by integrating an expert reasoning module in the evaluation step that infers potential operational problems in the activated sludge process from the simulation outputs. This module composed of knowledge-based flow diagrams, analyses influent and effluent compositions and operating conditions at each time interval, to detect favoring conditions for filamentous bulking, foaming and rising sludge problems. The diagrams were developed based on knowledge acquired from literature and the authors' expertise in this field. With the new approach, the user can not only evaluate different control strategies based on process removal efficiencies, sludge production, aeration and pumping energy, and controller performance, but also consider the propensity of microbiologically related solids separation problems that numerical models cannot describe.

The number of publications related with the application of KBS to wastewater treatment systems has being increasing through the last years. Although not exhaustively, a number of some successful examples of application will be referred in what follows. In the nineties, Sànchez-Marrè et al. [60] developed the DAI-DEPUR system. This system is based on an integrated multi-level architecture for WWTP supervision in real time including several knowledge bases that are connected for solving the global control task. Latter on Rodríguez-Roda et al. [61] presented an integrated system for supervision and control of an activated sludge process of a continuous flow reactor plant, which was based on a knowledge based part and an adaptative controller that was continuously adjusted according to the information given by KBS.

Expert systems (ES) have also been applied to the monitoring of anaerobic wastewater systems. Puñal et al. [62] developed an ES for the monitoring and diagnosis of the operation of an anaerobic WWTP treating highly polluted wastewater from a fiberboard factory. The monitoring system was applied to a hybrid UASB–UAF pilot scale reactor working in a real factory. The system was able to react to alterations from normal operating conditions as changes in the hydraulic residence time and organic load, which are among the most usual causes of instability in real plants. In both cases, the ES delivered valuable recommendations and, after following them, the system recovered its normal state in a short period of time. The rapid identification of the evolution of the variables enables the operator to take preventive or recovering measures in order to avoid the process to reach a serious ''destabilization'' state, very difficult to be recovered.

Comas et al. [63] successfully applied a knowledge-based approach in a full-scale plant for the supervision of the deflocculation problem in activated sludge processes. The methodology used integrated online, offline and heuristic information organized in three steps namely, the development of a decision tree involving knowledge acquisition and representation, the implementation of that knowledge into a rule-based system finally followed by a validation step. The authors pointed out that the advantage of the KB approach when compared to a classical control system is that, besides detecting turbidity at the effluent and actuate automatically, the ES is also, by itself, to detect the cause of the perturbation namely if the effluent turbidity is due to overloading of the clarifiers, to undesired denitrification in the clarifiers, or to bad formation of activated sludge flocs. Also important is the fact that, the examination of the temporal trends in certain significant variables enables the KBS to detect some problems before severe episodes occur in the plant mainly because if a severe situation of deflocculation or filamentous proliferation develops, recovering the proper activated sludge floc or re-directing the process to its normal state may take a long time.

Comas et al. [64] also expanded the KBS concept by applying it in the development and implementation of a Knowledge-Based Decision Support System (KBDSS) for the identification of adequate treatment for small communities taking into account difference knowledge sources to make up a comprehensive and accurate knowledge data base. By taking into account information about the community and landscape, the receiving environment, and the available wastewater treatment technologies as well information about removal efficiencies, space requirements and climatic, geological and hydrographical features, among others, the developed system was able to assist in the selection of the treatment level adequate to fulfill the target quality standards for the receiving environment and to select the specific type of treatment.

More recently, Martínez et al. [65] applied the methodology to a real episode of non-filamentous bulking using data from a real WWTP. The set of heuristic rules previously established identified a non-filamentous bulking situation as well as the possible cause for the problem. By means of the recalling algorithm, the system searched within 20 episodes of non-filamentous bulking and identified one as being the most similar to the current state of the biological system. The KBS retrieved the complete episode involving that day including control strategies, daily comments and lesson learnt in the past. The information was then analysed and adapted to the current situation by the operator, who, considering all this information, decided to apply the same control strategy as previously. The concept is being developed through the last years and lately Martínez et al. [66] proposed a dynamic knowledge-based decision support systems able of dealing with complex problems of long duration within the activated sludge process. Dynamic reasoning allows the system to understand how the process evolves, to react differently according to the process status trend and to propose the most suitable control strategies, based not only on the literature but also on the feedback derived from the evolution of the process. Instead of giving a generic solution, a specific long-term control strategy is determined to solve the situation, from beginning to end, and continuously updated according to the feedback and evolution of the process.

The dynamic KBS include three levels being the first level correspondent to data gathering (including both online and offline, numerical and qualitative data) as in the static model. Since activated sludge processes are dynamic systems, the optimal monitoring of the process must also include information about the evolution of the main variables. In order to faster and clearly detect sudden deviations and/or trends, some derivatives may be analysed instead of the real value of the variables. The second level, includes the reasoning module to infer the process status (problem diagnosis and detection of cause) in order to reach a reasonable proposal for a control strategy. This is accomplished through a knowledge-based expert system which, by means of dynamic reasoning and using data gathered at the first level and knowledge structured in heuristic rules, is able to reach a dynamic diagnosis, identify the potential cause, and suggest a long-term control strategy to restore the process to normal operation. Finally, the third level correspond to the decision support level, where the conclusions of the dynamic reasoning level are provided to the process operators, who refine and apply the control strategy. Once a new problem is identified and a new control strategy is started, dynamic reasoning only checks the required data to be aware of the evolution of the process, avoiding unnecessary reasoning that would lead to repetitive diagnoses (as in static reasoning systems).

4 Fuzzy Logic Systems

In a very simplistic way, Fuzzy Logic (FL) is an AI control technique that uses a set of pre-defined rules to relate inputs of the system to be controlled with outputs that want to be predicted, without any further knowledge of the system's behavior. FL uses the expertise accumulated through several hours of observation, for instance with the knowledge based systems mentioned before, to establish the set of rules that integrate the fuzzy logic control process [67]. What mainly distinguish this method from other AI methods is that FL can be regarded as a methodology for computing with words rather than numbers. Even if words are less precise than numbers, their use is closer to human intuition what facilitates the implementation and analyses of the model. The method assumes that precision is expensive and sometimes of relative importance. The guiding principle of soft computing techniques as fuzzy logic is to exploit the tolerance for imprecision, uncertainty and partial truth of uncertain and imprecise environments to achieve tractability, robustness, and low cost solution [68].

The empirical knowledge frequently used by trained WWTP operators is often expressed in a fuzzy rule form, such as: "if the levels of organic matter (BOD_5) are low and the levels of dissolved oxygen (DO) are high, then the expected water quality (WQ) is good". This can be translated to the form

If BOD_5 is low **and** DO is high **then** WQ is Good

where the 'if-part' of the rule "BOD_5 is low and DO is high" is called the antecedent or premise, while the 'then-part' of the rule "WQ is Good" is called the consequent or conclusion. Notice that the input (BOD_5 and DO) and the output (WQ) variables are defined in terms of low, high, good defining the fuzzy set (or universe of discourse) of each variable.

It is important to point out that if a system can be represented by a single equation, it should be modeled with a straightforward control technique. Traditional proportional, integral and derivative (PID) control systems remain an accurate and cost-efficient solution to control problems involving simple, linear systems. In general, fuzzy logic is best applied to nonlinear, time-variant and ill-defined systems. The application of the fuzzy logic methodology evolves three main steps namely, fuzzification, inference of the process rules and defuzzification:

1. Fuzzification

The first step in designing a fuzzy controller is to understand and define the system in terms of its input and output variables. Once the variables are known, they must be subdivided into membership functions. If X is the input space and its elements are denoted by x, then a fuzzy set A is defined as a set of ordered pairs

$$A = \{(x, \mu_A(x)), x \in X\}$$

where $\mu_A(x)$ is the membership function of x in A which expresses the extent to which x (input value) matches the category specified by A (low, medium, high, good, moderate, etc). A membership function (MF) is an arbitrary curve whose shape can be defined to suit from the point of view of simplicity, convenience, speed, and efficiency. Amongst the most usual are the triangular, trapezoidal, Gaussian and bell membership functions.

In the traditional two-valued logic, the control answers are of the type 'yes' or 'no' (or '1' and '0'). This is a somehow rigid concept for many systems where the variable being controlled can vary in a range, instead of having a fixed set-point without affecting the system's performance, or where it is more important to detect changes in the process then to quantify them. In the fuzzy sets theory established by Zadeh [69], the input is associated with a degree of membership, which will be a numeric value belonging to the interval [0, 1]. In other words, the fuzzy reasoning has the ability to reply to a 'yes-no' question with a 'not-quite-yes-or-no' answer. Taking into example a controller for the dissolved oxygen (DO) in a given system, the traditional 'yes/no' control system will signal '1' if the DO is 5 mg/L (set-point) and '0' if the DO measured is 4.8 mg/L. However, as shown in Figure 3, the fuzzy controller will evaluate this measurement according to the pre-determined membership functions (trapezoidal in this example) defining the intervals for which the DO is considered low, medium and/or high. A measured DO equal to 4.8 mg/L belongs to the medium fuzzy set within a degree equal to 0.9 in the [0, 1] interval.

The input is always a crisp numerical value limited to the universe of discourse of the input variable (in this case the interval between 0 and 10 mg/L) and the output is a fuzzy degree of membership in the qualifying linguistic set (always the interval between 0 and 1). The capacity to accept a gradation of states to describe a given system is one of the main advantages of fuzzy logic systems mostly if they are used to describe complex systems where a range of different conditions can come over requesting different control solutions. Since not all the variables have the same importance, sometimes it is necessary to evaluate the influence of each variable in the

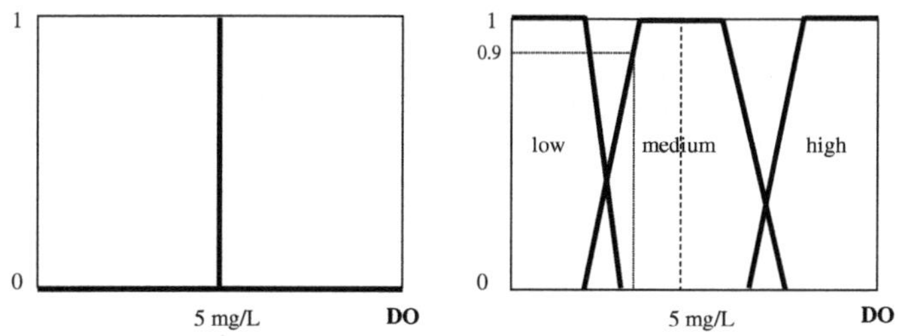

Fig. 3. Schematic representation of the dissolved oxygen (DO) control signal processed by a traditional 'yes/no' controller (left) and a fuzzy controller (b) when the desired set-point is 5 mg/L (right).

final score. This is done by giving a weight to every rule (a number between 0 and 1), which is applied to the number given by the antecedent.

2. Inference of the process rules

When the antecedent of a given rule has more than one part, a fuzzy operator is applied to obtain one number that represents the result of the antecedent for that rule. The input to the fuzzy operator is two or more membership values from fuzzified input variables. The output is a single truth value which will be used to shape the output fuzzy set. The most common fuzzy operators are the union (OR), intersection (AND) and additive complement (NOT). These operations are the ones known as standard and they can be considered generalizations of the conventional set operations. If two fuzzy sets *A* and *B* are defined on the universe *X*, for a given element *x* belonging to *X*, the following operations can be carried out:

Intersection, AND	$\mu_{A \cap B} = \min(\mu_A(x), \mu_B(x))$
Union, OR	$\mu_{A \cup B} = \max(\mu_A(x), \mu_B(x))$
Additive complement, NOT	$\mu_{\bar{A}} = 1 - \mu_A(x)$

After applying the fuzzy operators, the degree of support for the entire rule is obtained and used to calculate the output. The way the output is calculated depends on the fuzzy inference system (FIS) applied. Among the most common models are the Mamdani [70] and the Sugeno [71] type models, being the first the most usually applied. Mamdani-type inference model expects the output membership functions to be fuzzy sets. These fuzzy sets are represented by membership functions that are chosen to indicate the qualities of the consequent. If the antecedent is only partially true, having a value lower than '1', the output fuzzy set is truncated at this value. This procedure is called the minimum implication method [72].

Since decisions are based on the testing of all of the rules in an FIS, the rules must be combined in some manner in order to make a decision. Aggregation is the process by which the fuzzy sets, that represent the outputs of each rule, are combined into a single fuzzy set. The input of the aggregation process is the list of truncated output functions returned by the implication process for each rule. The output of the aggregation process is one fuzzy set for each output variable. The most used aggregation procedure is the maximum method [72], which consists in the union of all truncated output fuzzy sets. In the Sugeno-type inference model, the output is calculated as a weighted average of the result obtained from the implication step, for each rule. This procedure simplifies the computation required but is unable to correctly describe non-linear systems.

3. Defuzzification

Despite fuzziness helps the rule evaluation during the intermediate steps, the final desired output for each variable is generally a single number. The defuzzification step consists in transforming the fuzzy output into a non-fuzzy numerical value. The input for the defuzzification process is a fuzzy set (the aggregate output fuzzy set) and the output is a single number (note that this only apply in the case of the Mamdani-type FIS). The most popular defuzzification method is the centroid calculation, which returns the center of area under the curve formed by the output fuzzy set [72].

In order to make the whole process clearer, Figure 4 depicted the application of the Mamdani-type FIS to control a complex lab scale WWT system comprising anaerobic digestion and anoxic/aerobic modules for combined high rate biological nitrogen (N) and carbon (C) removal as reported by Pires et al. [73].

The main objective of the control system is to ensure low concentrations of nitrate, nitrite and COD in the plant effluent, actuating in the output variables of the fuzzy control system, which are the external recycle (R) and the bypass (B) flow rates. The COD/N ratio at the influent of the anoxic reactor and the ammonium concentration in

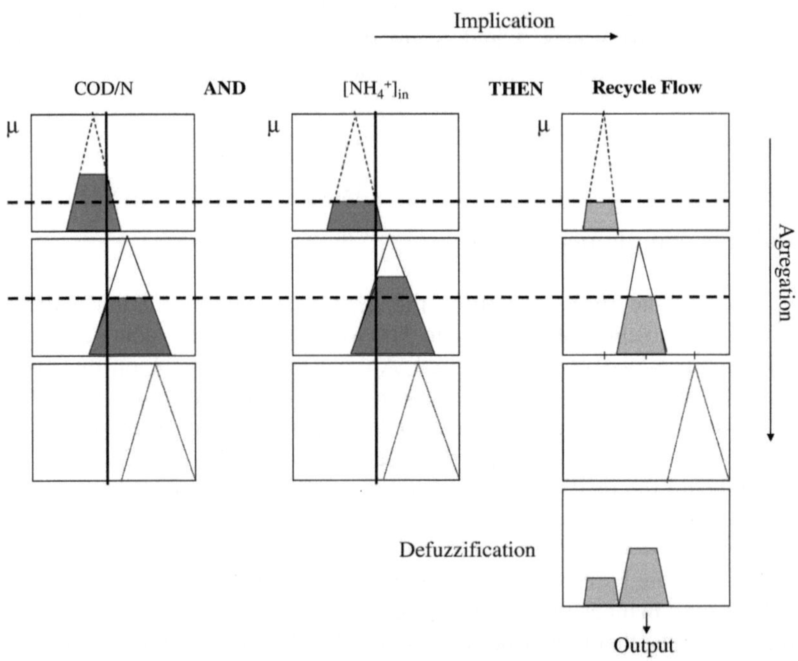

Fig. 4. Fuzzy inference diagram for the control of the recycle flow of a lab scale WW system with two input variables and three rules

the influent of the plant are the two input variables considered in the control system. In Figure 4, the columns refer to the input/output fuzzy sets, and rows are the fuzzy rules. The input fuzzy sets were defined as {very low, low, normal, high and very high} while the output fuzzy sets were defined as {null, minimum, normal, medium and maximum}. Linear membership functions were defined and a total of 25 rules were established between the input variables and each of the output variables. Notice that the figure is only reporting 3 rules for one of the outputs as it intends to be only an example to help to consolidate what was described before. The authors applied the model to react to imbalances deliberately imposed to the system as a step increase in ammonia concentration. With the help of the FIS, the recycle flow rate from the aerobic to the anoxic module and the bypass flow rate from the influent directly to the anoxic reactor were automatically changed when the new conditions were recognised by the expert system. The system proved to properly react in order to set adequate operating conditions that timely led to recover efficient N and C removal rates.

In recent years, fuzzy-logic-based methods have demonstrated to be appropriated to address uncertainty and subjectivity in environmental problems. The applications range from water quality index monitoring to optimization of the oxygenation level in batch reactors. For instance, Ocampo-Duque et al. [74] developed a methodology based on FIS to assess the water quality in a Spanish river. The authors evaluated 27 parameters divided in five main groups as inputs to predict a fuzzy water index quality. Trapezoidal membership functions were used to represent "low", "medium", "high", "poor", "average", and "good" fuzzy sets. The results obtained agreed with official reports and expert opinions about the pollution problems in the studied area, proving that the methodology is a suitable and alternative tool to be used in developing effective water management plans.

The online monitoring of DO, pH and ORP has proven to be an economic and useful technique for process control in sequencing batch reactors [75]. The ORP is primarily used to determine the end of the anoxic phase, by determining the so-called "nitrate knee" while pH sensors can be used to identify the so called "ammonia valley" and "nitrate apex". DO sensors have also been used to identify the end of the nitrification process, which in systems with constant aeration can be identified by an increase in DO. Having this in mind, several attempts have been reported in order to use this information to efficient control of SBR systems using fuzzy logic.

Traore et al. [76] presented a study where the main objective was to develop an online control action of the dissolved oxygen concentration, measured in a SBR pilot plant installed in a WWTP treating urban wastewater for organic matter and ammonia removal in which the DO had a high influence. The DO concentration had to be stable and as near as possible to a set point, fixed from expert knowledge of the pollution present in the process considered. The authors compared three kinds of DO control strategies namely simple on/off, PID and Mamdani fuzzy logic-based DO control. It was demonstrated that the fuzzy controller offered the possibility to easily define a non-linear control law starting from vague system knowledge while the traditional on/off and PID had much more difficulty to carry out an effective control.

Promising control performance were also achieved by Peng et al. [77] who studied a bench-scale anoxic–oxic (A/O) system to treat a synthetic starch wastewater for one year. The objectives were to improve nitrogen removal and reduce operating costs by using on-line sensors DO, pH and ORP. The authors found that the DO in the first aerobic zone could be used to estimate influent ammonia load under constant aeration and that the pH profiles and DO profiles along the length of aerobic zones give good indication of the nitrification. Further, ORP value at the end of aerobic zone showed a good correlation with the effluent ammonia and nitrate nitrogen concentration. This information was used to develop a FIS for online aeration, nitrate recirculation flow and external carbon dosage control.

Bae et al. [78] developed an automatic FL control and diagnosis based upon the operator's knowledge of the behavior of an SBR suitable for small-scale wastewater treatment. The input of the controller was DO lag-time, which has a strong relation to the loading rate. This approach can replace an expensive sensor, which measures the loading rate and ammonia concentration in the reactor, with a cheaper DO sensor. The proposed controller can assured optimal operation and prevented over-feeding problems.

Marsili-Libelli [79] proposed an inferential engine to control the timing of a SBR. Normally, this process requires a periodic switching between anaerobic and aerobic conditions being operated on a fixed time-schedule. The proposed strategy controls timing as the result of an inference process whereby indirect process indicators, represented by pH, ORP and DO, give information about the state of the process through a fuzzy clustering algorithm to decide whether each phase is about to end. This information, extracted form the noisy process signals through wavelet processing, constitutes the antecedent of a fuzzy inferential engine whose output is the switching decision variable. In this way the duration of each phase is limited to the time strictly necessary for the actual loading conditions. The training and test data were obtained from a pilot plant, on which the switching algorithm was then tested, showing that a significant time saving could be achieved with respect to conventional fixed timing. More recently, Marsili-Libeli et al. [80] applied the same methodology to monitor a lab-scale SBR treating nitrogen-rich wastewater from a sanitary landfill leachate with six months of uninterrupted operation, during which the process conditions varied widely. The monitoring system proved capable of adjusting the process operation, in terms of phase length and external COD addition, to the varying environmental and loading conditions, with a percentage of correct phase recognition of 95%. Furthermore, the system could be operated by a remote station using the platform native web publishing tools.

One of the critical point affecting classical biological reactors is the secondary settler efficiency. In some cases, the bacteria concentration or nature can lead to a deficient settling or to a bulking phenomenon [81]. Thus, a permanent follow-up of some physical and biological parameters influencing the sludge quality is needed. Actions to apply could vary from precise chlorination to manual removal of superficial foam, including some adjustments related to recycling flows. Traore et al. [82] studied the ability of fuzzy logic to integrate this human knowledge in order to maintain the secondary settler under stable conditions. The proposed approach is based on an easy methodology to control the sludge blanket by measuring simple online (influent flow) and offline (a daily measured Sludge Volumetric Index, SVI) parameters and manipulating the RAS recycle flow. The fuzzy controller was first optimized and tested using data generated by the COST simulation benchmark [36] and then successfully applied to a real activated sludge WWTP.

The control of anaerobic WWTP using FL systems has also been addressed and some successful applications are reported in the literature. These are very sensible and complex systems to model mainly due to the heterogeneous population of microorganisms and the many different phenomena and complex flow distributions that take place in the bioreactors. For instance, an overload can shut down the whole plant which further can lead to an unavailability of several weeks [83]. The usual parameters used for control of these systems are the hydrogen concentration, the pH, the gas production rate and the methane content in the biogas.

Murnleitner et al. [84] used a fuzzy logic expert system for modeling and control a two-stage anaerobic digestion system for which much qualitative knowledge was available. The system was designed to handle very strong fluctuations in the concentration of the substrate and the volumetric loading rate. Hydrogen concentration together with methane concentration, gas production rate, pH and the filling level of the acidification buffer tank were used as input variables for the fuzzy logic system. The manipulated variables were the flow rate from the acidification buffer tank into the methane reactor, the temperature and pH of both reactors, the circulation rate of the fixed bed reactor, back flow from the methane reactor into the acidification, and the control of the feed into the acidification buffer tank. The developed control system was successfully tested without using TOC, COD or equivalent measurements. The authors believe that though the control system was developed and tested with a fixed bed reactor, the control strategy presented can certainly be adapted to other reactor configurations, too.

Carrasco et al. [85] also used a fuzzy-logic-based diagnosis system for the determination of acidification states on an anaerobic plant processing wastewater from a fiber-board factory. The diagnosis system uses experts' knowledge to determine the acidification state of the process. The acidification state of the system within the anaerobic reactor was determined by means of evaluating the fuzzy inference system for the given inputs. The results obtained show the capacity of the system to determine anomalies in the operation of the bioreactor, establishing the current acidification state within a range between normal and overloaded, through intermediate degrees of acidification due to either hydraulic or organic overloads.

5 Artificial Neural Networks

In a very simplistic way, an Artificial Neural Network (ANN) is a mathematical technique that searches automatically for the best linear or non-linear relationships between cause (input) and effect (output). The ANN concept was developed as an analogy to the human brain behaviour. In this way, and unlike other expert systems where knowledge is made explicit in the form of rules, neural networks generate their own rules by learning from examples shown to them. An ANN combines the human association and the analytical power of a computer. Trying to simulate the brain's behaviour, an ANN can be defined as a network of artificial neurons, also called perceptrons, connected in such a way that information passing through the network, can be transmitted and processed along the course. The way the neurons are connected determines the shape of the network. As shown in Figure 5, the model of an artificial neuron is usually based on the following parameters [86]:

- Input connections ($x_1, x_2, \ldots x_n$): these inputs have weights ($w_1, w_2, \ldots w_n$) bound to them and one unitary weighted input (bias, x_0) linked to each neuron
- Input function (*f*): determines the aggregated net input signal to the neuron as

$$u = f(x, w). \tag{9}$$

where the function *f* is usually the summation function

$$u = \sum_{i=0}^{n} x_i w_i . \tag{10}$$

- Activation function (f_a): determines the activation level of the neuron and is normally typically a step function or a sigmoid function, that takes the argument *u* and produces the output f_o

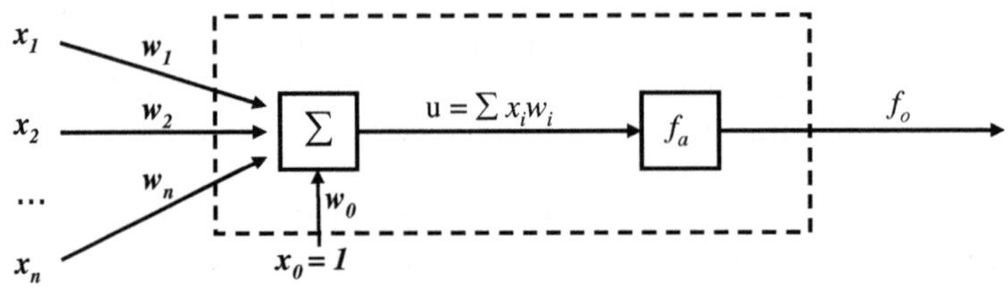

Fig. 5. Schematic representation of an artificial neuron

Note that w and x_0 are both *adjustable* scalar parameters of the neuron. The central idea of neural networks is that such parameters can be adjusted so that the network exhibits some desired or interesting behaviour which will be expressed by f_o. It is so possible to train the network by adjusting the weight or bias parameters to achieve some desired end.

Neural networks are information processing systems that demonstrate the ability to learn, recall, and generalize from training data. They usually have a large number of highly connected processing elements that can be connected in various topologies. The way the neurons are connected determines the shape of the network [87]. Generally, a NN has three functional layers namely the input, the hidden, and the output layer. The information enters the NN at the input layer and then all layers process these signals through the NN until they reach the output layer. Connection weights represent the strength of neural connections. Each input is weighted and the sum of the weighted inputs and the bias forms the input to the next transfer function. This transfer function can be a simple mathematical function like a sigmoid, hyperbolic tangent, sine or linear function. The output of the transfer function is passed directly to the output path of the neuron. A simple NN is shown in Figure 6. It contains four input, two intermediate (hidden), and one output neuron.

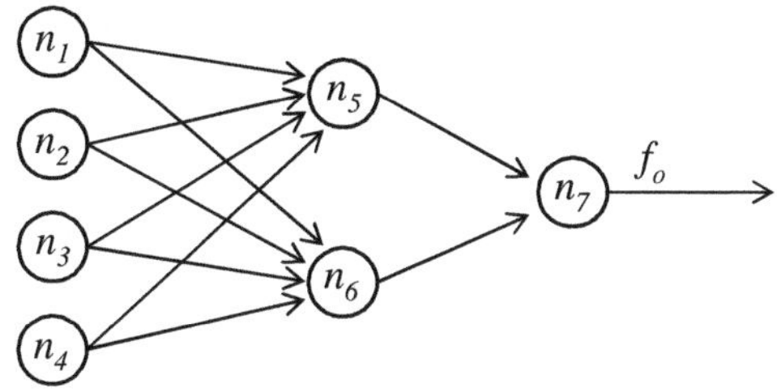

Fig. 6. Example of the structure of a simple neural network

The application of an ANN model to modelling purposes involves a sequence of two main steps which are the learning and the recall processes:

- Learning is the process of adapting or modifying the connection weights in response to training data. Based on the calculated differences between the derived or measured output of a certain process and the results obtained by ANN, the so-called residuals, parameters or weights are automatically adjusted in order to minimize the final total error. So an ANN is able to learn from its own mismatches or mistakes. The learning ability of a Neural Network is achieved by a learning (training) algorithm, which is usually classified as [87]:
  - Supervised: training examples include both the input vectors and the output vectors.
  - Unsupervised: only the input vectors are available, so the algorithm only learns some internal features of the input vectors
  - Reinforcement learning: is a combination of the above-mentionated algorithms based on presenting the input vectors and looking at the resulting output vector. If the result is considered as satisfactory the existing connection weights are increased, otherwise they are decreased.
- Recall refers to how the ANN performs when new but representative data are presented to the ANN (prediction). Before starting with the learning step it is very important to verify and screen the data to be sure the examples presented to the ANN are representative examples of the problem to be solved or modelled. According to Schulze et al. [88], Figure 7 schematizes the algorithm that describes the way an ANN works.

ANNs are normally very effective to capture the non-linear relationships that exist between variables in complex systems and can also be applied in situations where insufficient process knowledge is available to construct a white-box model to describe the system. The main requirements to define an ANN are representative data from of the system being studied, software and hardware with enough memory and capacity, knowledge of construction of an ANN and statistics to validate those models. The resulting network is robust against process noise or instrumentation bias and plant-specific behaviour can be automatically learned. Despite the fact that the main idea of the model is the establishment of relations between inputs and outputs using only data

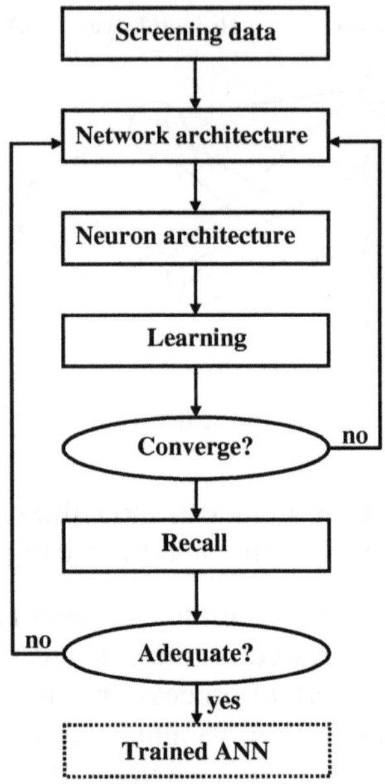

Fig. 7. A stepwise network operation, according to Schulze et al. [88]

measured from the system, a close cooperation with process experts is always advantageous and advised.

According to several authors, the advantages of using neural networks are that almost every relation between cause and effect can be modelled and no assumptions need to be made considering the nature of the relation Moreover, pre-processing of data is minimized and once calibrated or trained, an ANN can be transformed easily into program code so anyone can use it. Among the main disadvantages one should mention the large amount of data necessary to calibrate the ANN. The risk of over-parameterization is also frequent during the training stage. Additionally, once calibrated, the ANN is prepared to react accurately only to input data similar to the one used during the calibration step [88]. Extrapolation to other systems, for example with other types of activated sludge (other microbial decomposition) or other pollutants present in the wastewaters, may have many limitations and always should be performed with a great caution.

Due to its characteristics, ANN is being applied in many different applications including the prediction of water quality parameters like chloride, ammonium, and dissolved oxygen as well as for an accurate prediction of water pollution and water and energy consumption [88].

Pollution of beaches and loss of recreational water amenity caused by municipal effluent discharges from coastal primary WWTP have been a problem around the world. Oil and grease (O&G) are one of the significant pollutants in the effluent that impacts beaches and recreational water quality and have stringent regulatory discharge limits. Raha [89] developed a three-layer feed-forward back-propagation ANN model applying a supervised normal cumulative delta learning rule and hyperbolic tangent transfer function to predict the effluent O&G concentrations from a coastal primary and chemically assisted primary WWTP up to 3 days in advance. Results show that the models are able to predict effluent O&G with fair reliability with a RSMEP ranging between 3.6–4.5 mg/L. The tool provided water utility operators to take appropriate process control actions in sewerage transport and treatment operations, so as to enable the WWTP to produce effluent with O&G well below the license limits. Gontarski et al. [90] used a back-propagation neural network to predict the elimination of TOC in the effluent of an industrial treatment plant, using the delta-bar-delta algorithm [91] for the estimation of weights and the sigmoid function as the neuron transfer function. Different input parameters were considered during the ANN training as flow rations, pHs, DO, TSS, etc. A sensitivity analysis permitted to conclude that in this particular case, the liquid flow rate and the pH of the inlet stream were the most important variables to take into account in controlling the plant.

Sinha et al. [92] designed and trained an ANN based model for simulating the performance of an UASB reactor treating high-strength wastewater. The model inputs were organic loading rate, hydraulic retention time and influent bicarbonate alkalinity. The output variables were the effluent substrate concentration, the reactor bicarbonate alkalinity, pH and volatile fatty acid concentration, the average gas production rate and the percent methane content of the gas. Training of the NN model was achieved using a large amount of experimentally obtained performance data. Training was followed by validation using independent sets of performance data obtained from the same reactor. Simulations were performed using the validated ANN model to determine the impact of changes in parameters like influent COD and hydraulic retention time on the reactor performance. The simulation results permitted to get important insights into key variables that were responsible for influencing the working conditions of the UASB reactor under varying input conditions.

Strik et al. [93] developed an ANN to predict several gaseous trace compounds like hydrogen sulfide and ammonia in the biogas flow from an anaerobic digestion system. This highly energetic biogas applied into fuel cells will result in a significantly higher electrical efficiency and could contribute to an increase of renewable energy production. However it is necessary to guarantee that it is not contaminated with the mentioned gaseous compounds that will poison the fuel cell. Hydrogen sulfide and ammonia concentrations in biogas were modeled successfully using the MATLAB Neural Network Toolbox [94]. The results obtained by the authors showed that ammonia in biogas can indeed be present up to 93 ppm. With these kind of tool an operator is able to take every day, in a few minutes time, the right precautions being possible to foresee, control, reduce or even avoid the production of the toxic trace compounds for fuel cells.

Chen et al. [95] addressed the problem of the viability of secondary effluent wastewater reuse using an ANN model to facilitate the decision. This is a special concern in semi-arid and arid regions were it is being discussed the feasibility of reusing treated wastewater for aquifer recharge, agricultural irrigation, lawn sprinkling, toilet flushing, etc. The authors applied a neural network model as a forecasting tool to facilitate the decision-making process in effluent reuse applications. The proposed model was designed to use three online monitoring parameters (pH, ORP, and DO) in conjunction with three offline nutrient tests (BOD_5, NH_3, and NO_3) to construct the neural network model able to predict the concentration of nitrogen species in the final effluent. The possible reuse of the treated effluent depends on the values calculated for these variables. A rainfall index was also included to help identify the dynamic weather condition and provide the essential information on the operational status of a specific wastewater reclamation system.

Finally, and with the same purpose reported before, Fogelman et al. [96] presented a rapid oxygen demand measuring technique that uses UV–Vis multi-wavelength spectroscopy and artificial neural networks (ANNs) to indirectly determine chemical oxygen demand (COD) levels. The influence of turbidity and spectral range as input parameters to the ANN was analyzed and it was revealed the importance of the incorporation of the spectral range below 200 nm in the accuracy of the COD predictions given by the ANN model. It was also pointed out that there is a turbidity limit of approximately 150 NTU to use unfiltered samples. For higher turbidity values, the accuracy of the model decreases significantly.

6 Hybrid Model Schemes

As a final remark it is important to notice that often these models are used in association instead of independently. In fact, experts agree in that the powerfulness of AI methods can be largely enhanced when using combinations of different AI methods instead of individual methods as the ones presented in this chapter. An example of a hybrid model is for instance a model based on first engineering principles, where specific functionalities, as reaction kinetics or parameter estimation have to be predicted from process data. In such cases, artificial intelligent methods can be used to predict the missing/unknown data by taking into account only input and output data from the entire process or from a specific element of the process.

Poch et al. [97] presented a hybrid KBS integrating different AI methods (KBS and ANN) with numerical methods (classical control systems or models) capable of overcoming the limitations found when solving complex problems with a sole classical or AI technique. During the period of exhaustive validation, the environmental decision support systems handled with 123 different problem situations and suggested suitable action strategies, in most of the cases. The WWTP problem situations detected included foaming, rising, filamentous bulking, under loading, overloading, deflocculation (including possible toxic shock), hydraulic shock, mechanical fault, poor primary settler performance, non-biological origin problem on clarifier and influent nitrogen/organic matter shock. From those, about 80% were successfully identified (about one-third in advance and two-third the same day).

Molga et al. [98] developed and tested a hybrid first-principles neural network model where the trained neural network represented the unknown kinetics of biological decomposition of pollutants. The obtained information was implemented into the structure of the conventional model, which contains mass-balance equations. The net was trained with experimental data obtained during a one-year of performance of a full-scale biological treatment plant processing textile wastewater. Because of the use of the ANN in the hybrid system, the performance of the considered WWTP could be modeled based on the experimental data only without any additional information on the mechanism and the kinetics of biodegradation of treated contaminants required. The authors point out that this is a great advantage of neural modeling because time-consuming kinetic investigations can be avoided. However, they also refer that with such an approach, all properties of the modeled system, as well as all model simplifications, are included into the neural network parameters (weights) making the generalization of the model risked.

Another advantageous example of association of AI models is the use of MSV methods to pre-process data before it enters an ANN or FL structure. The objective is twofold: to uniform the weight of the input variables mostly when they represent dimensionally different variables of the process and eliminate correlated data selecting the minimum representative data that is sufficient to be used for the NN or FL model. PCA is among the most used pre-treatment method substituting in such a way the input variables selectivity analysis. The PCA technique helps the nonlinear ANN mapping by its orthogonal transformation of variables and reduction of system dimensionality.

Christos et al. [99] and Oliveira-Esquerre et al. [100] both used and association of PCA and ANN to predict the effluent BOD removal efficiency in a horizontal subsurface flow (HSF) constructed wetland and a biological WWTP from a pulp and paper plant, respectively. In both cases, the PCA was first applied for the proper selection of the parameters entering the ANN. The previous reduction of the number of input parameters permit to develop simpler nets, with a lower number of neurons and hidden layers, which significantly simplifies the control system, decreasing the computation time and permitting a faster access to the output controlled value.

Yoo et al. [101] and Yoo and Lee [102] applied a nonlinear fuzzy partial least squares (FPLS) algorithm for modeling nonlinear characteristics of biological WWTP. The FPLS model not only possesses nonlinear modeling ability, but also the robustness and interpretability of the PLS and fuzzy methods providing more stable estimations of output on extrapolation. FPLS embeds the nonlinear characteristics of FL modeling into the regression framework of the partial least squares (PLS) method. As a result, the FPLS model gives a more favorable modeling environment in which the knowledge of experts can be easily applied. The authors showed that the model gave good modeling performance and higher interpretability than any other nonlinear PLS modeling method.

Perendeci et al. [103] applied a neural fuzzy model based on ANFIS (Adaptive-Network Based Fuzzy Inference System) using online input variables parameters (CH_4 %, Q_{gas}, $Q_{anarecycle}$, $Q_{inf\text{-}bypass}$ and Q_{inf}) to estimate the effluent COD of a real scale anaerobic wastewater treatment plant of a sugar factory. ANFIS is a technique for automatically tuning first-order Sugeno type inference systems based on training data [104]. It consists of five key components: inputs and outputs, database and pre-processor, a fuzzy system generator, a fuzzy inference system, and an adaptive neural network representing the fuzzy system [105]. The ANFIS model was able to estimate the water quality discharge parameter with success for the case when only limited online variables were available without requiring the measurement of inlet COD which is an offline parameter to be determined experimentally. The information provided by the prediction procedure is sufficiently reliable for the plant monitoring and performance check.

7 Conclusion

The actual more severe environmental regulation rules will certainly lead to a more responsible and effective application of reliable prediction and control strategies into WWTP. The current believe is that real time monitoring and anticipative prediction of internal or external events that may affect the regular system behavior are the key points for research in the near-future.

The complexity that characterizes WWTS difficult the use of traditional models to obtain reliable descriptions as well as predictions and/or extrapolations of the system behavior. Moreover, the actual development of sensors significantly enlarge the amount of data that can be monitored for a given system. These has obvious advantages but simultaneously difficult the analysis and evaluation of the status of the system for a given period. Also, it is important to have in mind that, in the case of WWTS, fast and reliable prediction is usually more important than precise quantification. This can be considers one of the main reasons way WWTS became a challenging issue for artificial intelligence models development and validation.

Different AI models, including the ones that are being most extensively applied to WWTS, were synthetically described and reference works dealing with these models were reported. Without being exhaustive, the examples presented through this chapter intend to demonstrate that AI methods can indeed be used to WWTS monitoring and control. Depending on the system, on the amount of monitored data and the kind of information that need to be obtained, different AI models can be applied. However, the actual *state of the art* is focusing on the use of associations of the different AI methods to originate robust tools able to accurately online characterization of these complex systems.

List of Acronyms

ADM - anaerobic digestion model
AI - artificial intelligent
ANFIS - adaptive-network based fuzzy inference system
ANN - artificial neural networks
APCA - adaptive principal components analysis
ASM1 - activated sludge model No. 1
BOD - biological oxygen demand
BSM1 - benchmark simulation model 1
COD - chemical oxygen demand
DO - dissolved oxygen
E - error matrix
FIS - fuzzy inference system
FL - fuzzy logic
FPLS - nonlinear fuzzy partial least
FT-IR - fourier transform infra-red
KBDSS - knowledge-based decision support system
KBS - knowledge based systems
MF - membership function

MPCA - multi-way principal component analysis
MVS - multivariate statistical methods
NIPALS - nonlinear iterative partial least squares
ORP - oxidation-reduction potential
P - loadings matrix
PC - principal component
PCA - principal component analysis
PI - proportional-integral control
PID - proportional-integral-derivative control
PLS - partial least squares ,
X - data matrix
RAS - returned activated sludge
RMSECV - root-mean-square error of cross-validation
RMSEP - root-mean-square error of prediction
SPE - squared prediction error
SVI - sludge volume index
T - scores matrix
TA - total area
TL - total length
TOC - total oxygen demand
TSS - total suspended solids
VFA - volatile fatty acids
WWTS - wastewater treatment systems
WWTP - wastewater treatment plant

λ_i - eigenvalue
m - number of rows (of matrix M)
n - number of columns (of matrix M)

References

1. Poch, M., Comas, J., Rodriguez-Roda, I., Sanchez-Marre, M., Cortes, U.: Designing and building real environmental decision support systems. Environ. Mod. & Soft. 19, 857–873 (2004)
2. Jeppsson, U.: Modeling aspects of wastewater treatment processes. Ph.D. Thesis, Lund Institute of Technology, Sweden (1996)
3. Henze, M., Grady, Jr., C.P.L., Gujer, W., Marais, G.V.R., Matsuo, T.: Activated Sludge Model No. 1. IAWQ Scientific and Technical Report No. 1, London, UK (1987)
4. Henze, M., Gujer, W., Mino, T., Matsuo, T., Wentzel, M.C.M., Marais, G.V.R.: Activated Sludge Model No. 2. IWA Scientific and Technical Report No. 3, London, UK (1995)
5. Henze, M., Gujer, W., Mino, T., Matsuo, T., Wentzel, M.C.M., Marais, G.V.R.: Activated Sludge Model No. 2. IWA Scientific and Technical Report No. 3, London, UK (1995)
6. Olsson, G., Newell, B.: Wastewater Treatment Systems; Modeling, Diagnosis and Control. IWA Publishing, London (1999)
7. Dochain, D., Vanrolleghem, P.: Dynamical modeling and estimation in wastewater Treatment Processes. IWA Publishing, London (2001)

8. Gernaey, K.V., van Loosdrecht, M.C.M., Henze, M., Lind, M., Jørgensen, S.B.: Activated sludge wastewater treatment plant modelling and simulation: state of the art. Environ. Mod. & Soft. 19, 763–783 (2004)
9. Cinar, O., Daigger, G.T., Graef, S.P.: Evaluation of IAWQ activated sludge model no. 2 using steady-state data from four full scale wastewater treatment plants. Water Environ. Res. 70, 1216–1224 (1998)
10. Belanche, L.A., Valdes, J.J., Comas, J., Roda, I.R., Poch, M.: Towards a model of input–output behaviour of wastewater treatment plants using soft computing techniques. Environ. Mod. & Soft. 14, 409–419 (1999)
11. Pearson, K.: On lines and planes of closest fit to systems of points in space. Philosophical Magazine Series B 2, 559–572 (1901)
12. Hotelling, H.: Multivariate quality control illustrated by the testing of sample bombsights. In: Eisenhart, O. (ed.) Selected techniques of statistical analysis, pp. 113–184. McGraw-Hill, New York (1947)
13. Wise, B.M.: Adapting Multivariate Analysis for Monitoring and Modeling Dynamic Systems, Ph.D. Dissertation, University of Washington, Seattle (1991)
14. Wise, B.M., Ricker, N.L.: Recent Advances in Multivariate Statistical Process Control: Improving Robustness and Sensitivity. In: IFAC Symposium on Advanced Control of Chemical Processes, Toulouse, France (October 1991)
15. Qin, S.J., Dunia, R.: Determining the number of principal components for best reconstruction. J. Process Control 10(2-3), 245–250 (2000)
16. Kresta, J.V., MacGregor, J.F., Marlin, T.E.: Multivariate statistical monitoring of process operating performance. Can. J. Chem. Eng. 69, 35–47 (1991)
17. Ku, W., Storer, R.H., Georgakis, C.: Disturbance detection and isolation by dynamic principal component analysis. Chem. Intell. Lab. Syst. 30, 179–196 (1995)
18. Rosen, C., Lennox, J.A.: Monitoring wastewater treatment operation. Part I: Multivariate monitoring. Wat. Res. 35, 3402–3410 (2001)
19. Lennox, J., Rosen, C.: Adaptive multiscale principal component analysis for online monitoring of wastewater treatment. Wat. Sci. Tech. 45, 227–235 (2002)
20. Lee, J.-M., Chang, K.Y., Choi, S.W., Vanrolleghem, P.A., Lee, I.-B.: Nonlinear process monitoring using kernel principal component analysis. Chem. Eng. Sci. 59, 223–234 (2004)
21. Wold, H.: Soft modeling, the basic design and some extensions. In: Joreskog, K., Wold, H. (eds.) System under indirect observations. North Holland, Amsterdam (1982)
22. Wold, S., Esbensen, K., Geladi, P.: Principal component analysis. Chemom. and Intel. Lab. Sys. 2, 37–52 (1987)
23. Rosen, C.: A Chemometric Approach to Process Monitoring and Control with Applications to Wastewater Treatment Operation. Ph.D. Thesis, Lund Institute of Technology, Sweden (2001)
24. Geladi, P., Kowalski, B.R.: Partial least squares regression: A tutorial. Analytical Chimica Acta 185, 1–17 (1986)
25. Choi, S.W., Lee, I.-B.: Multiblock PLS-based localized process diagnosis. J. of Process Control 15, 295–306 (2005)
26. Wold, S., Kettaneh-Wold, N., Skagerberg, B.: Nonlinear PLS modeling. Chemometrics Intell. Lab. Syst. 7, 53–65 (1989)
27. Lee, D.S., Lee, M.W., Woo, S.H., Kim, Y.J., Park, J.M.: Multivariate Online Monitoring of a Full-Scale Biological Anaerobic Filter Process Using Kernel-Based Algorithms. Ind. Eng. Chem. Res. 45, 4335–4344 (2006)
28. Nomikos, P., MacGregor, J.F.: Monitoring batch processes using multi-way principal component analysis. AIChE J. 40, 1361–1375 (1994)

29. Villez, K.: Multivariate and qualitative data analysis for monitoring, diagnosis and control of sequencing batch reactors for wastewater treatment. Ph.D. Thesis, University of Gent, Belgium (2007)
30. Yoo, C.K., Lee, D.S., Vanrolleghem, P.A.: Application of multiway ICA for on-line process monitoring of a sequencing batch reactor. Water Research 38, 1715–1732 (2004)
31. Walczak, B., Massart, D.L.: Dealing with missing data Part I. Chemometrics Intell. Lab. Syst. 58, 15–27 (2001)
32. Walczak, B., Massart, D.L.: Dealing with missing data: Part II. Chemometrics Intell. Lab. Syst. 58, 29–42 (2001)
33. Åström, K.J., Wittenmark, B.: Computer controlled systems, theory and design, 3rd edn. Prentice Hall, Inc., Englewood Cliffs (1997)
34. Tomita, R.K., Park, S.W., Sotomayor, O.A.Z.: Analysis of activated sludge process using multivariate statistical tools-a PCA approach. Chem. Eng. J. 90, 283–290 (2002)
35. Le Bonte, S., Potier, O., Pons, M.N.: Toxic event detection by respirometry and adaptive principal components analysis. Environmetrics 16, 589–601 (2005)
36. Copp, J.: The COST Simulation Benchmark: Description and Simulator Manual. Office for Official Publications of the European Community: Luxembourg (2001)
37. Lee, C., Choi, S.W., Lee, I.-B.: Sensor fault diagnosis in a wastewater treatment Process. Wat. Sci. Tech. 53, 251–257 (2006)
38. Amaral, A.L., Ferreira, E.C.: Activated sludge monitoring of a wastewater treatment plant using image analysis and partial least squares regression. Analytica Chimica Acta 544, 246–253 (2005)
39. Gut, L., Płaza, E., Hultman, B.: Assessment of a two-step partial nitritation/Anammox system with implementation of multivariate data analysis. Chemom. & Intel. Lab. Sys. 86, 26–34 (2007)
40. Fuentes, M.J., Font, R., Gomez-Rico, M.F., Molto, J.: Multivariant statistical analysis of PCDD/FS in sewage sludges from different areas of the Valencian Community (Spain). Chemosphere 67, 1423–1433 (2007)
41. Steyer, J.P., Bouvier, J.C., Conte, T., Gras, P., Harmand, J., Delgenes, J.P.: On-line measurements of COD, TOC, VFA, total and partial alkalinity in anaerobic digestion processes using infra-red spectrometry. Wat. Sci. Tech. 45, 133–138 (2002)
42. Hansson, M., Nordberg, Å., Mathisen, B.: On-line NIR monitoring during anaerobic treatment of municipal solid waste. Wat. Sci. Tech. 48, 9–13 (2003)
43. Sousa, A.C., Lucio, M.M.L.M., Neto, O.F.B., Marcone, G.P.S., Pereira, A.F.C., Dantas, E.O., Cesar, W.D.M., Araujo, U., Galvão, R.K.H.: A method for determination of COD in a domestic wastewater treatment plant by using near-infrared reflectance spectrometry of seston. Analytica Chimica Acta 588, 231–236 (2007)
44. Dias, A.M.A., Moita, I., Páscoa, R., Alves, M.M., Lopes, J.A., Ferreira, E.C.: Activated sludge process monitoring through in situ near-infrared spectral analysis. Wat. Sci. Tech. 57, 1643–1650 (2008)
45. Langergraber, G., Fleischmann, N., Hofstädter, F.: A multivariate calibration procedure for UV/VIS spectrometric quantification of organic matter and nitrate in wastewater. Wat. Sci. Tech. 47, 63–71 (2003)
46. Lourenço, N.D., Chaves, C.L., Novais, J.M., Menezes, J.C., Pinheiro, H.M., Diniz, D.: UV spectra analysis for water quality monitoring in a fuel park wastewater treatment plant. Chemosphere 65, 786–791 (2006)
47. Rieger, L., Langergraber, G., Kaelin, D., Siegrist, H., Vanrolleghem, P.A.: Long-term evaluation of a spectral sensor for nitrite and nitrate. Wat. Sci. Tech. 57, 1563–1569 (2008)

89. Raha, D.: Comparative Study of Artificial Network Modeling in Predicting Effluent Oil and Grease from Coastal Primary and Chemically Assisted Primary Sewage Treatment Plants. Environ. Forensics 8, 371–381 (2007)
90. Gontarski, C.A., Rodrigues, P.R., Mori, M., Prenem, L.F.: Simulation of an industrial wastewater treatment plant using artificial neural networks. Comp. and Chem. Eng. 24, 1719–1723 (2000)
91. Jacobs, R.A.: Increased rates of convergence through learning, rate adaptation. Neural Networks 1, 295–307 (1988)
92. Sinha, S., Bose, P., Jawed, M., John, S., Tare, V.: Application of neural network for simulation of upflow anaerobic sludge blanket (UASB) reactor performance Biotechnol. Bioeng. 77, 806–814 (2002)
93. Strik, D.P.B.T.B., Domnanovich, A.M., Zani, L., Braun, R., Holubar, P.: Prediction of trace compounds in biogas from anaerobic digestion using the MATLAB Neural Network Toolbox. Environ. Mod. & Soft. 20, 803–810 (2005)
94. The MathWorks, MatLab Documentation Set: MATLAB Neural Network Toolbox (2002)
95. Chen, J.C., Chang, N.B., Shieh, W.K.: Assessing wastewater reclamation potential by neural network model. Eng. Appl. of Art. Intell. 16, 149–157 (2003)
96. Fogelman, S., Zhao, H., Blumenstein, M.: A rapid analytical method for predicting the oxygen demand of wastewater. Anal. Bioanal. Chem. 386, 1773–1779 (2006)
97. Poch, M., Comas, J., Rodríguez-Roda, I., Sànchez-Marrè, M., Cortés, U.: Designing and building real environmental decision support systems. Environ. Mod. & Soft. 19, 857–873 (2004)
98. Molga, E., Cherbanski, R., Szpyrkowicz, L.: Modeling of an Industrial Full-Scale Plant for Biological Treatment of Textile Wastewaters: Application of Neural Networks. Ind. Eng. Chem. Res. 45, 1039–1046 (2006)
99. Akratos, C.S., Papaspyros, J.N.E., Tsihrintzis, V.A.: An Artificial Neural Network Model and Design Equations for BOD and COD Removal Prediction in Horizontal Subsurface Flow Constructed Wetlands. Chem. Eng. J. (2008) (in press)
100. Oliveira-Esquerre, K.P., Mori, M., Bruns, R.E.: Simulation of an industrial wastewater treatment plant using artificial neural networks and principal component analysis. Braz. J. of Chem. Eng. 19, 365–370 (2002)
101. Yoo, C.K., Ho Bang, Y., Lee, I.-B., Vanrolleghem, P.A., Rosén, C.: Application of Fuzzy Partial Least Squares (FPLS). Modeling Nonlinear Biological Processes. Korean J. Chem. Eng. 21, 1087–1097 (2004)
102. Yoo, C.K., Lee, I.-B.: Integrated framework of nonlinear prediction and process monitoring for complex biological processes. Bioprocess Biosyst. Eng. 29, 213–228 (2006)
103. Perendeci, A., Arslan, S., Celebi, S.S., Tanyolac, A.: Prediction of effluent quality of an anaerobic treatment plant under unsteady state through ANFIS modeling with on-line input variables. Chem. Eng. J. (2008) (in press)
104. Jang, R.J.S.: ANFIS: Adaptive - network - based fuzzy inference system. IEEE Trans. On Systems, Man and Cybernetics 23, 665–685 (1993)
105. Tay, J.H., Zhang, X.: A fast predicting neural fuzzy model for high-rate anaerobic wastewater treatment systems. Water Res. 34, 2849–2860 (2000)

6

Multiobjective Genetic Algorithms for the Optimisation of Wastewater Treatment Processes

Benoît Beraud[1], Cyrille Lemoine[1], and Jean-Philippe Steyer[2]

[1] Anjou Recherche, Chemin de la Digue, BP 76,
78603 Maisons Laffitte, France
{benoit.beraud,cyrille.lemoine}@veolia.com
[2] INRA, UR50, Laboratoire de Biotechnologie de l'Environnement, Avenue des Etangs,
11000 Narbonne, France
steyer@supagro.inra.fr

Abstract. The combination of multiobjective genetic algorithms with wastewater treatment plant (WWTP) models provides an efficient framework for the evaluation, optimisation and comparison of WWTP control laws. This chapter presents a methodology developed for this efficient combination. Existing models and simulation software are used. They are combined with NSGA-II, a multiobjective genetic algorithm capable of finding the best tradeoffs (Pareto front) among multiple opposed objectives. Long term evaluations of the optimized solutions are proposed to check their robustness. An application of the methodology on the Benchmark Simulation Model 1 is presented and illustrates the benefits of the methodology.

1 Introduction

The recent advances in the development of reliable online sensor for the monitoring of wastewater treatment plants (WWTPs) are leading to the development of new control strategies. The comparison of these control laws on real processes is difficult due to the conjunction of three factors.

First, large timescales are needed for an unbiased evaluation of a new control law performances due to the low growth rate of bacteria used for the pollution removal. This is due to the required adaptation of bacteria to the new operating conditions induced by the new control law.

Second, the continuous change of the WWTP influent characteristics (flowrate, loading, temperature) is also inducing bias in the comparison of the various control laws. A solution would consist in the measurement of these perturbations to correct the performances values. This is however very difficult due to the possible adaptation of the bacteria to these perturbations.

Third, the nonlinearities of the processes used in WWTPs induce difficulties in the manual search of optimal parameters for a given control law and sub-optimal parameters can induce bias in the comparison of the control law.

As a consequence of these three factors, it is impossible to evaluate and optimize control laws on real processes. It is therefore necessary to develop a new methodology. The solution proposed in this chapter consists in using the WWTPs models for

M. do Carmo Nicoletti, L.C. Jain (Eds.): Comp. Intel. Tech. for Bio. Mod., SCI 218, pp. 163–195.
springerlink.com

the evaluation of control laws performances. These models are combined with an optimisation technique, the multiobjective genetic algorithm NSGA-II, to find the best control laws parameters.

This chapter presents the new methodology developed and its application on a case study. The characteristics of the WWTP influent and processes are described in section 2. State of the art models are detailed and a review of their use for the optimisation of WWTP functioning is presented.

Section 3 presents the challenges of the optimisation of WWTP models and existing optimisation solutions. As there is currently no existing solution solving all challenges, the new methodology proposed in this chapter is detailed, based on existing models and multiobjective optimisation.

NSGA-II, the multiobjective genetic algorithm used in this chapter, is presented in section 4, preceded by an introduction to genetic algorithms.

The methodology is finally applied in section 5 on the Benchmark Simulation Model 1, a literature case study.

Conclusions are then drawn in section 6 on the benefit of the methodology developed, on its strengths and weaknesses and on the perspective of its application on real industrial processes.

2 Wastewater Treatment Processes and Their Models

The choice of the optimisation methodology to use must be based on the characteristics of the system under study. The diverse processes used for the treatment of wastewater in conventional activated sludge systems, their non-linearity and the high level of disturbances induced by the incoming wastewater are typically limiting the range of reliable optimisation techniques. This section will highlight these characteristics together with a description of state-of-the-art models of the WWTP processes.

2.1 Main Characteristics of Wastewater and Treatment Processes

Wastewater. Wastewater is the collection of all reject water from households and small industries not equipped with their own treatment plant. It is composed of many pollutants. This pollution is usually referred to in terms of chemical oxygen demand (COD), total suspended solids (TSS), total nitrogen (NGL) and ammonia. The chemical oxygen demand is a measurement of the quantity of oxygen required to reduce all pollutants using a chemical way. It is a reliable measurement of organic pollution. The concentration of total suspended solids represents the amount of particulate compounds, including bacteria and inert compounds.

The most important characteristic of wastewater is that it is strongly influenced by daily and weekly variations induced by human activities. Long term modifications are also occurring. This includes seasonal events like holidays (either an increase or decrease of the population) but also long term demographic changes.

Typical variations of the influent flow rate at the wastewater treatment plant are illustrated in Fig. 1 (the flow rate corresponds to a typical daily production per habitant in the absence of industry).

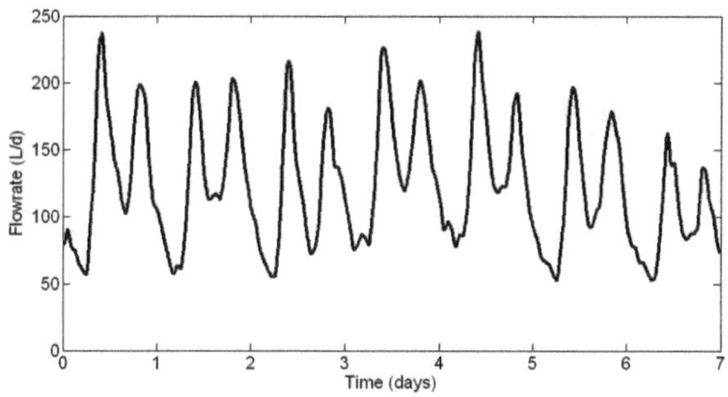

Fig. 1. Typical variations of influent flow rate

Similar variations occur for pollutants concentrations, but a delay between the peak of flowrate and the peak of pollutants concentrations is usually observed (the flowrate peak is located one and two hours before the peak of pollutants concentrations).

The pattern of these variations is influenced by:

- the sewer topology,
- the state of the pipe (infiltration, hole, etc.),
- rain fall events in case of combined sewer (collecting rain and wastewater).

These variations induce big perturbations on the processes of the wastewater treatment plant. Traditionally, these perturbations are limited by the high volumes of the process reactors. Nowadays, more compact systems are constructed and advanced control laws and actuators capable of handling these perturbations are required.

Treatment Processes. In order to treat this wastewater, traditional municipal WWTPs use two main processes: the activated sludge and the clarification-thickening.

The activated sludge process was discovered in 1913 in the UK with experiments regarding the treatment of wastewater in a draw-and-fill reactor (or sequenced batch reactor). First, wastewater was filled in a reactor and oxygen was supplied. Pollution was hence removed from water thanks to biodegradation. Then a phase of decantation was performed. Finally, clean water was removed and sludge was kept in the reactor. These actions were repeated many times. The experiments showed that the water quality was enhanced cycle after cycle. Scientists first thought that the sludge was activated (in a manner similar to activated carbon) and the process was therefore named the activated sludge process. The name of the process stayed even when it was realized later that there was no activation of the sludge but a concentration of bacteria and a selection of the best performing organisms.

Three main reactions are occurring in an activated sludge process concerning the removal of pollutants: the oxidation of organic compounds (transformation in carbon dioxide), the nitrification (transformation of ammonia in nitrate) and the denitrification (transformation of nitrate in gaseous nitrogen). The two first reactions require the presence of oxygen while the last one occurs only in the absence of oxygen. This is the main reason why activated sludge are usually intermittently aerated, even if advanced control laws are developed that could lead to simultaneous reactions by providing the exact amount of oxygen required [1, 2].

The process of clarification-thickening consists in separating solid particles from water and soluble compounds. Its double name is due to the fact that it can be seen from the two sides: clarification as the goal is to have only water in the overflow; thickening as the goal is to keep the sludge (which contains the bacteria) in the system with as less water as possible in order to reintroduce it in the activated sludge unit.

Usually, these two processes are placed in series in order to have continuous operation of the plant as illustrated in Fig. 2. The tank containing the activated sludge is aerated intermittently to provide oxygen to the bacteria consortium. The settler is then used to separate the cleaned water from the particulate compounds. The sludge, which contains the bacteria, is reintroduced into the activated sludge unit. Some sludge is also wasted out of the settler in order to keep a constant mass in the system.

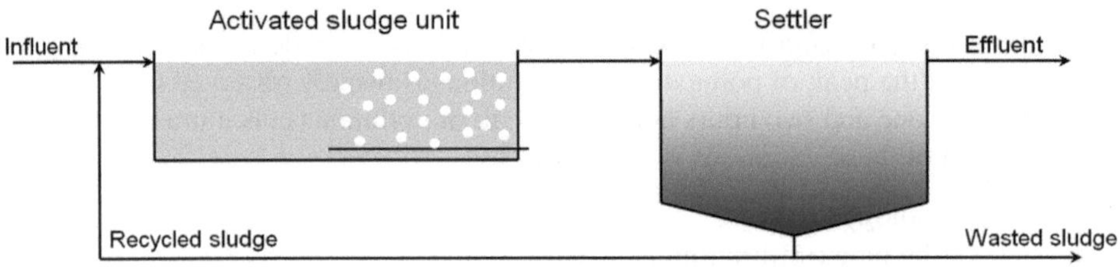

Fig. 2. Typical layout of a wastewater treatment plant

Three very different time scales should be considered in a wastewater treatment plant: a short one (typically one minute) corresponding to biochemical reactions, a medium one (typically one day) corresponding to the hydraulic residence time and a longer one (typically one week to one month) corresponding to the evolution of the sludge. This is an important factor to keep in mind for the simulations and optimisations of a wastewater treatment plant.

2.2 Mathematical Models of the Activated Sludge Process

In order to simulate and better understand the activated sludge process, four models for activated sludge processes were published during the last twenty years by the IWA (International Water Association), formerly IAWQ (International Association on Water Quality). These models are known in the literature as the ASM models (i.e., the activated sludge models ASM1, ASM2, ASM2d and ASM3). These are nowadays the most used models to represent the behaviour of activated sludge processes, even if various minor modifications have been proposed and used locally.

All these four models are based on the same principle of description. First, the mixed liquor (activated sludge and wastewater) is finely detailed into components that stand for different compounds and organisms (ranging from 13 for ASM1 to 19 for ASM2 and ASM2d). This step is named the fractionation. Then, processes of components transformations are introduced together with kinetic laws to describe the evolution of the system. The reaction rates of these different processes are dependant from the concentration of some compounds and organisms in the mixed liquor by means of linear relations or Monod activation terms.

One key issue in all these models is that most components of the fractionation can not be measured directly, especially organic compounds and bacteria. They have to be expressed as a fraction of another global parameter like the COD or the TSS.

ASM1. ASM1 is the first model developed by Henze et al. [3]. It is a dynamic model and the main processes occurring in an activated sludge reactor are included (oxidation of organic compounds, nitrification and denitrification) as well as the decay of the biomass, the reaction of ammonification of organic nitrogen and the hydrolysis of the particulate substrates. These processes are illustrated in Fig. 3. The fractionation of the mixed liquor is based on 13 components: particulate and soluble substrate (X_S and S_S), particulate and soluble inert compounds (X_I and S_I), autotrophic and heterotrophic organisms (X_A and X_H), particulate inert compounds arising from biomass decay (X_P, similar to X_I), oxygen (S_O), ammonium and ammonia nitrogen (S_{NH}), nitrate and nitrite (S_{NO}), soluble and particulate organic nitrogen (S_{ND} and X_{ND}) and alkalinity (S_{ALK}). ASM1 is the simplest model of the four IWA but some processes are missing.

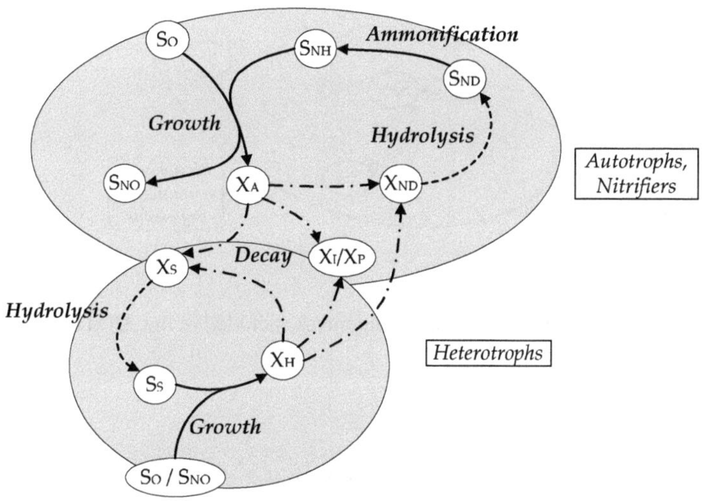

Fig. 3. Processes and components included in the ASM1

ASM2 and ASM2d. In ASM2 [4], the processes of biological phosphorus removal and chemical phosphorus coagulation are added. This induce a finer fractionation of the mixed liquor with new components corresponding to these processes: soluble phosphorus (S_{PO4}), phosphorus accumulating organisms (X_{PAO}) and their internal cell storage of organics (X_{PHA}) and poly-phosphate (X_{PP}), metal-hydroxides (X_{MeOH}) and metal-phosphate (X_{MeP}). Soluble organic substrate (S_S) of ASM1 is divided into fermentable substrate (S_F) and fermentation product (S_A), with a new process of fermentation describing the transition between these two compounds. The processes of ASM2 are illustrated in Fig. 4. Effect of the temperature on the model parameters is also included in this model. The difference between ASM2 and ASM2d [5] is the denitrification capacity of phosphorus accumulating organism which is considered only

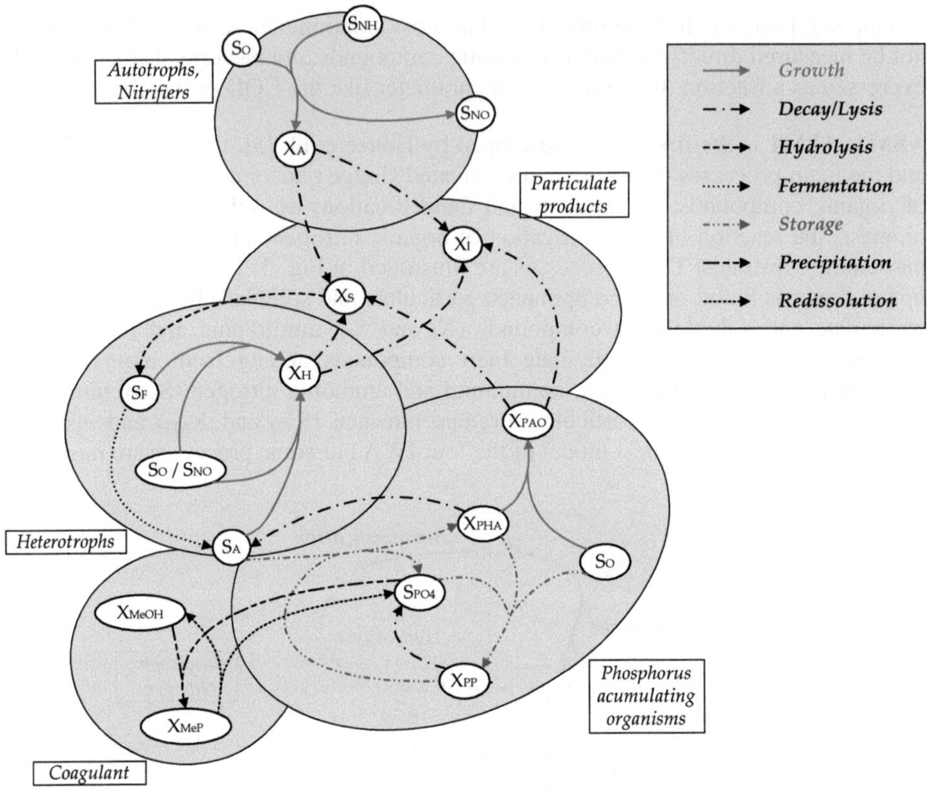

Fig. 4. Processes and components included in the ASM2

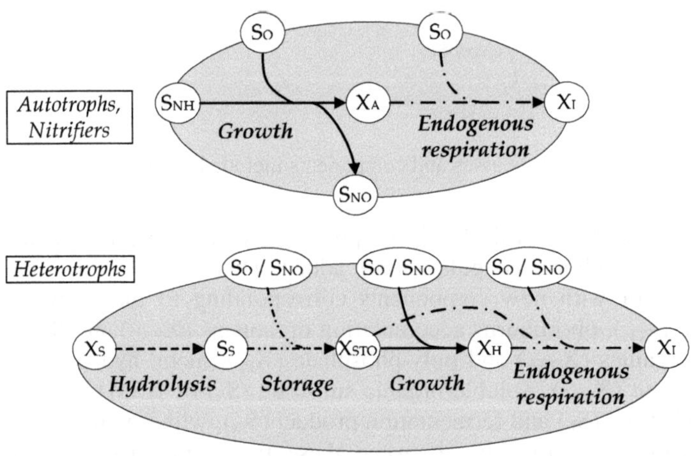

Fig. 5. Processes and components included in the ASM3

in ASM2d. In each of these models, the main drawback compared to ASM1 is an increased number of parameters to calibrate.

ASM3. ASM3 [6] is a modification of the ASM1. The concept of death/lysis is replaced with endogenous respiration and internal cell storage of heterotrophic bacteria is added. A new component corresponding to this storage (X_{STO}) is added to the fractionation. This is meant to allow a separation of autotrophic and heterotrophic biomass (see Fig. 5 compared to Fig. 3). The addition of internal cell storage also allows a better representation of batch configurations (draw-and-fill reactors). Soluble and particulate organic nitrogen are removed from the fractionation as well as the corresponding processes of hydrolysis and ammonification. This is related to the very difficult measurement of these two components as well as the difficult estimation of the associated kinetics. Finally inert compound X_I and inert compound arising from biomass decay X_P are grouped in a single component X_I.

2.3 Mathematical Models of the Clarifiers

The other main process of wastewater treatment required with activated sludge processes is the clarifying of water. Many different kinds of models are available for settlers, ranging from the simplest 0-dimensionnal models to the most complicated 3-dimensionnals models.

0-Dimensionnal Conceptual Models. Since the overflow rate concept was introduced by Hazen [7], conceptual 0-Dimensionnal models are utilised for the design of settling tanks. However, the traditional model of Hazen is only applicable to primary settling tanks because the settling rate is assumed to be dependent only on the particle size. This condition is more likely to be fulfilled in primary than secondary settling tanks. More recently, Krebs et al. [8] introduced a conceptual model for secondary settling tanks, which is based on a linear solids concentration profile in the solids blanket. The main limitation of these models is that they can not easily be applied for control and operation of a WWTP, but mostly only for its design.

0-Dimensionnal Flux Models. The limiting flux theory is based on the consideration that the settling velocity of particles is related to their current concentration as a gradient of concentration is always present in the vertical dimension and concentrations of particles below a given point are higher and limiting the possible velocity. The classical flux theory is introduced by Kynch [9] and returns a hindered solids settling velocity. The main limitation of these models is that the theory only provides an underflow solids concentration.

1-Dimensionnal Layered Flux Models. 1-Dimensionnal models layered models are based a discretization of the settler in a set of horizontal layers. In each layer, the concentration of sludge is assumed to be constant and only the transfer between the layers is described. The theory of the limiting flux is used in these models. Two fluxes are therefore making the sludge concentration evolve (see Fig. 6). First, a settling flux represents the decantation of the sludge using the hindered solids settling velocity of the limiting flux theory. Second a transport flux (or advection flux) takes into account

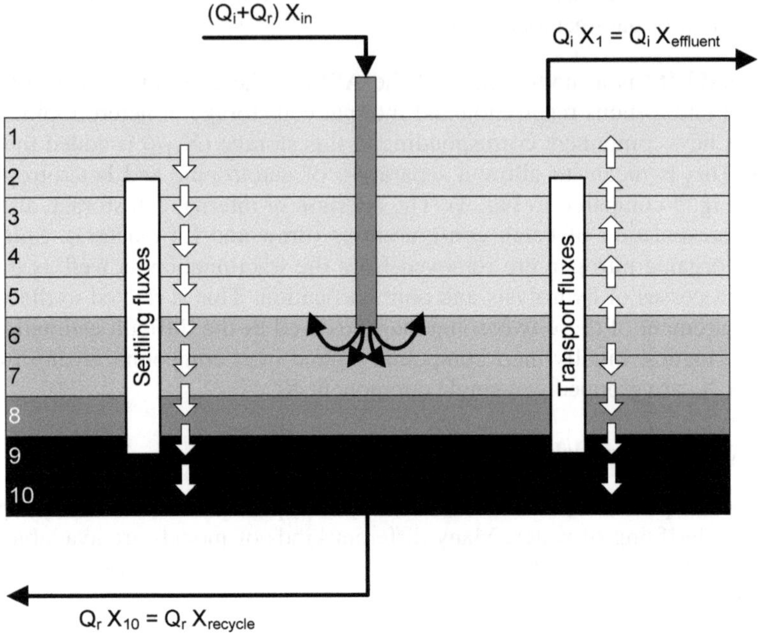

Fig. 6. 1-dimensionnal modelling of clarifiers

the water velocity across layers. Many models are available in this area and the review of Grijspeerdt et al. [10] showed that the one of Takács et al. [11] is the most reliable for dynamic simulations of secondary settlers.

Adaptations of this model have been proposed to include biological reactions occurring in secondary clarifiers in this model [12]. Two solutions are studied: inclusion of an activated sludge model for each layer of the secondary clarifier or addition of a model for depletion of oxygen and a defined fraction of nitrate in the underflow flux (this reaction is mostly occurring at the bottom of the clarifier where there is enough activated sludge). The second solution is found to be the best compromise between computation time and model accuracy.

The main advantage of the layered models are their limiting computer complexity and their possible coupling The main limitation of the layered models is that they have to be calibrated with more complex model (2D or 3D). The choice of the number of layer has a particular importance on the parameters of the model. Another limitation is the low accuracy of the estimation of particles concentrations in the effluent.

1-Dimensionnal Advection-Dispersion Models. Instead of using a limited number of layers for settling tank modelling, the 1D advection-dispersion equation can be solved numerically. This partial differential equation is [13]

$$\frac{\partial X}{\partial t} + u\frac{\partial X}{\partial y} + \frac{\partial v_S X}{\partial y} - D_C\frac{\partial^2 X}{\partial y^2} = 0. \qquad (1)$$

where X is the solids concentration, y the vertical coordinate, u the bulk liquid velocity and v_S the settling velocity of solids. To account for non-ideal flow behaviour, dispersion is introduced into the model by the dispersion coefficient D_C.

The main limitation of this model is that it requires solving a partially differential equation. This limit avoids its combination with activated sludge models for dynamic simulations.

2 and 3-Dimensionnal Models. In a real plant there are many factors influencing the performance and the capacity of the settling tank; many boundary and flow conditions cannot be reflected in 1D models. There are four categories of unconsidered influences [14]:

- geometry, e.g. shape of the basin, inlet and outlet arrangements, and baffles
- flow, e.g. density effects causing non-uniform velocity profiles, potentially resulting in turbulence and/or short-circuits from the inlet to the outlet
- solids removal mechanism, which results in many unsteady effects
- environmental conditions, e.g. wind shear, air and inlet water temperature

The prediction of the settling tank performance with 0D and 1D models is therefore a matter of calibration. Hence, more advanced models are needed. 2D and 3D models have the potential to describe the internal flow pattern and the appropriate solids and solutes transport phenomena. Their application is mainly related to the evaluation of internal design changes like the addition of baffles, on the simulation of rectangular settlers and on the calibration of 0D and 1D models [15].

The disadvantage of these models is that their use is computationally very demanding and, therefore, they cannot be used for control purposes.

2.4 Benchmark Models for the Evaluation of Control Laws

The availability of these models of secondary treatment and the recent advances in computing power have made the simulation of plant models possible on desktop computers. Long simulations of many weeks are now feasible. From another perspective, advanced control laws are developed and published but the tests are made on specific WWTP models with specific objectives. This does not allow an unbiased comparison of the performances of these various control laws. Advanced criteria and a common model are therefore necessary.

In the framework of a COST/IWA task group, a Benchmark Simulation Model No 1 (BSM1) has been proposed by international scientists to assess these topics [16, 17, 18, 19]. This model is composed of five activated sludge tanks in series and a secondary clarifier (see Fig. 7). The activated sludge tanks are modelled with ASM1 [3] and the secondary clarifier uses the model of Takács et al. [11]. This benchmark model allows researchers to have a common framework for the evaluation of their control law performances.

Many criteria are included with this model for the evaluation of the process performances (effluent quality, aeration energy, pumping energy, sludge production, cost of addition of external carbon, etc.) as well as criteria for the evaluation of the performances of the controller itself (integral of squared or absolute error, maximum deviation from set point, etc.).

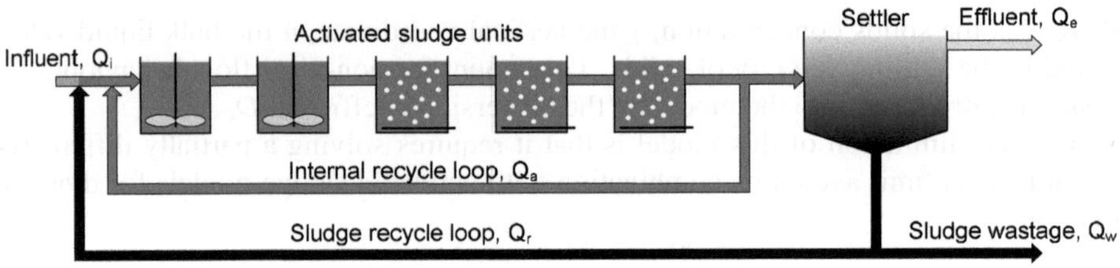

Fig. 7. Plant layout used in the Benchmark Simulation Model No 1

Together with this model, three different influent datasets are provided, representing respectively dry weather, rain weather and storm weather, each with a duration of fourteen days. Mean values are also provided for the computation of steady states of the model.

Additional models to describe sensors and actuators behaviour and a long term influent dataset are also proposed in the literature [20]. They allow the evaluation of control laws long term performances. This long term influent dataset is generated thanks to a phenomenological model capable of representing typical disturbances occurring at the outlet of a sewer network [21].

Finally, a Benchmark Simulation Model No 2 (BSM2) has been recently proposed [19, 22, 23]. This is a plant wide model which includes a thickener, an anaerobic digester and a dewatering unit as well as a primary clarifier in addition to the activated sludge tanks and clarifier of BSM1. With this model more advanced control strategies can be tested and their impact is evaluated on the real whole plant model. This is necessary due to return fluxes that can have huge impact on the treatment performances. This also allows the optimisation of other objectives related to the sludge treatment like the biogas production of the digester or the real total sludge production.

2.5 Conclusion

This section presented the typical models used for representing WWTPs and for their simulation. The Benchmark Simulation Models are the typical application of such models for development of control laws. Their simulation time is quick enough to allow their use for theoretical optimisation of their control laws. The challenges of such optimisations, existing solutions and their drawbacks as well as the new methodology developed are presented in the next section.

3 Optimisation of WWTP Control Laws

The availability of the precedent models is sufficient to allow on-desk optimisation of WWTP processes and control laws. However, their theoretical optimisation is challenged by mainly two problems.

3.1 Challenges Encountered and Existing Solutions

The two main problems challenging the theoretical optimisation are (i) the continuous non-steady state operation of the WWTP and (ii) the amount of possibly opposed objectives that have to be considered. The first problem is the consequence of continuous and huge variations in the incoming pollutant loads previously presented in section 2.1. The opposed objectives of the second problem are for instance the effluent quality, the energy consumption, the sludge production, the biogas production, the risk of malfunction of the plant, the maintenance cost, etc. Two kinds of solutions are proposed in the literature to provide an optimisation tool with such problems [24].

Optimisations Based on Steady-State Simulations and Perturbation Analysis. In a first category of optimisation attempts [25, 26], the authors are considering two objectives: one is the economical optimality of the process, the other takes into account the dynamical performances around a nominal point. These attempts are based only on steady state computations, which are complimented with a perturbation analysis.

In Steyer and Harmand [25], a controller is optimized for the nitrification process. The objective is to minimize the "noise to signal" ratio of sensors (hence inducing a better control of set points) by manipulating the coefficients of the control law under constraints of operational performances. In Vera *et al.* [26], the optimisation of the design and control law of a WWTP bioprocess is performed. The objective is to minimize the error of the controller and the investment costs corresponding to the chosen design. A step perturbation is considered for the evaluation of the controller performance, the goal being to reject this perturbation to stay as close as possible to a fixed set point. Constraints are added to ensure a proper operation of the process.

The main lack of these optimisation attempts is that it does only take into account the average performances and the rejection of a single perturbation. The extension of the results to continuously changing perturbations that arise at the WWTP inlet is not considered and may be at risk.

Optimisations Based on Dynamic Simulations with a Single Objective. In the second category of optimisation attempts, authors focus only on a single economical performance index which allows the use of dynamic simulations for the optimisation. Operational objectives are transformed into constraints that have to be satisfied. Such techniques have the drawback that they do not provide clear insight about the trade-off occurring between the economical and the operational objectives. Examples are the optimisation of (i) the control of the aeration [27, 28, 29] and (ii) the global plant design and control [30].

In Balku and Berber [27], the objective is to minimize the energy consumption by manipulating the aeration sequences under constraints on the effluent quality. Dynamic simulations are performed for each candidate solutions. The main lack of this method is that the optimal aeration profile found for the perturbations used in the simulation may not be adequate for other perturbations. As the control law is acting in open loop (no measurements of incoming pollution or current process states are made), no correction is possible. Another lack is that the impact of the optimal aeration profiles on the long term is not assessed and it may lead to problems like the washout of bacteria from the system for instance [31].

In Fikar *et al.* [28], the objective is to minimize the mean total nitrogen concentration in the effluent also by manipulating aeration sequences. This goal is meant to also minimize the aeration energy and to be the most critical objective of the studied case. A Non Linear Programming (NLP) solver dedicated to dynamic optimisation is used in this work. To address the potential risk of washout, a constraint is added that initial and finals states must be the same, these values being included in the optimisation variables. A control law based on measurement of nitrate and oxygen is derived from the optimal aeration profile found. This allow a closed-loop functioning and help to reject perturbations. The main limitation is that only one day of dry weather is considered for the simulations. It limits the optimality of the solution as rain weather is not considered.

In Holenda *et al.* [29], the objective is to minimize the effluent quality, considering that the aeration energy will be optimized at the same time. A genetic algorithm is used and 10 days of dynamic simulations are performed for each candidate solution but the same daily variations are used each day. Once more the main limitation is that the solution is specific to the perturbation considered. Very good performances are obtained for this specific case but they may degrade quickly when the perturbation change which is always the case in WWTPs. Another limitation is the choice of the initial states. It is assumed to be related to the number of cycle per day. This assumption has a low accuracy. For instance, initials states are obviously not the same if the aeration is on 10% or 90% of time in average.

In Rivas *et al.* [30], steady-state and dynamic optimisations of the design and operation of a complex WWTP layout are performed. Different optimisations are performed in sequence with various objectives and manipulated variables. This allows having an insight into the potential performances. The main limitation is that all objectives and all potentially manipulated variables are not considered at the same time. The assumption behind it is that the problems are uncoupled, which may not always be the case. Another limitation is that no clear vision is of the potential compromise between the potentially conflicting objectives is possible.

Conclusion. The challenges identified in the optimisation of the WWTP control laws have not yet been fully addressed in the literature. Promising solutions have been proposed and will serve as a basis for the development of a methodology addressing all challenges. This methodology is presented in the next section.

3.2 New Methodology Proposed

In order to address the two challenges identified before (i.e. the continuous non-steady state operation of the WWTP and the amount of possibly opposed objectives) and the limitations of existing work considering our application, we propose a new methodology based on multi-objective genetic algorithms and repeated dynamic simulations.

Motivations for the Choice of Genetic Algorithms. The choice of using a genetic algorithm is based on the observation that mainly two categories of algorithms are used for optimisations of nonlinear systems: techniques based on the theory of Non-Linear Programming (NLP) and techniques based on the theory of Genetic Algorithms (GA).

NLP techniques are supposed to require fewer evaluations to reach the optimum but they require information about the derivatives of the objective function (which may require a large number of simulations when their exact formula cannot be expressed like it is usually the case for WWTP optimisation). Another major drawback is that they can be stuck in a local optimum.

On another hand, GA techniques usually require large number of evaluations but no information about the derivatives is needed and they are capable to find the global optimum which is very interesting from an application point-of-view. But the main advantage of genetic algorithms for the optimisation of wastewater treatment processes is that they can use simulations of a whole model (or even interconnected models) inside the objective functions. These simulations can even be based on commercial packages, without any need to have a complete view of all equations inside the models. Closed loop controllers that can be used for the monitoring and supervision of the process operation can also be included in the simulator and optimised. This combination of GAs with dynamic simulations of WWTPs is the selected solution.

Combination of Genetic Algorithms and Models Simulations. An application combining an optimisation technique and dynamic simulations is called dynamic optimisation. In our case, it is based on three main components (see Fig. 8): an optimisation algorithm, a simulator and a performance model.

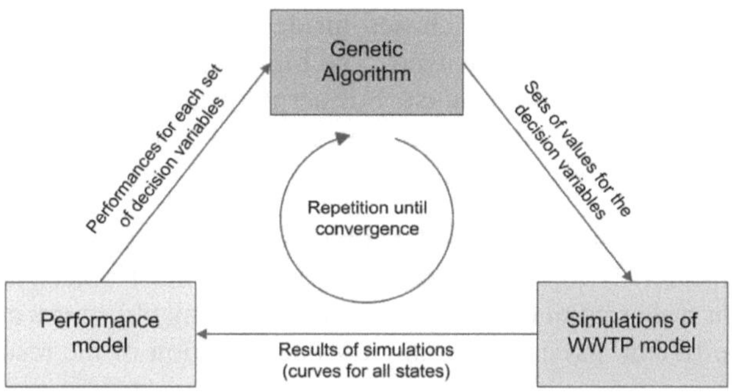

Fig. 8. Flowchart for the dynamic optimisation of WWTP models

The genetic algorithm is the optimisation technique chosen in our study. Its main characteristic is that it proposes new solutions to test. These solutions contain different values of the decisions variables that are manipulated by the genetic algorithm. In our application these decisions variables are set points, bounds or internal parameters of WWTP control laws.

For each solution, an evaluation of objective(s) and sometime constraints has to be made. In our case, this is done via dynamic simulations of the WWTP model which have been modified according to the decision variables of the current solutions. More details about these simulations are given in the next subsection.

The results of the simulations are values of all model states varying with the simulation time. These have to be transformed into objectives and constraints values. This is done with the performance model(s). The performance criteria (objectives and constraints) are typically based on averages, integrals, minimum or maximum values, etc.

The whole procedure is repeated for each generation of the genetic algorithm until the convergence of the problem and the termination of the optimisation.

Benefits of a Multiobjective Approach. As previously mentioned, many opposite objectives are usually encountered in WWTP optimisation.

Traditional optimisation approaches can only handle one single objective. A way to aggregate all objectives in a single one is a weighting scheme as in Eq. 2 (obj_i are the various objectives, α_i are weights and *obj* is the aggregated objective value).

$$obj = \alpha_1 \cdot obj_1 + \alpha_2 \cdot obj_2 + \alpha_3 \cdot obj_3 + \ldots \tag{2}$$

This technique is convenient as it is the only way to use traditional algorithms. However, the choice of the weights α_i is subjective to the knowledge of the analyst and has influence on the final solution obtained when objectives are opposite. The decision maker which will be provided with the solution is usually not involved in this choice and a frequent question is the sensitivity of the solution to a slightly different weighting scheme [32].

Another approach consists in adapting existing optimisation algorithms to search for what is called the Pareto front which is the set of best compromises between all opposite objectives. Best compromises are solutions for which there is no other solution which have better values for all objectives. An illustration is provided in Fig. 9 where candidate solutions are in light and dark grey and the Pareto front is in dark grey.

The global solution can then be chosen inside this set of best solutions, having a clear insight into the trade-offs involved. The big advantage of this technique is that no more choice is made by the analyst. Furthermore, the decision maker has all information he needs to make his choice. This situation is almost the contrary of what happens when using a weighting scheme.

Choice of the Objectives for the Optimisation. The main advantage of multi-objective optimisation is to allow separate consideration of objectives in agreement with the decision maker point of view. However, too many objectives can make the optimisation less efficient and can prevent easy interpretation of the results. The second

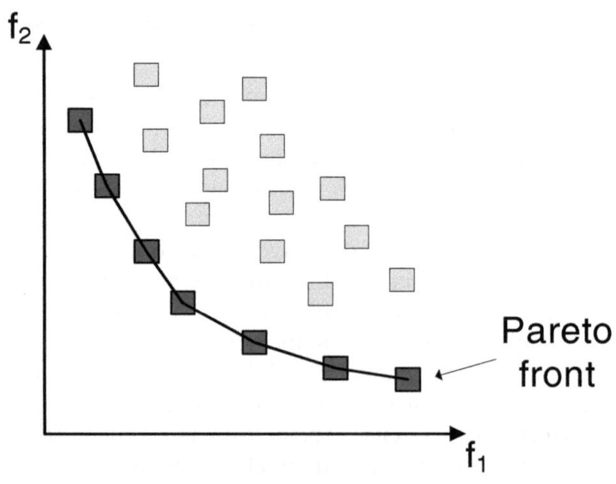

Fig. 9. Example of a set of potential solutions (light and dark grey) and the associated Pareto front (dark grey)

point is very important to take into account since even the best techniques may not be applied in practice if the interpretation of the results is too difficult. A solution can be to use techniques for multicriteria decision making (MCDM). A complete review of these techniques can for instance be found in the book by E. Triantaphyllou [33].

In the case of the optimisation of wastewater treatment processes, the objectives are pretty easy to identify. The first one is usually the effluent quality. Depending on the local legislation and on the problem of optimisation considered, it may be necessary to divide this objective into smaller ones corresponding to each pollutant or group of pollutants (organic pollution, nitrogen, etc.). The second objective is usually the operational cost. It may include the various energy consumptions (aeration, pumping, mixing, etc.) or the cost of external carbon addition, etc.

Many other objectives may be considered like the biogas production issued from the sludge treatment, the cost of construction/modification of the WWTP, the risk of malfunctioning of the WWTP, etc. The use of such objectives depends on the inclusion of additional models in the simulator and/or the development of specific performance indexes.

Simulation Procedure. In order to have pertinent evaluations of the WWTP simulations, it is important to ensure that the plant model is stabilized before starting to evaluate performance criteria. This is necessary as the biomass is for instance evolving very slowly and a single week of simulation is not enough to reach stabilisation.

A first solution could have been to use traditional techniques of steady state computations before the simulation. These steady-states computation require a constant input and the corresponding values may not be representative of the stabilized state of the plant under dynamic conditions. This is due to the strong variations of the incoming wastewater (see section 2.1), the slowly evolving biomass and the non-linearity of WWTP models. Another limitation is that steady-states are taking much time to compute when stiff control laws are used (like on/off controls of the aeration).

To address this problem, an adaptation could be to find a pseudo steady-state for slowly evolving states (like the biomass concentration) under dynamic conditions. Such techniques are still under development but could provide interesting results with a limited computing time [34].

We propose in this chapter another solution adapted from pseudo steady-state computations which consist in repeating a week of dynamic simulations until the difference of slowly evolving states between the beginning and the end of the simulation is negligible. This is very computing demanding as many weeks of simulations are typically required but it can be reduced thanks to the repetitive aspect of genetic algorithms.

As many simulations are done, it is possible to have good guesses of initial states before the first week of simulation based on previous simulations. A memory is used to store the stabilized states of previous simulations together with the decision variable values associated. For a given simulation, the final states of the nearest previous simulation (in term of decision variables values) are used as initial values and quick convergence to a stabilized functioning is usually obtained after few weeks of simulation.

The final process is illustrated in Fig. 10. For each solution, the initial states of the simulation are set to the stabilized ones of the nearest previous simulation. One week of simulation is then performed with dynamic influent datasets. Convergence to stabilized functioning is then checked for slowly evolving states. If the convergence has

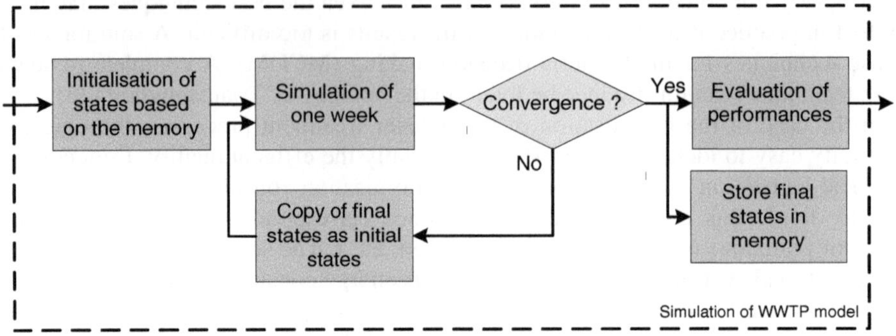

Fig. 10. Flowchart of the simulation procedure used for unbiased evaluations

not yet been reached, a new week a simulation is performed with its initial states set to the final values of the previous simulation. When the convergence is reached, performances can be computed and the final states values obtained are stored together with the current decisions variables values for next simulations.

Evaluation of the Long Term Performances and Robustness. The simulation procedure presented before allows having a stabilized functioning for the evaluation of the performances. However, the number of perturbations encountered in the simulation is rather limited. In reality, plenty of perturbations arise every year at a WWTP and it is necessary that the WWTP is capable of handling these events in an optimal way.

Using simulations with long time horizon in the optimisation is not reliable due to the very long computing time associated. However, the robustness can be checked a posteriori, as a proof for the application of the optimisation results. This is reliable in term of computing time as only a limited amount of simulations corresponding to the final population have to be made. If the results show that some solutions are not reliable on the long term, they may be removed from the set of optimal solutions. Another possibility is to analyze these wrong solutions in order to make conclusions about necessary modifications of the control law or some additional constraints that have to be placed on the optimisation problem.

This step is really important in order to prevent wrong results that will behave too poorly in the reality and which may put discredit on the optimisation method (which is probably reliable in fact, only the formulation of the problem is not adequate).

Chachuat *et al.* [31] showed for instance that their previous results based on an optimisation for 24h functioning of an activated sludge unit are not reliable for long term operation due to a washout of bacteria to the effluent. Additional constraints were simply required and sufficient to provide reliable results.

Two other examples are provided in Fig. 11. For both figures, only the influent datasets used for the evaluation period changed but the simulation model is exactly the same. The short term performance is the one obtained during the optimisation based on the final evaluation of one week. The long term performance is obtained by simulating the WWTP model with each solution of the optimisation with a long influent dataset. This dataset is composed of 609 days and illustrates typical variations of the influent of a WWTP. It is part of BSM1_LT, an extension of the BSM1 for the

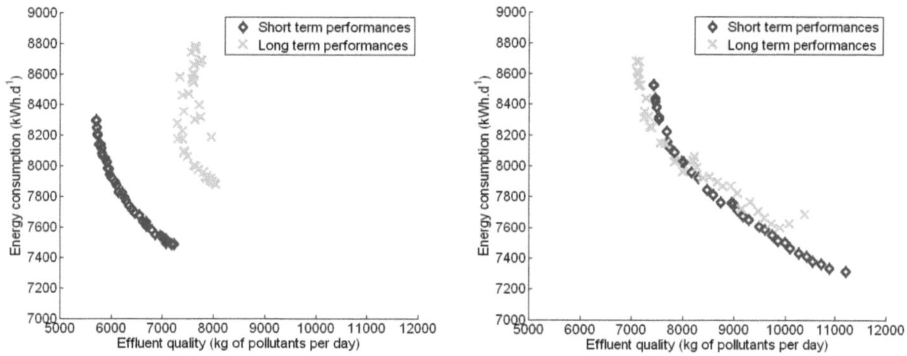

Fig. 11. Comparison of long term performances: unstable (left) or stable (right)

assessment of long term performances [20, 21]. Performances are median values of the 609 days of simulation.

On the left part of Fig. 11, two problems can be observed: first, there is a strong deviation between the short term performances and the long term performances. This may be of major importance considering for instance legislation aspects. The second problem is that the points located on the upper part of the figure are not anymore Pareto-optimal on the long term because better Pareto-optimal points with lower effluent quality and energy consumption. The results on the right part of the figure are reliable as only very little deviation is observed between short term and long term performances. Moreover, the best median effluent quality achieved with solutions of the right part of the figure is better than the best one of the left part.

This figure clearly shows the benefit of using long term simulations for a good analysis of optimisation results and also to have opportunities to enhance the methodology by identifying lacks.

3.3 Conclusion

The two main problems challenging the theoretical optimisation have been presented in this section. Existing solutions are not jointly addressing the two aspects. A new methodology is hence proposed, based on the combination of existing WWTP models and multiobjective genetic algorithms. A simulation procedure for reliable evaluation of the WWTP performances has been proposed, based on a repetition of a week of dynamic data. For the evaluation of the robustness of optimized solutions, long term simulations have been proposed based on 609 days of simulation. Finally, the benefits of the multiobjective approach have been presented together with typical objectives to consider. The remaining point to present is the multiobjective genetic algorithms NSGA-II which is used for the optimisations.

4 Genetic Algorithms and Their Multiobjective Adaptations

In this section, focus is put on the optimisation technique. First, a general presentation of genetic algorithms is made, followed by a review of main multiobjective adaptations of genetic algorithms. Details are finally given about NSGA-II, the algorithm used in this study.

4.1 Presentation of Genetic Algorithms

Genetic algorithms are quickly presented in this section to provide insight into the core ideas of their functioning. Our aim is not to provide a full review of all their subtleties and variations but only to give the reader a better understanding of the functioning of these algorithms, their strengths and weaknesses. For those wishing to have more details, plenty of chapters and books are available in the literature. A good start can be the chapter of Reeves [35].

Introduction to Genetic Algorithms. The origin of genetic algorithms (GAs) is attributed to John Holland who published a book exposing the root of this technique in 1975 [36]. It was also that year that Ken De Jong, a graduate student of Holland, completed his doctoral thesis providing a clear insight into the capabilities of GAs for optimisation [37]. However, at that time, only very few real applications existed, mainly due to the important computing power required. The real interest of the scientific and practitioner communities took off few years later, in 1989, when David Goldberg [38], another graduate student of Holland, published a very influential book that was a real catalyst for the application of the GA theory. The real take off beginning this year is also probably linked with the increase of computing power that began to be available in many research centres at that time.

GAs are based on two key principles used for the generation of new candidate solutions: mutation and recombination. Mutation consists in altering the current solution to see what could be achieved in his neighbourhood. It is a pretty old technique and can be seen as a simple modification of traditional hill-climbing techniques. Recombination consists in taking two good solutions and exchanging some of their characteristics. The idea behind it is to mimic what happens in the evolution of species, when some genes are exchanged and could lead to better characteristics of the individuals, only the best fitted individuals been capable to survive on the long term. The recombination is assumed to provide fast convergence to the optimum (compared to mutation) and to avoid the stop of the algorithm in a local extremum. The main drawback is that only near optimal solutions are found.

Core Principles of Genetic Algorithms. Genetic algorithms were first developed for the optimisation in a discrete search space and we will therefore first focus only on this case to present the main concepts. Extensions to continuous search space will be presented in the last paragraph.

With a discrete search space B^n

$$B^n = \{0,1\}^n \tag{3}$$

and a function f

$$f : B^n \to \Re \tag{4}$$

the general optimisation problem is to find

$$\min_{x \in B^n} f \ . \tag{5}$$

In this expression, f is the objective function and x, named an individual, is a vector of decision variables composed of n binary values. If the problem consists in maximising an objective function, the modification to transform it into a minimisation problem is obvious.

The flowchart of the operations of a genetic algorithm is presented in Fig. 12. As in all optimisation algorithms, an iterative procedure is used. First, the GA is initialized and a random initial population is generated. Objective values are evaluated for this first population. Then, cyclic repetitions are performed based on five steps:

- evaluation of the termination criteria; if it is satisfied, end the optimisation,
- selection of parents in the previous population
- recombination of parents to create the new population
- mutation of the population
- evaluation of the objective values of each solution of the population

As stated in the first step, the repetition ends when the termination criteria is satisfied. Each step will now be detailed.

Initialisation of the Genetic Algorithm. Before the initial population can be built, its size has to be chosen. This is one of the drawbacks in the use of GAs: no general rule is available for this choice and it is mainly based on the experience of the GA user. A general guideline about this choice is that the size of the population has to be related

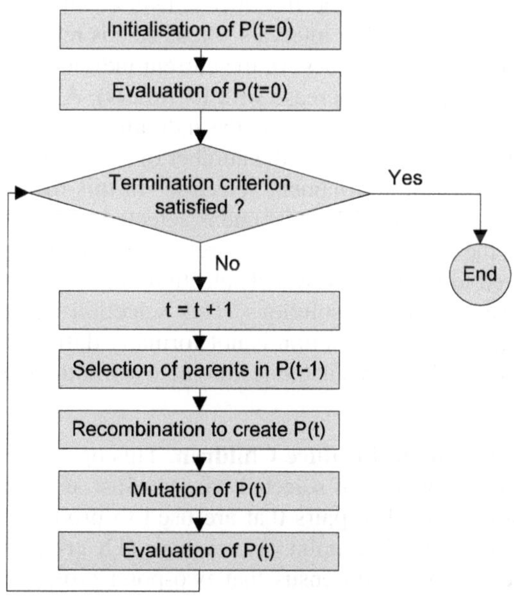

Fig. 12. Flowchart of a genetic algorithm

to the size of the search space. Some theoretical propositions in this area are presented in Reeves [39]. They conducted to the conclusion that both values (0 or 1) of the decision variables should be found in at least one individual of the initial population and therefore a population size of *O(log n)* is sufficient to cover the search space.

The next problem is the choice of a method for the generation of this initial population. A traditional method consists in using random values. This method has the advantage to be very simple, but it is obvious that it may happen quite often that the randomly chosen population does not cover the whole search space. Another solution therefore consists in using a model for the generation of this population. A simple and efficient method in this topic is the generalization of the Latin hypercube, which ensure good distribution of the population among the search space. This type of model is especially profitable in the case of the search in non-binary spaces.

A final important point to consider is the inclusion of a known high quality solution into the initial population. This may help in having a quick convergence of the genetic algorithm, but there is risk of premature convergence, where the algorithm will be stuck in the local optimum near this initial solution. This may happen by a loss of the diversity of the population as the known solution is really good compared to other potential solutions randomly generated and the GA will therefore go very fast in this direction without having necessarily other candidate solutions [35]. It is therefore a risky choice, with benefits and drawbacks.

Selection of Parents. This step consists in choosing the individuals (which will be called parents) among the current population. They are used to generate the children for the next population. The selection of the parents should be related to the objective function values in order to promote better solutions. For this step, the roulette-wheel selection and the tournament selection are the two most popular mechanisms.

The roulette wheel selection is the first mechanism developed. In this method, the space between 0 and 1 is divided in intervals whose size is related to the fitness (measure of goodness of the objective values) of the current individuals (the size of the interval associated to each individual increase with its fitness). A random number between 0 and 1 is then generated and the parent is the individual whose interval contains the random number. This is repeated unless the number of parents required is reached.

Another mechanism is the tournament selection. In this method, k individuals are randomly chosen and compared. The best one is selected as a parent. This selection is also repeated unless the required number of parents is reached. The main advantage of this method compared to the roulette wheel selection is that it only needs an operation of comparison between potential solutions. This selection can therefore even work when the value of the objective function is not formally defined but only comparison between solutions is possible. This is the technique used in the multi-objective genetic algorithm NSGA-II.

Recombination of Parents to Produce Children. This operation consists in generating new candidate solutions out of selected parents. Just as in natural evolution, the parents are randomly grouped in pairs that are used to produce two children for the new generation. A single random point of crossover (1X crossover) is the traditional operation, but there is now a consensus that two-point crossover (2X) is preferable [40]. Illustration of these two mechanisms is provided in Fig. 13. 1X crossover consists in randomly choosing a position in the chromosomes of the two parents and

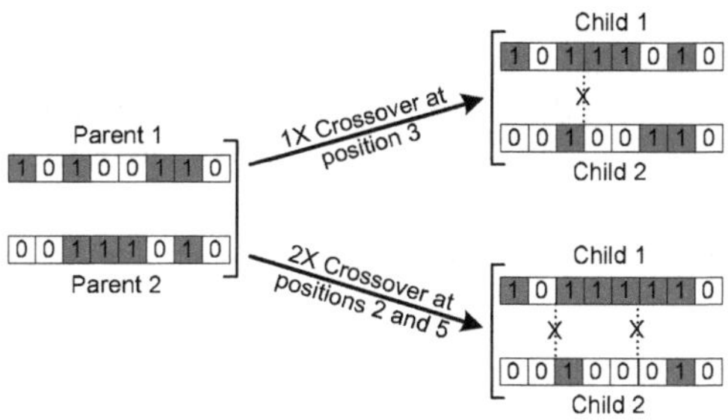

Fig. 13. Operations of 1X crossover (top) and 2X crossover (bottom) for individuals composed of 8 bits. The white squares indicate a value of 0 and the grey 1.

exchanging the values of the two chromosomes after this position. 2X crossover consists in randomly choosing two positions and exchanging the values of the two chromosomes in between these two positions.

Mutation of Children. The mutation consists in randomly altering one decision variable value. This operation is usually based on a random probability of occurrence. The main objective of this operation is to be a safeguard to always introduce diversity in the population in order to avoid to be stuck in a sub-optimal region and also to explore the whole search space. The mutation rate (probability of occurrence of this operation) is a compromise between the exploration of the search space and the number of function evaluations before reaching the optimum.

Termination of the Optimisation. The choice of the criteria for the termination of the GA is probably the most difficult one. As GA is a global and stochastic optimisation procedure, it is difficult to know when the algorithm has reached the optimum. The solution usually consists in using a combination of a maximum number of generation without any change in the best solution found, a maximum total number of function evaluations and sometimes a criteria on the population minimum diversity (when the diversity is too low, the algorithm can stop as we may assume that no more enhancement is possible).

Adaptation of Binary Genetic Algorithms to Real Search Spaces. The introduction to GAs of the previous paragraphs is based on the assumption that we are optimizing binary variables. In real life problems, optimisation variables are more usually in integer or real search spaces, and the direct transformation of these numbers in their binary coding is not the most adequate one (the operation crossover and mutation are typically very badly performing when such a simple transformation is used).

For integer variables, Gray coding can be used and has been proven to provide higher quality results than binary coding [41, 42]. The key principle of this coding is that only one bit has to be changed to move from one integer value to the next one (i.e. plus one or minus one).

For real variables, the main problem is the size of the search space. When the precision required on the decision variable is high, the dimension of the problem when with a binary alphabet (either binary code or Gray code) can be very high. It has been the main motivation for the developments of real-coded genetic algorithms. Instead of using bits as genes in the individuals, real values are used. Then, the operations of crossover and mutation are redefined to provide similar performance and functioning of the optimisation. Many variants are available in the scientific literature [43, 44].

4.2 Multiobjective Genetic Algorithms

Adaptations of the traditional genetic algorithms presented before are available in the literature to solve the problem of searching for the Pareto front of a given multiobjective problem. Such GAs are named Multi-Objective Genetic Algorithms (MOGAs). The three most common ones that are widely accepted and working on a wide range of problems are the Strength Pareto Evolutionary Algorithm 2 [45], the Pareto Envelope-based Selection Algorithm II [46] and the Non-dominated Sorting Genetic Algorithm II [47]. Other very simple algorithms are also available like PAES [48] or NPGA [49].

Choice of the Multiobjective Genetic Algorithm. PAES and NPGA are mainly used due to the criticism that NSGA-II, PESA-II and SPEA 2 are time consuming in their operation as each solution has to be compared to all other ones to sort them. In fact, this criticism is of absolutely no concern in the application of this study as the computation time required for the internal logic of the GA is very small compared to the time required for the evaluation of a single solution which implies the run of a model simulation. An order of 100 is usually observed between the two operations (internal logic of the GA and evaluation of the solutions). As GAs performances are far more important for our study than their algorithmic complexity, these two algorithms are not selected for our study.

The most important criterion to select a GA is in our case the number of function evaluations required to reach the optimum. For all common problems, the average performance level of SPEA2, PESA-II and NSGA-II is quite similar [46, 45] and the choice of using one instead of the others depends on the problem studied. No criteria have been indeed found to help us in the choice of the algorithm to use, probably mainly due to the high variety of problems that can be solved thanks to evolutionary multi-objective algorithms. Moreover, it is difficult to transform most real world problems into a problem used for algorithms benchmarking. Finally, as the performances are very tight for all three algorithms, the best one can change from one WWTP optimisation problem to another one.

The choice of the algorithm for this study is therefore mostly based only on *a priori*. It is the algorithm NSGA-II that is chosen based on the number of successful and real applications found using this algorithm [50, 51, 52] and on the simple practical consideration that its implementation on Matlab is freely and already available (which is not the case for other algorithms).

The Non-Dominated Sorting Genetic Algorithm. The Non-Dominating Sorting Genetic Algorithm is one of the first adaptations of GAs for the search of the Pareto front proposed by Srinivas and Deb [53]. The second version (NSGA-II) was proposed by Deb et al. [47] to answer three main criticism on NSGA-I: the high computational complexity of $O(MN^3)$ for the GA internal computations (M is the number of objectives and N is the population size), lack of elitism (elitism consists in keeping the best Pareto front found so far to use latter in the GA, it was proven to provide better results) and the use of a technique for diversity preservation subject that is sensitive and based on a parameter that has to be user-defined. All these criticisms are solved with the NSGA-II.

Before describing the internals of NSGA-II, two concepts introduced in this algorithm have to be defined: the non-dominated sorting and the crowding distance computation.

The non-dominated sorting consists in assigning a 'rank' to each solution of the population. The procedure to compute this rank is the following (see Fig. 14 for an illustration). First, the solutions belonging to the Pareto front of the current population are assigned a rank 1 (F1). They are then temporary removed from the population and the solutions belonging to the new Pareto front are assigned a rank 2 (F2). They are then temporary removed and the procedure is repeated until all solutions have a rank. The solutions with the lower rank are of course the best ones and will be favoured during the operation of selection.

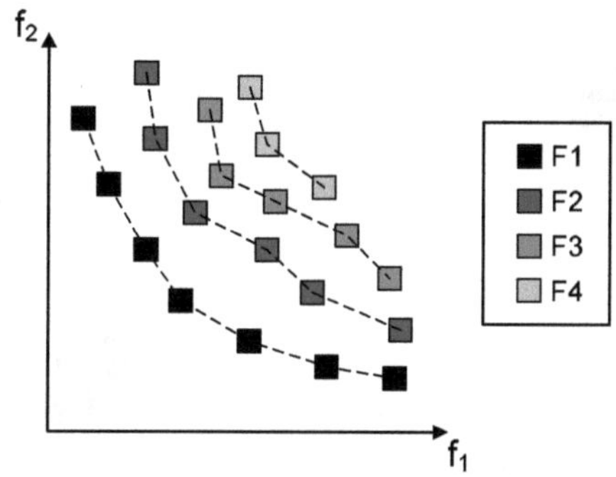

Fig. 14. Example of the non-dominated sorting of a set of solutions

It is then necessary to differentiate solutions with the same rank. This is made with the second concept of crowding distance. The main goal of a MOGA is to preserve the diversity of the solutions in order to have a good coverage of the Pareto front. For a given solution, the crowding distance is therefore the measure of the sum of the distances between the two nearest solutions along each objective. Solutions with big crowding distance are important to preserve while solutions with small crowding distance can be discarded as they have other solutions near them. Solutions at the boundaries of the Pareto front (worst and best solutions for each objectives) are assigned an infinite crowding distance as they are very interesting ones considering the preservation of diversity.

The execution of NSGA-II is then based on these two concepts as illustrated in Fig. 15 that will now be detailed.

After the classical initialisation of the first population (1) and its evaluation (2), the solutions are sorted according to their non-domination rank and crowding distance (3). This set forms the first population P(t=0). An iterative procedure is then performed:

- tournament selection on the current population P(t) based on the sorting of the population previously made (4)
- classical recombination and mutation to produce children Q(t) (5)
- evaluation of objective values for each child (6)
- mixing of current population P(t) and children Q(t) in combined population R(t) (7)
- non-dominated sorting of R(t) and computation of crowding distance (8)
- selection of the new population P(t+1) (9) :
  - first fronts whose summed number of solutions is lower than the population size are automatically selected
  - the best solutions of the next front are selected according to their crowding distance
  - other solutions of this next front and of following fronts are rejected
- evaluation of the termination criteria, if it is not satisfied, perform a new generation (10)

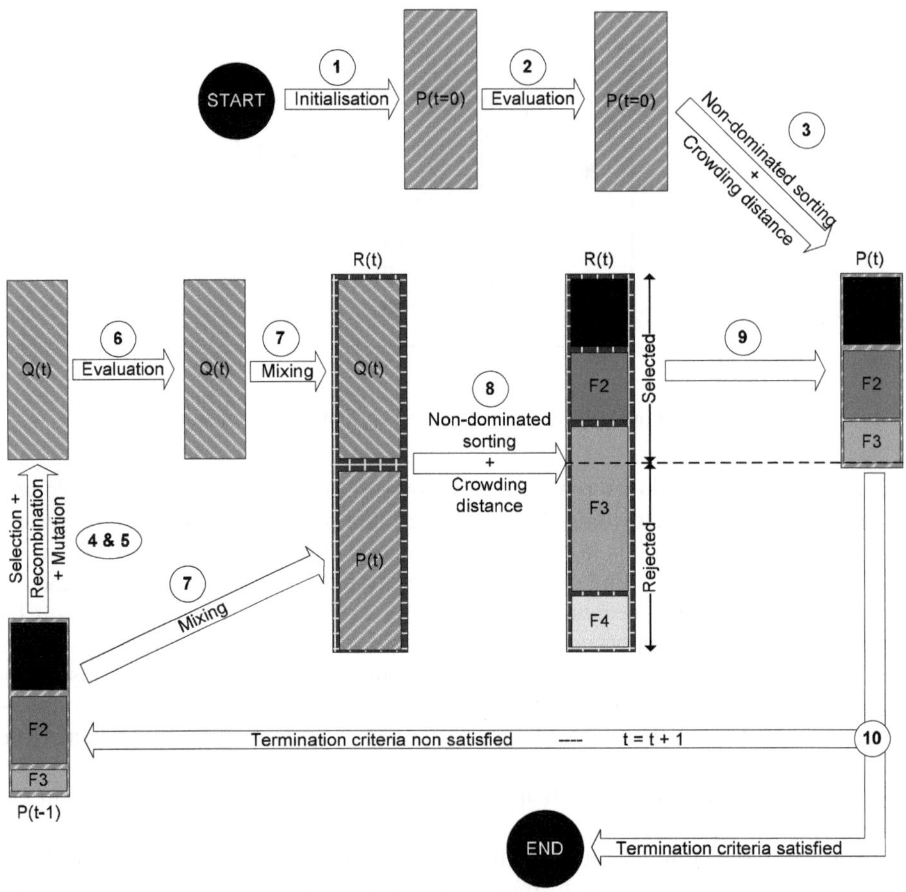

Fig. 15. Flowchart of NSGA-II operations

This algorithm is very competitive with SPEA-2 and PESA-II and its main quality is the very good diversity of the solutions proposed. Another important point is the robustness of its default parameters which are assumed to provide good performances for most problems. The only parameters that have to be adjusted on all problems are the population size and the termination criteria. This algorithm is therefore perfectly suited for the dynamic optimisation of combined models of WWTPs and their control laws.

Setting of NSGA-II Parameters. In order to assess which are the best values of these two main NSGA-II parameters, the easiest way is to find a similar problem that has already been solved and use the same parameters. If no similar problem is available (or if it is not sure what a similar problem is), the only solution is to run the optimisation algorithm with various parameters and to check with some indexes which are the best parameters. It is important to use repetitions of the optimisation for the same parameters because GAs are stochastic. It may hence happen that they produce bad results in some occasion.

In order to assess the quality of the various tests performed, two very good indexes are the convergence (distance between the current Pareto front and the optimal Pareto front) and the diversity (distance between points of the Pareto front). The convergence index has to be minimal while the diversity index has to be maximal.

Two measures for these performance indexes are proposed by Deb *et al.* [54]. They are running metrics, which therefore allow a comparison of the different settings during the whole optimisation. This allows comparing the various alternatives at each generation, not just after a given amount of function evaluations.

Conclusion. This section briefly introduced the main concepts of genetic algorithms and their multiobjective adaptations. Five MOGAs were proposed, among which NSGA-II was selected for its robustness and ease of use. Finally metrics for the assessment of MOGAs performances were presented.

5 Application of NSGA-II for the Optimisation of Wastewater Treatment Processes

Now that the methodology of simulation and optimisation has been clearly defined and the genetic algorithm chosen, an application on a case study is developed. It is based on a simple control scheme for the BSM1. Two objectives are considered and four decision variables have to be optimized.

5.1 Definition of the Case Study

The original BSM1 layout is used. A new control law is proposed, composed of three PI controllers (see Fig. 16). Each control acts on the aeration of each aerated units by controlling the dissolved oxygen concentration and manipulating the oxygen transfer coefficient (which is linked to the air flow rate in real WWTP but the inclusion of such a link is not necessary in this application). A fixed set point is used for each controller and will be optimized.

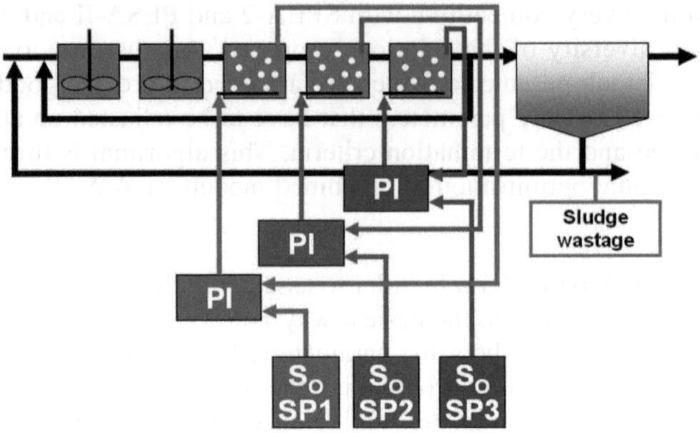

Fig. 16. Controller layout for BSM1 based on dissolved oxygen measurements

In this application we assume that the internal settings of the controllers (gains and time constants) have already been adjusted to provide good performances for all kind of dissolved oxygen set point (the range studied is quite limited and robust parameters are available).

Only four parameters have therefore to be optimized: the three set points and the wastage flow rate. The three first allow minimizing the energy consumption while ensuring good effluent quality. The last one allows adjusting the quantity of bacteria in the system. The ranges of each decision variable are summarized in Table 1. These are the bounds used by the genetic algorithms.

Table 1. Design parameters for the optimisation problem

Variable	Unit	Min	Max
Set point 1 S_O	$g.m^{-3}$	0	8
Set point 2 S_O	$g.m^{-3}$	0	8
Set point 3 S_O	$g.m^{-3}$	0	8
Wastage flow rate	$m^3.d^{-1}$	0	400

In this optimisation, two objectives are chose: the effluent quality (Eq. 6) and the energy consumption (Eq. 7 and Eq. 8), as defined in the BSM1 [16].

The effluent quality is defined in the BSM1 as a weighted sum of the effluent loads of total suspended solids (TSS), chemical oxygen demand (COD), biochemical oxygen demand (BOD), total Kjeldahl nitrogen (TKN) and NO_x nitrogen (S_{NO}) in the effluent:

$$EQ = \frac{1}{1000 \cdot T} \cdot \int_0^T \begin{pmatrix} 2 \cdot TSS(t) + 1 \cdot COD(t) \\ + 2 \cdot BOD(t) + 20 \cdot TKN(t) \\ + 20 \cdot S_{NO}(t) \end{pmatrix} \cdot Q_e(t)dt \ . \tag{6}$$

The aeration energy is very linked to each plant, depending on the aeration technique chosen as well as the performances of the material used. The formula proposed in the BSM1 is an empirical approximation valid only for tanks of 1333 m^3. Its computation is based on the oxygen transfer coefficient (K_La) in each aerated reactor, T is the duration of the evaluation (7 days):

$$AE = \frac{1}{T} \cdot \int_0^T \sum_{i=3}^{i=5} 0.4032 \cdot K_La_i(t)^2 + 7.8408 \cdot K_La_i(t)dt \,. \tag{7}$$

In BSM1, the pumping energy is directly related to the integral of the sum of the internal recycle flow rate (Q_a), the return sludge flow rate (Q_r) and the waste sludge flow rate (Q_w). The formula proposed is the following:

$$PE = \frac{0.04}{T} \cdot \int_0^T Q_a(t) + Q_r(t) + Q_w(t)dt \,. \tag{8}$$

5.2 Assessment of the GA Parameters

As explained in section 4.2, only two parameters of NSGA-II are really necessary to adapt to each problem: the population size and the termination criteria. In this application the termination criteria is only a number of generation to complete.

The two indexes of convergence and diversity of Deb *et al.* [54] are used to search for the best values of the two parameters on our case study. Five populations size have been chosen to be tested (12, 20, 48, 100 and 200 individuals). For each population size, the same number of function evaluation is performed (6000). It means that 500 generations are made for the population of 12 individuals, and only 30 generations for the 200 individuals. As GAs are stochastic, five repetitions of the optimisation are performed for each population size. The mean performance indexes obtained during the five repetitions are presented in Fig. 17.

The conclusion for this test case is that a population size of about 48 or 100 is adequate due to the good convergence and diversity obtained. The choice between these

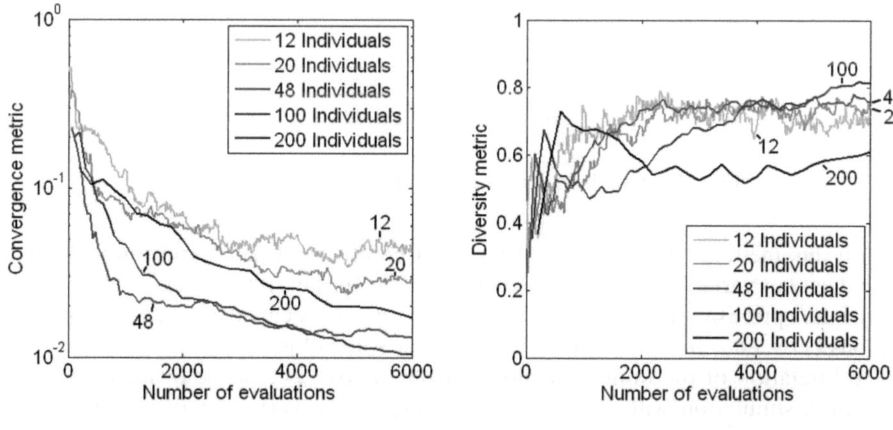

Fig. 17. Mean convergence and diversity metrics for the various population sizes

two parameters is difficult as the smaller size ensures a quicker convergence and diversity but the bigger size ensures better performances on the long term. However, in the case of the optimisation of WWTPs, there is big uncertainty about the modelling and the controller is usually not sensitive to small parameters errors. Therefore, the absolute convergence and diversity is not very important. On the contrary, a wish of the end user is always to limit the computing time. Therefore, quick convergence and diversity is preferable. A population size of 48 individuals and a number of 4000 function evaluations therefore seems to be reasonable for our case study.

The investigation performed here can of course not be made for all optimisation problems due to the large computing time required (typically one or two weeks for the five repetitions on a desktop PC.

NSGA-II is however very robust and it is therefore advised to look inside such details only when there is large uncertainty about the possible best parameters. When a similar problem can be found with known parameters, it is reliable to use these settings.

5.3 Results of the Optimisation

After the optimisation, the Pareto front obtained can be compared to the performances proposed in BSM1 in the open loop and closed loop configurations, as shown in Fig. 18. Huge improvements are observed. At constant effluent quality, it is possible to reduce the energy consumption from 10% to 20% compared to the original BSM1 performances. At constant energy consumption, the effluent quality is reduced by about 25%.

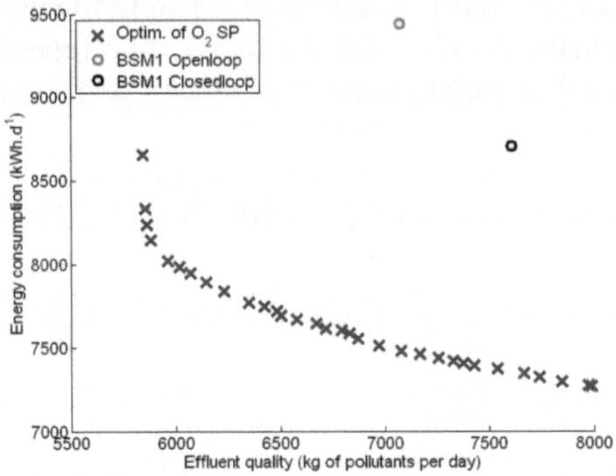

Fig. 18. Optimised performances of the O_2 control scheme

With this method each solution of the final Pareto front has corresponding values of the decision variables. This can then be used for further analysis or to apply the results of the optimisation on the real process. These values can also be used to discover relations between some decision variables and/or some objectives.

The limitation of the improvements presented above is that the evaluation is only based on a simulation with a dry weather influent. The variations of this dataset are important but not representative of a rain or a storm event. Therefore, the next step consists in simulating the WWTP with the settings issued from the optimisation for a long time period.

5.4 Evaluation of the Robustness of Results

In the case of the control law based on oxygen set points previously optimised, the long term performances obtained are illustrated in Fig. 19. 609 days of simulation are performed and the median performances are computed. The reduction of energy consumption observed on the short term results is still observed on the long term simulations but the best enhancement of the effluent quality is not as high as previously mentioned (only 9 % instead of 25%). Some solutions are even not Pareto-optimal anymore. This is probably linked with the choice of the dataset used for the evaluation period. Additional constraints may also be required.

This is a practical illustration showing that these long term simulations are really necessary before providing the results to the decision maker. They are very valuable to check the performances of the simulation procedure chosen.

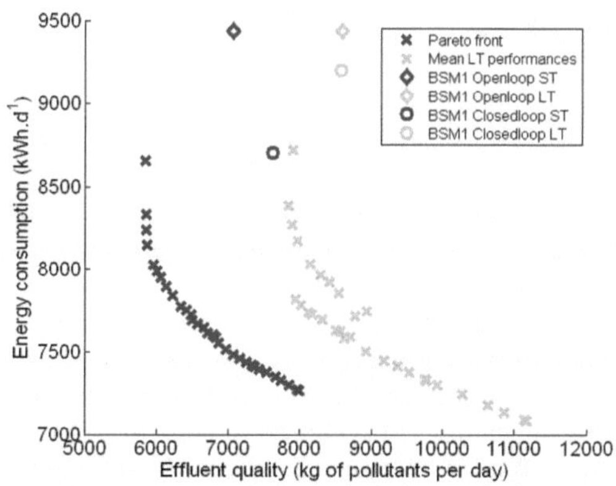

Fig. 19. Comparison of short term and long term performances of the oxygen control scheme

5.5 Conclusion

This application showed the benefits of the multiobjective optimisation methodology proposed in this chapter. It allows a clear insight into the tradeoffs between two conflicting objectives. Long term evaluations allow to better analyse the robustness of the results and to provide only relevant solutions to the decision maker. Enhancement of the problem formulation can even be detected with these long term evaluations (addition of constraints, test of new influent datasets for the evaluation period, etc.).

6 Conclusion and Perspective of Application at the Industrial Scale

This chapter illustrates the benefits of combining the available wastewater treatment plant models with the multiobjective genetic algorithm NSGA-II. The multiobjective approach allows having clear insight in the tradeoffs arising between opposed objectives. The genetic algorithm allows using state of the art models and can solve problems with many local minima. Addition of long term evaluations of the final results allow to check the robustness of the results obtained after the optimisation with much more perturbations.

The use of this optimisation procedure allows simple implementation on industrial problems. The only requirement is an accurate model of the WWTP and the knowledge of the adequate criteria to optimize (for instance local legislation on the effluent quality, price of energy, price of reactants, etc.). A simple link between the genetic algorithm and the state of the art modelling software used in the specific industries is then only required to implement the optimisation. Typical performance indexes can be made available that can simply be tuned to each local case of study.

Potential fields of application are far wider than the case of WWTP optimisation presented in this chapter. In the field of water, it can for instance concern the optimisation of the design, renewal and online control of water and wastewater networks, but it can also concern the optimisation of the design of any industrial application made of many processes that have to be integrated together, or the automated synthesis of an optimal control scheme.

References

1. Thaure, D., Lemoine, C., Daniel, O., Moatamri, N., Chabrol, J.: Optimisation of Aeration for Activated Sludge Treatment with Simultaneous Nitrification Denitrification. Water Science and Technology 58(3), 639–645 (2008)
2. Olsson, G., Nielsen, M., Yuan, Z., Lynggaard-Jensen, A., Steyer, J.P.: Instrumentation, Control and Automation in Wastewater Systems. IWA Publishing, London (2005)
3. Henze, M., Grady Jr., C.P.L., Gujer, W., Marais, G.v.R., Matsuo, T.: Activated Sludge Model No. 1. IAWPRC Scientific and Technical Report No. 1, IAWQ, London (1987)
4. Henze, M., Gujer, W., Mino, T., Matsuo, T., Wentzel, M.C., Marais, G.v.R.: Activated Sludge Model No. 2. IAWPRC Scientific and Technical Report No. 3, IAWQ, London (1995)
5. Henze, M., Gujer, W., Mino, T., Matsuo, T., Wentzel, M.C., Marais, G.v.R., van Loosdrecht, M.C.M.: Activated Sludge Model No. 2d. Water Science and Technology 39(1), 165–182 (1999)
6. Gujer, W., Henze, M., Mino, T., van Loosdrecht, M.C.M.: Activated Sludge Model No. 3. Water Science and Technology 39(1), 183–193 (1999)
7. Hazen, A.: On Sedimentation. Transactions of the American Society of Civil Engineers 53, 45–71 (1904)
8. Krebs, P., Armbruster, M., Rodi, W.: Numerische Nachklärbecken-Modelle. KA-Wasserwirtschaft, Abwasser, Abfall 47(7), 985–999 (2000)
9. Kynch, G.J.: A theory of Sedimentation. Transactions of the Faraday Society 48, 166–176 (1952)
10. Grijspeerdt, K., Vanrolleghem, P., Verstraete, W.: Selection of One-Dimensional Sedimentation: Models for On-Line Use. Water Science and Technology 31(2), 193–204 (1995)
11. Takács, I., Patry, G.G., Nolasco, D.: A Dynamic Model of the Clarification-Thickening process. Water Research 25(10), 1263–1271 (1991)
12. Gernaey, K.V., Jeppsson, U., Batstone, D.J., Ingildsen, P.: Impact of Reactive Settler Models on Simulated WWTP Performance. Water Science and Technology 53(1), 159–167 (2006)

13. Ekama, G.A., Barnard, J.L., Günthert, F.W., Krebs, P., McCorquodale, J.A., Parker, D.S., Wahlberg, E.J.: Secondary Settling Tanks: Theory, Modelling and Operation. IAWQ Scientific and Technical Report No. 6, IAWQ, London (1997)
14. De Clerq, B.: Computational Fluid Dynamics of Settling Tanks: Development of Experiments and Rheological, Settling and Scraper Submodels. PhD thesis, Ghent University, 338 p. (2003)
15. Weiss, M., Plosz, B., Essemiani, K., Meinhold, J.: Sedimentation of Activated Sludge in Secondary Clarifiers. In: Fifth World Congress on Particle Technology (2006)
16. Copp, J.: The COST Simulation Benchmark - Description and Simulator Manual. Office for Official Publications of the European Communities, Luxembourg (2002)
17. IWA Task Group on Benchmarking of Control Strategies for WWTPs, `http://www.benchmarkwwtp.org`
18. Jeppsson, U., Pons, M.N.: The COST 624 Benchmark Simulation Model - Current State and Future Perspective. Control Engineering Practice 12(3), 299–304 (2004)
19. Nopens, I., Benedetti, L., Jeppsson, U., Pons, M.N., Alex, J., Copp, J.B., Gernaey, K.V., Rosen, C., Steyer, J.P., Vanrolleghem, P.A.: Benchmark Simulation Model No 2 - Finalisation of Plant Layout and Default Control Strategy. In: IWA World Water Congress 2008, Vienna (2008)
20. Rosen, C., Jeppsson, U., Vanrolleghem, P.A.: Towards a Common Benchmark for Long-Term Process Control and Monitoring Performance Evaluation. Water Science and Technology 50(11), 41–49 (2004)
21. Gernaey, K.V., Rosen, C., Jeppsson, U.: WWTP Dynamic Disturbance Modelling - an Essential Module for Long-Term Benchmarking Development. Water Science and Technology 53(4-5), 225–234 (2006)
22. Jeppsson, U., Rosen, C., Alex, J., Copp, J., Gernaey, K.V., Pons, M.N., Vanrolleghem, P.A.: Towards a Benchmark Simulation Model for Plant-Wide Control Strategy Performance Evaluation of WWTPs. Water Science and Technology 53(1), 287–295 (2006)
23. Rosen, C., Vrecko, D., Gernaey, K.V., Pons, M.N., Jeppsson, U.: Implementing ADM1 for Plant-Wide Benchmark Simulations in Matlab/Simulink. Water Science and Technology 54(4), 11–19 (2006)
24. Sakizlis, V., Perkins, J.D., Pistikopoulos, E.N.: Recent Advances in Optimisation-Based Simultaneous Process and Control Design. Computers and Chemical Engineering 28(10), 2069–2086 (2004)
25. Steyer, J.P., Harmand, J.: Economical Design of Biological Systems: New Tools for Advanced Integrated Process and Control Design. Laboratory report PUB020003014009255З INRA Narbonne, France (2000)
26. Vera, J., Torres, N.V., Moles, C.G., Banga, J.: Integrated Nonlinear Optimisation of Bioprocesses via Linear Programming. AIChE Journal 49(12), 3173–3187 (2003)
27. Balku, S., Berber, R.: Dynamics of an Activated Sludge Process with Nitrification and Denitrification: Start-up Simulation and Optimisation using Evolutionary Algorithm. Computer and Chemical Engineering 30(3), 490–499 (2006)
28. Fikar, M., Chachuat, B., Latifi, M.A.: Optimal Operation of Alternating Activated Sludge Processes. Control Engineering Practice 7(13), 853–861 (2005)
29. Holenda, B., Domokos, E., Rédey, A., Fazakas, J.: Aeration Optimisation of a Wastewater Treatment Plant using Genetic Algorithm. Optimal Control Applications and Methods 28(3), 191–208 (2007)
30. Rivas, A., Irizar, I., Ayesa, E.: Model-based Optimisation of Wastewater Treatment Plants Design. Environmental Modelling and Software 23(4), 435–450 (2008)

31. Chachuat, B., Roche, N., Latifi, M.A.: Long-Term Optimal Aeration Strategies for Small-Size Alternating Activated Sludge Treatment Plants. Chemical Engineering and Processing 44(5), 591–604 (2005)
32. Savic, D.: Single-Objective vs. Multiobjective Optimisation for Integrated Decision Support Integrated Assessment and Decision Support. In: First Biennial Meeting of the International Environmental Modelling and Software Society (2002)
33. Triantaphyllou, E.: Multi-Criteria Decision Making Methods: A comparative Study. Kluver Academic Publishers, Dordrecht (2000)
34. Platte, F., Kuzmin, D., Fredebeul, C.H., Turek, S.: Novel Simulation Approaches for Cyclic Steady-State Fixed-Bed Processes Exhibiting Sharp Fronts and Shocks. International Series of Numerical Mathematics 151, 207–223 (2005)
35. Reeves, C.R.: Genetic Algorithms. In: Glover, F., Kochenberger, G.A. (eds.) Handbook of Metaheuristics. International Series on Operations Research and Management Science, pp. 65–82. Kluwer Academic Publisher, Dordrecht (2003)
36. Holland, J.H.: Adaptation in Natural and Artificial Systems. University of Michigan Press, Ann Arbor (1975)
37. De Jong, K.A.: An Analysis of the Behavior of a Class of Genetic Adaptive Systems. Doctoral Dissertation, University of Michigan, 266 p. (1975)
38. Goldberg, D.E.: Genetic Algorithms in Search, Optimisation and Machine Learning. Addison-Wesley, Reading (1989)
39. Reeves, C.R.: Using Genetic Algorithms with Small Population. In: Forest, S. (ed.) Proceedings of Fifth International Conference on Genetic Algorithms, pp. 92–99. Morgan Kaufman, San Mateo (1993)
40. Eshelman, L.J., Caruana, R.A., Schaffer, J.D.: Biases in the Crossover Landscape. In: Schaffer, J.D. (ed.) Proceedings of 3rd International Conference on Genetic Algorithms, pp. 10–19. Morgan Kaufman, San Mateo (1989)
41. Mathias, K.E., Whitley, D.: Transforming the Search Space with Gray Coding. In: IEEE Conference on Evolutionary Computation, vol. 1, pp. 513–518 (1994)
42. Whitley, D.: A Free Lunch Proof for Gray versus Binary Encodings. In: Genetic and Evolutionary Computation Conference (GECCO 1999), pp. 726–733 (1999)
43. Deb, K.: Multi-Objective Optimisation Using Evolutionary Algorithms, 518 p. John Wiley & Sons, New York (2001)
44. Herrera, F., Lozano, M., Verdegay, J.L.: Tackling Real-Coded Genetic Algorithms: Operators and Tools for Behavioural Analysis. Artificial Intelligence Review 12, 265–319 (1998)
45. Zitzler, E., Laumanns, M., Thiele, L.: SPEA2: Improving the Strength Pareto Evolutionary Algorithm for Multiobjective Optimisation. In: Giannakoglou, K.C., Tsahalis, D.T., Periaux, J., Papailiou, K.D., Fogarty, T. (eds.) Evolutionary Methods for Design, Optimisation and Control with Application to Industrial Problems (EUROGEN 2001), pp. 95–100. International Center for Numerical Methods in Engineering (2002)
46. Corne, D.W., Jerram, N.R., Knowles, J.D., Oates, M.J.: PESA-II: Region-Based Selection in Evolutionary Multiobjective Optimisation. In: Genetic and Evolutionary Computation Conference (GECCO 2001), pp. 283–290 (2001)
47. Deb, K., Agrawal, S., Pratap, A., Meyarivan, T.: A Fast Elitist Non-Dominated Sorting Genetic Algorithm for Multi-Objective Optimisation: NSGA-II. In: 6th International Conference on Parallel Problem Solving from Nature, pp. 849–858. Springer, New York (2000)
48. Knowles, J.D., Corne, D.W.: Approximating the Non-Dominated Front using the Pareto Archived Evolution Strategy. Evolutionary Computation 8(2), 149–172 (2000)

49. Horn, J., Nafpliotis, N., Goldberg, D.E.: A Niched Pareto Genetic Algorithm for Multiobjective Optimisation. In: Proceedings of the First IEEE Conference on Evolutionary Computation, IEEE World Congress on Computational Intelligence, vol. 1, pp. 82–87. IEEE Service Center, Piscataway (1994)
50. Bekele, E.G., Nicklow, J.W.: Multi-objective Automatic Calibration of SWAT using NSGA-II. Journal of Hydrology 341(3-4), 165–176 (2007)
51. Majumdar, S., Mitra, K., Raha, S.: Optimized Species Growth in Epoxy Polymerization with Real-Coded NSGA-II. Polymer 46(25), 11858–11869 (2005)
52. Reed, P., Minsker, B.: Striking the Balance: Long-Term Groundwater Monitoring Design for Conflicting Objectives. Journal of Water Resources Planning and Management 130(2), 140–149 (2004)
53. Srinivas, N., Deb, K.: Multi-Objective Function Optimisation Using Non-Dominated Sorting Genetic Algorithms. Evolutionary Computation 2(3), 221–248 (1995)
54. Deb, K., Jain, S.: Running Performance Metrics for Evolutionary Multi-Objective Optimisation. Kanpur Genetic Algorithms Laboratory Report 2002004 (2002)

7

Data Reconciliation Using Neural Networks for the Determination of K_La

Nilesh Patel and Jules Thibault

Department of Chemical and Biological Engineering, University of Ottawa, Ottawa (Ontario), Canada K1N 6N5
niltel@yahoo.com, Jules.Thibault@uottawa.ca

Abstract. The oxygen mass transfer coefficient (K_La) is of paramount importance in conducting aerobic fermentation. K_La also serves to compare the efficiency of bioreactors and their mixing devices as well as being an important scale-up factor. In submerged fermentations, four methods are available to estimate the overall oxygen mass transfer coefficient (K_La): the dynamic method, the stationary method based on a previous determination of the oxygen uptake rate ($Q_{O2}X$), the gaseous oxygen balance and the carbon dioxide balance. Each method provides a distinct estimation of the value of K_La. Data reconciliation can be used to obtain the most probable value of K_La by minimizing an objective function that includes measurement terms and oxygen conservation models, each being weighted according to their level of confidence. Another alternative, for a more rapid determination of K_La, is using a neural network which has been previously trained to predict K_La from the series of oxygen conservation models. Results obtained with this new approach show that K_La can be predicted rapidly and gives values that are equivalent to those obtained with the conventional data reconciliation algorithm.

1 Introduction

The supply of oxygen is a critical factor in all aerobic fermentations. Indeed, aeration is one of the bottlenecks when operating aerobic fermentation because of the low solubility of oxygen and the constraint on the intensity of mixing due to the limiting shear stress that the cells can sustain. An insufficient oxygen transfer leads to a decrease of microbial growth and product formation. To assess if particular equipment would be able to supply oxygen at a non-limiting rate, it is essential to have a good estimate of the overall oxygen mass transfer coefficient (K_La). In submerged fermentations, K_La serves to compare the efficiency of bioreactors and their mixing devices to provide the right level of oxygenation. K_La is also one of the most important scale-up factors [1,2].

Many methods have been proposed for the determination of K_La for submerged fermentations. The majority of the investigations have however been performed with water and other model fluids, in an attempt to mimic as closely as possible conditions encountered in fermentation systems. These investigations are very useful because conditions are well defined and can be rigorously controlled, providing relatively

M. do Carmo Nicoletti, L.C. Jain (Eds.): Comp. Intel. Tech. for Bio. Mod., SCI 218, pp. 197–214.
springerlink.com

good estimates of K_La that can be used in design calculations. It is nevertheless preferable to determine K_La *in situ* under actual operating conditions since microorganisms, substrates, metabolites, viscosity and antifoam have an impact on the oxygen transfer rate [3]. In fact, K_La values in fermenters may differ substantially from values predicted for oxygen absorption into water or simple aqueous solutions even when differences in liquid physical properties such as viscosity and diffusivity are taken into account [4].

The two most common methods for K_La determination during the course of fermentation are the dynamic method [5] and the gaseous oxygen mass balance [6,7]. Using the information of these two methods, two other methods can be easily defined: the stationary method and the carbon dioxide production rate method [8,9]. For the dynamic method, and subsequently the stationary method, only a fast response dissolved oxygen probe is required to obtain the necessary data. The two gaseous methods, which require oxygen and carbon dioxide sensors, use global oxygen and carbon dioxide balances in the gas phase across the bioreactor, respectively.

K_La is obviously independent of the method employed to estimate its value. However, the four methods will invariably give four different estimates of K_La due to errors in measurements [10] and inaccuracy in the mass balance equations. Therefore, data reconciliation techniques could be used with advantage to obtain the most probable value of K_La where both the reliability of data measurements and the accuracy of each estimation method are taken into consideration. Data reconciliation essentially consists of writing and minimizing an objective function that considers the level of confidence on the various measurements and the mass conservation models. Pouliot et al. [8] defined an objective function comprised of the weighted contribution of twelve measurements and four mass conservation models to estimate the value of K_La at various fermentation times during the production of *Saccharomyces cerevisae*. Patel and Thibault [9] used the same technique to determine K_La during the production of citric acid using *Aspergillus niger*. Data reconciliation involves the minimization of an objective function while taking into account the accuracy of each mass balance equation and each measurement involved in the mass balance equations [11].

In the present investigation, it is proposed to evaluate K_La using a feedforward neural network that is trained strictly using the theoretical mass balance models prevailing during the growth phase of typical aerobic fermentations. The mass conservation models include the dynamic method, the stationary method and the two gas balance methods. It is desired to define a series of simple variables of a fermentation that could be used jointly to estimate rapidly the value of K_La. Because the feedforward neural network uses input data based on the simulation of the four conservation mass balances, data reconciliation is implicitly imbedded and used to determine a single value of K_La.

Cruz et al. [12] have used a neural network to estimate K_La during the growth phase of *Cephalosporium acremonium* whereas Rao and Kumar [13] used a neural network to model the mass transfer rate in an unbaffled surface aeration tank. Djebbar and Narbaitz [14] used successfully a neural network to analyze the mass transfer characteristics in air stripping towers and to simulate K_La. To fit the neural network model, they assembled a large database of pertinent air stripping towers. All these applications used past process data to fit a neural network that was used to estimate K_La in subsequent experiments.

In this investigation, the neural network is used as a metamodel to encapsulate four fermentation mass balance equations that are simulated over a wide range of operating conditions and K_La values such that it is not specific to one type of fermentation. The training of the neural network is performed without resorting to actual experimental data but only using data that were generated from basic principles over a wide range of K_La values. Most variables used for the training of the neural network are defined from simulated fermentations. When the artificial neural network has been defined, it can be used directly without requiring optimization to estimate K_La values.

This chapter is divided as follows. A brief description of the experimental system is presented, followed by a brief review description of the different methods for measuring K_La during fermentation and the associated data reconciliation technique. This is followed by a description of the proposed neural network methodology to estimate K_La and, finally, the main results are presented, compared and discussed.

2 Materials and Methods

The performance of data reconciliation for the determination of K_La using artificial neural networks and its comparison to the more conventional data reconciliation method will be evaluated for three different microorganisms: *Saccharomyces cerevisiae*, *Aspergillus niger* and *Trichoderma reesei*. The three different microorganisms were used at different time over a period of ten years such that conditions of fermentation and measurement equipment differed. Nevertheless, K_La determination was performed using the same methods. Since the general K_La neural network estimator derived in this investigation is not dependent on the strain, these differences should not matter. On the contrary, despite these differences, the validity of the reconciliation method proposed in this investigation will be reinforced.

2.1 Microorganisms and Fermentation Mediums

Saccharomyces cerevisiae - The culture medium composition was: 0.5 g peptone, 3 g yeast extract, 1.25 g KH_2PO_4, 1.25 g K_2HPO_4, 0.45 g $MgSO_4{\cdot}7H_2O$, 5.33 g $(NH_4)_2SO_4$, and 1.1 g glucose per litre of water. Two bags of 8 g of Fleischmann's quick-rise yeast were used as inoculum and were added to an Erlenmeyer flask containing 750 mL of the medium given above. The cells were incubated on an orbital shaker at 25°C for 1.5 h before being added to the fermenter. Glucose was used as thc carbon source. The growth behaviour of *S. cerevisiae* is strongly influenced by glucose concentration. To avoid the Crabtree effect, glucose solution (200 g/L glucose) was fed using a predetermined exponential feeding scheme in order to maintain a low concentration within the bioreactor. Fedbatch operation is important since one of the methods used to estimate the K_La value is based on the carbon dioxide production rate because the Crabtree effect would induce a systematic error in the K_La evaluation [15]. The objective of the feeding strategy was to maintain the respiratory quotient (RQ) around unity [16,17].

Aspergillus niger (ATCC 1015) - The freeze-dried culture was rehydrated and grown on Petri dish and subsequently transferred to an agar slant (SIGMA, Potato

Dextrose Agar, P-2182). It was then used to inoculate a 50 mL of culture medium. The composition of the culture medium was composed of the following constituents [18]: Sucrose, 140 g/L; NH_4NO_3, 2.5 g/L; KH_2PO_4, 2.5 g/L and $MgSO_4{\cdot}7H_2O$, 0.25 g/L. Three days before the start of experiment, the microorganisms were transferred into two 730-mL Erlenmeyer flasks of culture medium. At the start of each experiment, one Erlenmeyer was transferred into the bioreactor containing about 17 L of culture solution. Each experiment ran for a period of 12-13 days.

Trichoderma reesei RUT C-30 (ATCC 56765) - Stock cultures were supplied by Iogen Corporation, Ottawa. The glycerol stock solutions of spores were maintained at -80°C and were transferred on potato dextrose agar plates. New plates were prepared every month and kept at 4°C. The culture contained: glucose, 13 g/L; $(NH_4)_2SO_4$, 1.4 g/L; KH_2PO_4, 2.0 g/L; $MgSO_4 \cdot 7H_2O$, 0.6 g/L; $CaCl_2 \cdot 2H_2O$, 0.3 g/L; $FeSO_4 \cdot 7H_2O$, 5.0 mg/L; $MnSO_4 \cdot 7H_2O$, 1.6 mg/L; $ZnSO_4 \cdot 7H_2O$, 1.4 mg/L; $CoCl_2 \cdot 6H_2O$, 2 mg/L, Peptone, 2 g/L and Yeast Extract, 0.5 g/L. The pH of the medium was initially adjusted to 5.5 using 10N NaOH. Medium was autoclaved for 20 min at 121°C. Shake flask cultures were performed with a volume of 500 mL in a 1-L Erlenmeyer flask with three baffles. A spore solution in sterilized water was prepared from the plates.

2.2 Experimental System

Two types of bioreactors as shown in Fig. 1 were used: a stirred tank bioreactor (STB) and a reciprocating plate bioreactor (RPB). These bioreactors, built in our laboratories, are identical except for the mixing mechanism. The STB was used for fermentations performed with the three types of microorganisms whereas the RPB was also used for fermentations with *A. niger*.

The STB has a total volume of 22.5 L and is made of two concentric stainless steel columns. The bioreactor has an inner diameter of 228 mm and a column height of 550 mm. The outer tube has an internal diameter of 236 mm that leaves an annular gap of 3.5 mm to form a jacket where water, at an appropriate temperature, is continuously circulated to maintain constant the temperature of the fermentation broth. The top cover plate of the bioreactor has ports for sampling, feeding, and to hold a dissolved oxygen probe, pH probe and a thermocouple. Compressed air is fed at the bottom of the bioreactor after passing through a rotameter, a mass flow meter and a sterile gas filter. The gas sparger at the bottom of the bioreactor contains one hundred uniformly distributed holes, 1 mm in diameter. The gas flow rate is controlled by a mass flow meter. The exit and inlet gas streams were dehumidified before analyzing their oxygen and carbon dioxide compositions with a gas analyzer (Maihak, Multor 610) or a mass spectrometer (Amatek, Proline). The mixing device consists of three identical Rushton turbines mounted on the central shaft. Each turbine consists of 6 blades mounted on the periphery of a 50 mm diameter disk. Each blade is 25 mm long, 15 mm high and 1.5 mm thick. Four baffles were placed inside the mixing vessel.

The RPB is identical to the STB except for the reciprocating plate stack used as the mixing device. The plate stack consisted of 6 perforated stainless steel plates, 221 mm in diameter and 1.25 mm thick. Each plate was spaced 50 mm apart from one another. The perforations have a diameter of 19 mm and holes are distributed on an equilateral triangular pitch. The plate fractional free area, including the 3.5 mm annular space between the plate edge and bioreactor wall, is 0.357. The driving unit consists of a

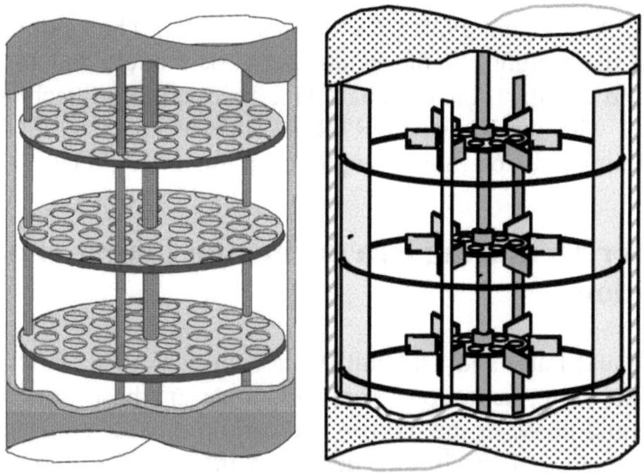

(a) (b)

Fig. 1. Schematic diagram of the Reciprocating Plate Bioreactor (a) and the Stirred Tank Bioreactor (b)

connecting rod, which imparts the reciprocating motion, a tenfold reducing speed transmission and a variable speed motor controlled by a microcomputer. An aluminum disc, containing 100 uniformly distributed perforations and mounted on the output shaft of the reducing transmission, is used in conjunction with an infrared optical switch (HOA-2001, Honeywell) to measure and control, with a microcomputer, the frequency of reciprocation by manipulating the power to the motor. A summary of the operating conditions for each fermentation is provided in Table 1.

Table 1. Summary of operating conditions and some instrumentation

Variables	*S. cerevisiae*	*A. niger*	*T. reesei*
Temperature (^{o}C)	30	30	28
Air flow rate (L/min)	10	10	7
Broth volume (L)	15 - 19	17	10
Agitation speed:			
STB (RPM)	400	100 - 400	350 - 400
RPB (Hz)	-	0.25 - 1.00	-
Gaseous O_2 measurement	Paramagnetic	Paramagnetic	Mass spectrometer
Gaseous CO_2 measurement	Infrared	Infrared	Mass spectrometer

2.3 Computer/Software Platform

All computer programs for the conventional data reconciliation, the development of artificial neural network models and for the simulation of the oxygen mass transfer in well-mixed bioreactors were written by the authors in FORTRAN programming language (COMPAQ Visual FORTRAN Version 6.1). The optimization algorithm used for finding the weights of the neural network models and for minimizing the objective function of the conventional data reconciliation technique was quasi-Newton available with the Visual FORTRAN software package. All programs were run on an IBM Thinkpad X60 microcomputer running at 1.83 GHz.

3 Methods for Measuring K_La during the Course of a Fermentation

Pouliot et al. [8] and Patel and Thibault [9] have used four methods to estimate K_La during the course of fermentation. These methods are based on the oxygen mass balance in the liquid phase or the oxygen and carbon dioxide concentration in the gas phase. These methods will be briefly described in turn. The nomenclature defining symbols and subscripts used in equations and figures is presented at the end of the chapter.

3.1 Dynamic Method

The dynamic oxygen mass balance within the fermenter is given by the following equation:

$$\frac{dC_L}{dt} = K_L a \left(C_L^* - C_L \right) - Q_{O_2} X \tag{1}$$

Equation (1) states that the rate of change of the dissolved oxygen in the fermenter is equal to the rate of oxygen mass transfer from the gas to the liquid phase minus the rate of oxygen uptake by the microorganisms.

In the dynamic method, first reported by Taguchi and Humphrey [5], the oxygen uptake rate (OUR or $Q_{O2}X$) and K_La are determined using the following procedure. As illustrated in Fig. 2, the gas supply and the agitation are stopped momentarily to cut the oxygen supply to the liquid phase so that the rate of decrease of dissolved oxygen is caused entirely by the OUR. The decrease in dissolved oxygen is usually linear and the slope of the plot of C_L as a function of time provides a direct estimate of the oxygen uptake rate. The underlying hypothesis is that the rate of oxygen uptake is unaffected by the absence of air bubbling and agitation, and lower dissolved oxygen concentration. Before the dissolved oxygen concentration reaches its critical lower limit, aeration and agitation are resumed and the dissolved oxygen concentration normally returns to its initial level. K_La can be estimated using Equation (1) using a graphical method or solving the equation using a finite difference method. The advantage of the latter is the possibility to easily include the dynamics of the dissolved oxygen probe (τ_P), which could induce an important bias for higher values of K_La. In that case Equation (2), which describes the dynamic response of the dissolved oxygen probe, is solved simultaneously with Equation (1).

$$\tau_P \frac{dC_P}{dt} = \left(C_L - C_P \right) \tag{2}$$

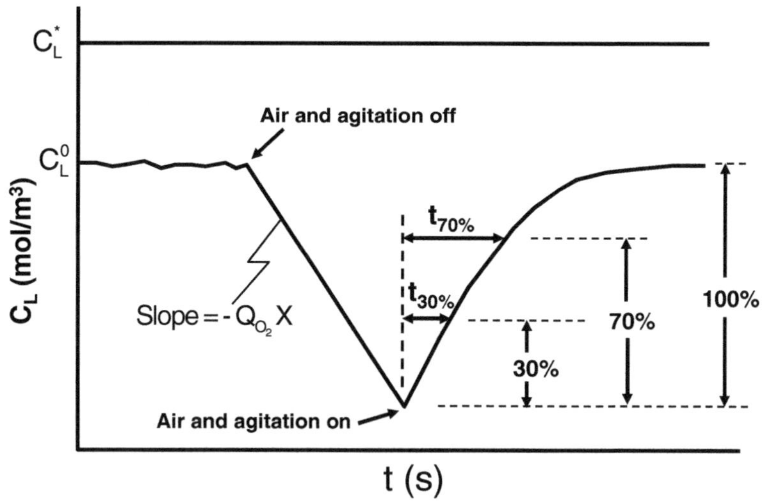

Fig. 2. Dissolved oxygen probe response during the dynamic method

It is important to point out that the estimate of K_La using the dynamic method does not depend on the estimation of $Q_{O2}X$. Indeed, based on Equation (1), $Q_{O2}X$ only affects the steady-state value of the dissolved oxygen and not the rate at which the steady-state dissolved oxygen is achieved, thus the estimation of K_La using the dissolved oxygen response on the reoxygenation of the medium is not affected the actual value of $Q_{O2}X$.

A small variant of this method has been used whereby aeration was cut momentarily whereas the agitation was reduced to 30 RPM in order to prevent cell sedimentation. This small modification does not affect the estimation of K_La [19].

3.2 Stationary Method

Once the estimation of $Q_{O2}X$ is available, it is possible to use Equation (1) along with the information prevailing in the pseudo-stationary phase preceding the dynamic test to directly calculate K_La.

$$K_La = \frac{Q_{O2}X}{C_L^* - C_L^0} \tag{3}$$

3.3 Oxygen Gas Balance Method

Under pseudo-stationary conditions, the oxygen deficit of the gas stream across the fermenter is equal to the oxygen uptake rate (OUR). The OUR in Equation (3) can be replaced by the gaseous oxygen overall mass balance to estimate K_La:

$$K_La = \frac{\frac{1}{V_L}\left(\frac{P_1}{RT_1} Q_{1,G}\, y_{1,O_2} - \frac{P_2}{RT_2} Q_{2,G}\, y_{2,O_2}\right)}{(C_L^* - C_L^0)} \tag{4}$$

3.4 Carbon Dioxide Gas Balance Method

The gaseous carbon dioxide production rate can also be used to estimate K_La provided that an estimate of the respiratory quotient (RQ) is available. Similar to Equation (4), K_La is obtained as follows:

$$K_La = \frac{\frac{1}{RQ}\frac{1}{V_L}\left(\frac{P_2}{RT_2}Q_{2,G}\,y_{2,CO_2} - \frac{P_1}{RT_1}Q_{1,G}\,y_{1,CO_2}\right)}{(C_L^* - C_L^o)} \qquad (5)$$

Good estimate of RQ is available for a large number of fermentations or can also be estimated from past fermentations. All these four methods have their own strengths and limitations at different stages of the fermentation process. For example, the dynamic method can be used with relatively higher precision during the initial stages of fermentation as the oxygen consumption is low and the dissolved oxygen concentration in the medium is high. Because of the low oxygen consumption, the difference between the inlet and exit gas concentrations is consequently very small at that stage of fermentation, and the gas balance methods for K_La estimation lack accuracy. As the fermentation progresses, the growth of biomass leads to an increase in oxygen consumption and to a decrease in the dissolved oxygen concentration. As a result, the gas balance methods become more accurate. Furthermore, at one point, the dissolved oxygen concentration may become too low to allow using the dynamic method and one has to rely strictly on the gas balance methods. Fortunately, it is under these circumstances that the gas balance methods provide the most accurate estimates of the values of K_La. Because the relative accuracy of the various methods constantly varies throughout the fermentation, averaging the values of K_La will not lead to the best estimate. To overcome this problem and to take into account the relative accuracy of all methods, a data reconciliation technique was used. This technique, in addition to considering the precision of each method, takes into account the reliability of all measurements involved in the estimation of K_La values.

4 Data Reconciliation Technique

4.1 Conventional Data Reconciliation Technique

Since it is now common for fermentation systems to be equipped with an O_2/CO_2 monitor or mass spectrometer [20], up to four methods are available to determine K_La, thereby leading to four different estimates. A simple average of the four values could be taken to give a unique and more precise value of K_La. However, as some methods are more accurate than others at different stages of fermentation, averaging K_La is not the best method to achieve a more precise value. Instead, a data reconciliation technique can be used with advantage to resolve this problem. This technique, briefly reviewed in the next paragraphs, considers the precision of each measurement and each estimation method to provide the best estimate of the K_La value. A more detailed description may be found in Pouliot et al. [8].

Equations (1) to (5) involve twelve measured or estimated process variables : the pressure (P), the temperature (T), the time constant of the dissolved oxygen probe

(τ_P), the saturated dissolved oxygen concentration (C_L^*), the dissolved oxygen concentration (C_L), the liquid volume (V_L), the respiratory quotient (RQ), the gas flow rate (Q_G), the oxygen mole fraction in the inlet gas ($y_{1,O2}$), the oxygen mole fraction in the outlet gas ($y_{2,O2}$), the carbon dioxide mole fraction in the inlet gas ($y_{1,CO2}$), and the carbon dioxide mole fraction in the outlet gas ($y_{2,CO2}$). These variables are generally subject to random and non-random errors. Data reconciliation can be used to minimize the impact of these errors in order that the adjusted or reconciled values of the process measurements obey mass conservation laws as well as other constraints. It is important to point out that, to perform process data reconciliation, the measured data must be redundant; that is, there exist more measured data than are necessary to satisfy system mass balances [21,22]. For the estimation of K_La, any of the four mass balance equations can be used such that data redundancy exists. In this case, data reconciliation is an appropriate methodology to converge to a unique K_La value while taking into account measurement and modeling errors.

In addition, the mass balance models are never perfect representations of the underlying behavior of the process due to the many modeling assumptions that have to be made. Some of these assumptions are: the liquid and gas phases within the bioreactors are perfectly mixed, K_La is uniform at all locations within the bioreactor and the oxygen uptake rate is identical before and after the in situ dynamic test. As a result, data reconciliation has to be performed by taking into account both the measurement errors and the process modeling errors. To achieve this dual objective, general criteria for data reconciliation can be defined to take simultaneously into consideration all measured variables and all conservation models, written in the form of a weighted summation, whereby each term is being affected by a weight that corresponds to the level of confidence that one has in each measurement and each conservation model. This dual objective has been used successfully for many applications [11,23-25].

The data reconciliation objective function for the estimation of K_La was formulated with (1) the twelve process variables associated with their respective level of accuracy and (2) the four mass conservation balance equations with an estimate of their level of confidence. This objective function is minimized to determine a unique value of K_La. The dynamics of the dissolved oxygen probe, represented as a first order system (Equation (2)), was also incorporated. In the objective function, the weighting factor for each measurement term was set equal to the inverse value of the variance associated to that measurement. Weighting factors of conservation models were determined considering the individual precision of each measured variable using a Monte Carlo simulation method [9]. The resulting objective function is relatively complex and was not reproduced in this chapter for lack of space. The reader is referred to previous literature to gain a better understanding of the conventional data reconciliation method [8,9]. To simplify the estimation of K_La, a new method is proposed in the next section.

4.2 Data Reconciliation via a Feedforward Neural Network

The conventional data reconciliation algorithm requires mathematical artillery that is relatively extensive and that must be used for all estimations of K_La. In order to simplify the procedure, it is proposed to use a feedforward neural network (FNN) to

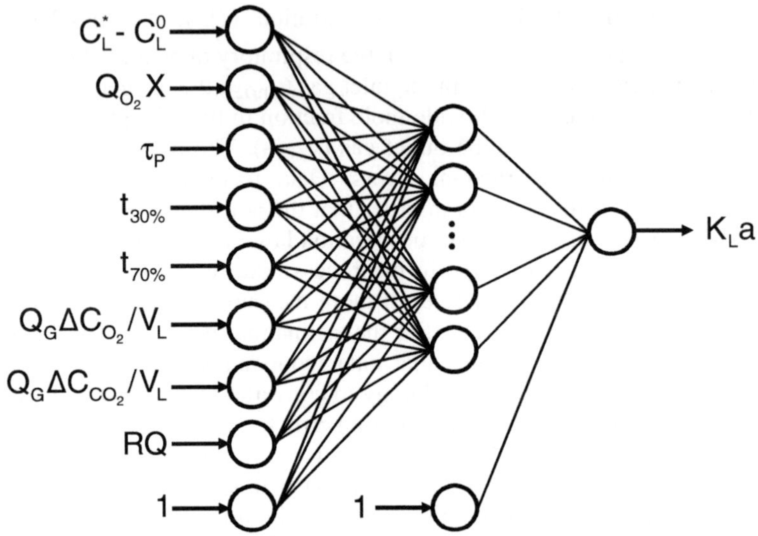

Fig. 3. Architecture of the feedforward neural network used for the estimation of K_La

predict K_La using process information that can be easily obtained from dynamic experiments and gaseous oxygen and carbon dioxide concentrations. A feedforward neural network draws its analogy from human brain neuronal system which is able to learn by modifying the synaptic connections. It consists of a layer of input neurons, a layer of output neurons and one or more hidden layers.

In this investigation, a three-layer FNN was used, that is a network with a single hidden layer as shown in Fig. 3. A FNN can simply be viewed as a general nonlinear model that relates a set of independent variables (inputs) and dependent variables (outputs). The inputs and outputs of the neural network are usually scaled into the range of 0 to 1. The neurons in the input layer simply store the coded information of the input variables and fan out this information to the functional neurons of the hidden layer. The hidden layer contains functional neurons (except for the bias in the hidden layer) which perform a nonlinear transformation of the weighted sum of the outputs of the neurons of the previous layer. The output neurons perform a similar operation whereby it transforms the weighted sum of the outputs of the hidden neurons to calculate the normalized output variables. To adequately represent the underlying phenomenon that the neural network is trying to model, it is necessary to train the model with representative data. In the training phase, an optimization algorithm is used to find a suitable set of parameters (connecting weights) that will minimize the prediction errors for the entire training data set. For a more comprehensive description of all aspects of neural networks including the underlying equations, the reader is referred to the introductory paper of Lippmann [26] and Thibault et al. [27].

It is assumed that the system of mass conservation Equations (1-5) applies for all three fermentation systems such that it is desired to derive a single neural network that will implicitly perform data reconciliation to calculate the most probable value of K_La. To train the neural network to perform this task, the dynamic method was simulated with different values of K_La, $Q_{O2}X$ and of the other twelve process

Table 2. Ranges of variation of each process variable to generate the training data set

Variable	Range	Variable	Range
P (Pa)	99 300 – 103 300	Q_G (10^4 m^3/s)	1.00 – 1.67
T (K)	299 - 303	$y_{1,O2}$	0.1992 – 0.2092
τ_P (s)	5 - 15	$y_{2,O2}$	0.189 – 0.205
C_L^* (mol/m^3)	0.23 – 0.25	$y_{1,CO2}$ (10^4)	4.0 – 6.0
C_L^0 (mol/m^3)	0.001 – 0.24	$y_{2,CO2}$ (10^4)	5.72– 229.3
V_L (m^3)	0.008 – 0.017	$Q_{O2}X$ (10^4)	14.3 – 991.5
RQ	0.80 – 1.50	K_La (s)	0.001 – 0.10

variables. A series of 500 simulations was performed with all these process variables chosen randomly, some calculated from other process variables, over a relatively wide, yet realistic, range of variation. The range of variation of these variables in the simulation is presented in Table 2. For each simulation, some process parameters were directly selected or defined to be used as inputs to the neural network for the estimation of K_La. This set of parameters was then used to train the feedforward neural network of Fig. 3 as a predictor for K_La. This neural network can be viewed as a software sensor of the oxygen mass transfer coefficient [28-30].

The neural network of Fig. 3 has eight process inputs that were defined from the simulation of Equations (1) - (5) and were selected because they are correlated with the value of K_La that prevails at any point during the fermentation. These variables are (see also Fig. 2):

1. C_L^* - C_L^0 is the difference between the dissolved oxygen (DO) concentration at saturation and the pseudo-stationary DO concentration prior to the initiation of the dynamic test.
2. $Q_{O2}X$ is the oxygen uptake rate (OUR) that is determined from the slope of the DO response when agitation and aeration are stopped. This value is not influenced by the dynamics of the DO probe so that the linear portion of the curve can be used directly for its evaluation.
3. τ_P is the time constant of the DO probe that was determined separately by fitting a first-order model to the response of the probe after its rapid transfer from a nitrogen-sparged medium to an oxygenated solution.
4. $t_{30\%}$ is the time required for the measured DO concentration to regain 30% of the range between the lowest DO concentration and the pseudo-stationary DO concentration. This time is influenced directly by K_La and the time constant of the DO probe. It is important to mention that the estimation of K_La is independent of the value of $Q_{O2}X$. Indeed, $Q_{O2}X$ only affects the final pseudo-stationary DO level but not the rate of change of the DO concentration while the system is returning to this final value after agitation and aeration have been resumed.
5. $t_{70\%}$, similar to $t_{30\%}$, is the time required for the measured DO concentration to regain 70% of the range of variation of the DO concentration. These two times are obviously directly impacted by the value of K_La as they represent the rise to the steady DO value when aeration and agitation are resumed.

6. $Q_G \Delta C_{O2}/V_L$ is the product of the gas flow rate and the change in concentration between the inlet and the exit of the fermenter divided by the liquid volume. This term represents the rate of oxygen consumption per unit volume of the fermentation broth.
7. $Q_G \Delta C_{CO2}/V_L$ is similar to previous term for the concentration of carbon dioxide. It represents the rate of carbon dioxide production per unit volume of the fermentation broth. When the value of the respiration quotient (RQ) is known, these two expressions would ideally be equivalent.
8. RQ is the respiration quotient and is usually fairly constant during a major part of the fermentation. Values of past experiments are usually used as an estimate.

This neural network performs data reconciliation using information pertaining to each mass conservation models to estimate a unique value of K_La. Once the neural network has been trained, it can be used rapidly to estimate K_La from the determination of the eight parameters. In addition to the nine input neurons (eight process inputs plus bias), the neural network used 7 hidden neurons and one output neuron. The number of input neurons was chosen when the sum of squares of both the training and validation data sets did not decrease significantly. Training was performed using a quasi-Newton optimization algorithm [31]. The parity plot of the theoretical (or simulated) K_La versus the predicted K_La using the neural network is presented in Fig. 4. As can be observed, the neural network is able to represent very

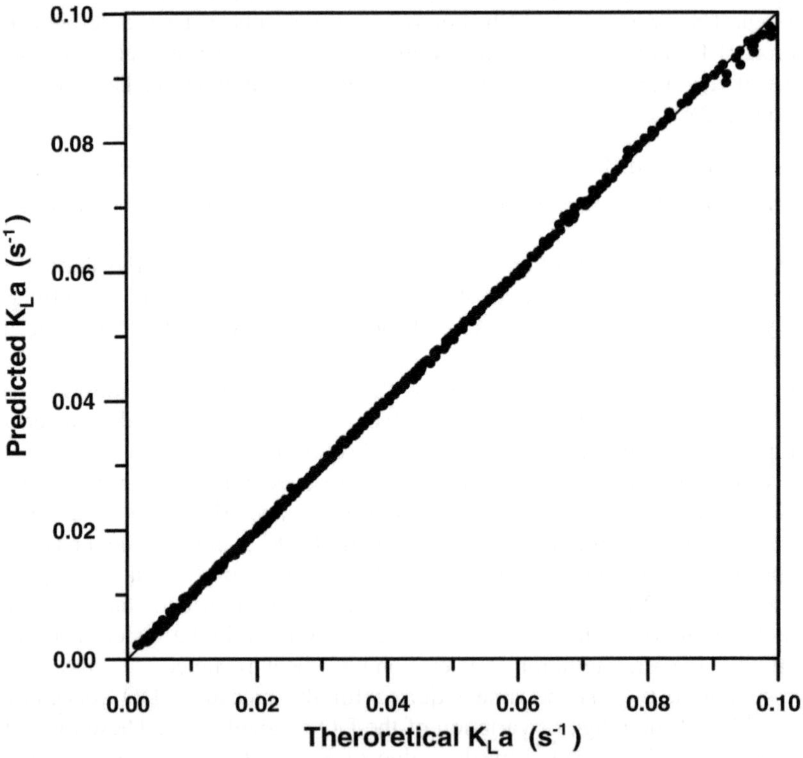

Fig. 4. Theoretical K_La versus predicted K_La for 500 simulations

well the 500 values of K_La that were determined over a relatively wide range of the numerous process variables. The neural network has been trained with 400 values taken randomly and the remaining 100 values were used for validation.

The quality of measured data affects the quality of the estimation of K_La. Alves and Nascimento [32] proposed an analysis of process data via first principle models and neural network models to detect outliers and systematic errors prior to perform data reconciliation. The use of the neural network developed in this investigation could eventually be extended to also detect outliers and systematic errors.

5 Results and Discussion

The data obtained during the various dynamic K_La tests performed during the fermentations of *S. cerevisiae*, *A. niger* and *T. reesei*, were used to test the ability of the neural network of Fig. 3 to predict K_La values. The neural network predictions were compared with the K_La values obtained with the conventional data reconciliation method. These results are presented in Fig. 5. The range of K_La for the three fermentation systems is different. For the production of baker's yeast, the fermentation has a relatively low viscosity and K_La is higher. As the viscosity of the fermentation broth increases, K_La decreases as observed for the fermentations with the two filamentous microorganisms, *A. niger* giving by far a more viscous fermentation broth. The average difference in the estimation of K_La between the two methods lies well within the two dotted lines that represent the ±20% variation in K_La prediction. The scatter in the difference of K_La for *S. cerevisiae* and *A. niger* is greater than for *T. reesei*. This difference is mainly attributed to the precision of the equipment used to measure the gas phase concentration. For *T. reesei*, a mass spectrometer was used which provided a significantly greater accuracy than the O_2 paramagnetic and CO_2 infrared measurements for fermentations performed with the other two microorganisms.

The real value of K_La is obviously unknown and the four different methods discussed above and given in Equations (1 - 5) provide different estimates of K_La. It is only possible to compare predictions among themselves. It is assumed in this investigation that the conventional data reconciliation method provides the most probable value of K_La because it takes into account each of the twelve measurements as well as the four conservation models along with their respective accuracy [8,9]. In the case of the conventional data reconciliation, the accuracy of each model was determined by performing a Monte Carlo simulation at the point of operation for each dynamic test in order to find the weighting factor to be used in the objective function. On the other hand, for the neural network, the underlying relationship that is developed depends only on the set of eight input variables of Fig. 3 that were derived from simulations. Therefore, it is not possible to change, as in the case of the conventional data reconciliation, the weighting factor of a particular input variable to adapt to its current relative accuracy. Despite this limitation, the results of Fig. 5 clearly show that the predictions of K_La using the artificial neural network are similar to those obtained by the conventional data reconciliation method.

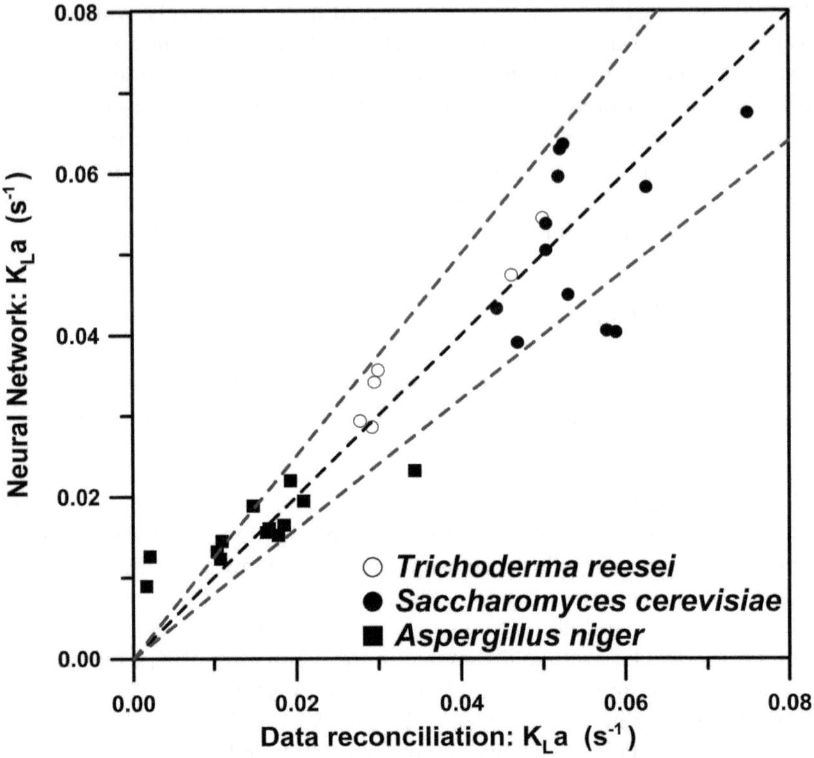

Fig. 5. Plot of the K_La values obtained with the neural network versus values obtained with the conventional data reconciliation technique. The two dotted lines from each side of the 45° line represent the range of ±20%.

The points on the parity plot of Fig. 5 appear to be relatively well distributed around the 45° line and there is no apparent systematic bias for any of the three fermentations. The level of accuracy observed in this comparison is relatively good and well within typical errors that are commonly seen in K_La estimation [14,33]. These results are very interesting, since a unique neural network was used for estimating K_La for the three types of microorganisms. These results may be considered surprising in view that a single model applies equally well to the three types of microorganisms. However, the neural network does not model the biochemical reaction but only the oxygen mass transfer coefficient from variables that are related to the fate of oxygen within the system. The level of dissolved oxygen is function of two factors, the oxygenation rate and the oxygen uptake rate, as described by Equation (1). When the fermentation medium is more viscous, as in the case of *A. niger*, the rate at which oxygen is transferred from the gas phase to the fermentation broth is significantly lower whereas is it higher for *S. cerevisiae*. On the other hand, the OUR for *S. cerevisiae* is significantly higher compared to *A. niger*.

To increase the sensitivity of the estimation, it could be possible to use training data generated over a more restricted range of operation that would be encountered in actual fermentation. The input vector of the neural network could also be reduced in the case of a specific application. For example, if the time constant of the DO probe was known, a constant value could be used while generating the data and its value would be removed from the input vector of the neural network. It would also be advantageous to perform a correlation analysis between a series of inputs and K_La in the training data set to determine the inputs that have the most impact on the estimation of K_La, thereby allowing choosing the best input vector to construct the neural network.

Different versions of the neural network could be used depending on the operating points. For example, at the beginning of the fermentation, the differences of oxygen and carbon dioxide concentrations between the inlet and the outlet gas streams are generally low. The accurate estimation of K_La by these two methods strongly relies on the precision of the O_2 and CO_2 sensors, and of the mass flow meter. In this case, K_La could be estimated very reliably with the dynamic and stationary methods such that a neural network without the two gaseous terms and RQ value (last three elements of the input vector) could be constructed and an enhanced prediction could be obtained. The reverse is true when the dissolved oxygen is very low and it is not possible to perform a dynamic test. In this case, the differences in gaseous concentrations would be closed to their maximum values and a greater accuracy in the estimation of K_La could be achieved considering only the gaseous terms. In that particular case, a neural network could be constructed with only four inputs because it is impossible to evaluate input variables $Q_{O2}X$, τ_P, $t_{30\%}$ and $t_{70\%}$. As a result, the latter four inputs could be safely eliminated from the network of Fig. 3. For the simulations performed to generate the data to train the neural network, values of $Q_{O2}X$, $t_{30\%}$ and $t_{70\%}$ were set to zero whenever the stationary DO level was less than 20% and a dynamic method could not be performed in practice. The neural network was nevertheless able to integrate this information.

6 Conclusion

In this investigation, the *in situ* estimation of K_La in fermentation using a neural network trained strictly on theoretical mass conservation equations was examined and compared with the estimation obtained with the more conventional data reconciliation algorithm. It was found that the neural network has the ability to estimate the value of K_La from process variables defined from the response of the dynamic method of Taguchi and Humphrey [5] and the gas balance methods. The values obtained were close to those obtained by the more conventional data reconciliation method.

The results obtained clearly showed that a single artificial neural network could be used for the estimation of K_La for different fermentation systems. Dedicated neural networks could be developed over a narrower range of K_La and for specific situations where not all the conservation models could be used. This would improve the estimation accuracy of K_La. However, results show that, given the typical estimation error normally associated with K_La determination, the unique artificial neural network of Fig. 3 is adequate.

Nomenclature

C_L	dissolved oxygen concentration (mol/m^3)
C_L^0	pseudo-steady-state dissolved oxygen concentration recorded at the initiation of the dynamic method (mol/m^3)
C_L^*	dissolved oxygen concentration in equilibrium with mean gaseous oxygen concentration (mol/m^3)
C_P	dissolved oxygen concentration recorded by the DO probe (mol/m^3)
DO	acronym for dissolved oxygen
K_La	overall oxygen mass transfer coefficient (s^{-1})
OUR	oxygen uptake rate (mol/m^3 s)
P	pressure (Pa)
Q_G	gas flow rate (m^3/s)
$Q_{O2}X$	oxygen uptake rate (mol/m^3 s)
R	gas constant (8.306 Pa m^3/(mol K))
RQ	respiratory quotient
t	time (s)
T	temperature (K)
V_L	liquid volume in the fermenter (m^3)
y	gaseous mole fraction

Greek letters

τ_P	dissolved oxygen probe time constant (s)

Subscripts

1	inlet stream
2	outlet stream
CO_2	carbon dioxide
G	gas
O_2	oxygen

Acknowledgement

The authors are grateful to the Natural Science and Engineering Research Council (NSERC) of Canada for their financial support.

References

1. Jarai, M.: Factors Affecting the Scale-Up of Aerated Fermentation Processes. Int. Chem. Eng. 19, 701–708 (1979)
2. Moo-Young, M., Blanch, H.W.: Design of Biochemical Reactors – Mass Transfer Criteria for Simple and Complex Systems. Adv. Biochem. Eng./Biotechnol. 19, 1–69 (1981)

3. Yagi, H., Yoshida, F.: Oxygen Absorption in Fermenters: Effects of Surfactants, Antifoaming Agents, and Sterilized Cells. J. Ferment. Technol. 52, 905–916 (1974)
4. Gauthier, L., Thibault, J., LeDuy, A.: Measuring K_La with Randomly Pulsed Dynamic Method. Biotechnol. Bioeng. 37, 889–893 (1991)
5. Taguchi, H., Humphery, A.E.: Dynamic Measurement of the Volumetric Oxygen Transfer Coefficient in Fermentation Systems. J. Ferment. Technol. 14, 881–889 (1966)
6. Siegel, S.D., Gaden Jr., E.L.: Automatic Control of Dissolved Oxygen Levels in Fermentations. Biotechnol. Bioeng. 4, 345–356 (1962)
7. Shuler, M.L., Kargi, F.: Bioprocess Engineering – Basic Principles. Prentice Hall PTR, New Jersey (1992)
8. Pouliot, K., Thibault, J., Garnier, A., Acuña Leiva, G.: K_La Evaluation During the Course of Fermentation Using Data Reconciliation Techniques. Bioprocess Biosystems. Engineering. 23, 565–573 (2000)
9. Patel, N., Thibault, J.: Evaluation of Oxygen Mass Transfer in Aspergillus Niger Fermentation Using Data Reconciliation. Biotechnol. Prog. 20, 239–247 (2004)
10. Brown, D.E.: Bioprocess Measurements and Control. Chemistry & Industry, 678–681 (September 16, 1991)
11. Hodouin, D., Bazin, C., Makni, S.: On-line Reconciliation of Mineral Processing Data. In: Proc. of the AIME/SME Symposium – Emerging Computer Techniques for the Mineral Industry, Reno, Nevada, pp. 101–110 (1993)
12. Cruz, A.J.G., Silva, A.S., Araujo, M.L.G.C., Giordano, R.C., Hokka, C.O.: Estimation of the Volumetric Oxygen Transfer Coefficient (K_La) From the Gas Balance and Using a Neural Network Technique. Braz. J. Chem. Eng. 16, 179–183 (1999)
13. Rao, A.R., Kumar, B.: Predicting Re-aeration Rates Using Artificial Neural Networks in Surface Aerators. Int. J. Appl. Environ. Sci. 2, 155–166 (2007)
14. Djebbar, Y., Narbaitz, R.M.: Neural Network Prediction of Air Stripping K_La. J. Environ. Eng. 128, 451–460 (2002)
15. Yamane, T., Shimizu, S.: Fed-batch Techniques in Microbial Processes. Adv. Biochem. Eng./Biotechnol. 30, 147–194 (1984)
16. Copella, S.J., Dhurjati, P.: A Detailed Analysis of Saccharomyces cerevisiae Growth Kinetics in Batch, Fed-batch, and Hollow-fiber Bioreactors. Chem. Eng. J. 41, B27–B35 (1989)
17. Wang, H.Y., Cooney, C.L., Wang, D.I.C.: Computer-Aided Baker's Yeast Fermentations. Biotechnol. Bioeng. 19, 69–86 (1977)
18. Atkinson, B., Mavituna, F.: Biochemical Engineering and Biotechnology Handbook, 2nd edn. Stockton Press, New York (1991)
19. Gagnon, H., Lounes, M., Thibault, J.: Power Consumption and Mass Transfer in Agitation Gas-Liquid Columns: A Comparative Study. Can. J. Chem. Eng. 76, 379–389 (1998)
20. Heinzle, E.: Present and Potential Applications of Mass Spectrometry for Bioprocess Research and Control. J. Biotechnol. 25, 81–114 (1992)
21. Crowe, C.M.: Observability and Redundancy of Process Data for Steady State Reconciliation. Chem. Eng. Sci. 44, 2909–2917 (1989)
22. Hodouin, D., Everell, M.D.: A Hierarchical Procedure for Adjustment and Material Balancing of Mineral Process Data. Int. J. Miner. Process. 7, 91–116 (1980)
23. Liebman, M.J., Edgar, T.F., Lasdon, L.S.: Efficient Data Reconciliation and Estimation for Dynamic Processes Using Nonlinear Programming Techniques. Comput. Chem. Eng. 16, 963–986 (1992)
24. Mah, R.S.H.: Chemical Process Structures and Information Flows. Batterworths, Boston (1990)

25. Makni, S., Hodouin, D., Bazin, C.: A Recursive Node Imbalance Method Incorporating a Model of Flowrate Dynamics for On-Line Material Balance of Complex Flowsheets. Miner. Eng. 8, 753–766 (1995)
26. Lippmann, R.P.: An Introduction to Computing With Neural Nets. IEEE ASSP Magazine 4(2), 4–22 (1987)
27. Thibault, J., Van Breusegem, V., Chéruy, A.: On-line Prediction of Fermentation Variables Using Neural Networks. Biotechnol. Bioeng. 36, 1041–1048 (1990)
28. Koprinkova-Hristova, P., Patarinska, T.: Neural Network Software Sensors Design for Lysine Fermentation Process. Applied Artificial Intelligence 22, 235–253 (2008)
29. Mangesh, M.G., Jana, A.K.: A comparison of Three Sets of DSP Algorithms for Monitoring the Production of Ethanol in Fed-Batch Baker's Yeast Fermenter. Measurement 41, 970–985 (2008)
30. Lawrynczuk, M.: Modelling and nonlinear predictive control of a yeast Fermentation biochemical reactor using neural networks. Chem. Eng. J. 145, 290–307 (2008)
31. Powell, M.J.D.: Some Global Convergence Properties of a Variable Metric Algorithm for Minimization without Exact Line Searches. In: ASM/SIAM Symp. on Non-linear Programming, New York (1975)
32. Alves, R.M.B., Nascimento, C.A.O.: Analysis and Detection of Outliers and Systematic Errors in Process Plant Data. Chem. Eng. Comm. 194, 382–397 (2007)
33. Ruchti, G., Dunn, I.J., Bourne, J.R.: Comparison of Dynamic Oxygen Electrode Methods for the Measurement of K_La. Biotechnol. Bioeng. 23, 277–290 (1981)

8

A Computational Intelligent Based Approach for the Development of a Minimal Defined Medium: Application to Human Interleukin-3 Production by *Streptomyces lividans 66*

Keyvan Nowruzi, Ali Elkamel, Jeno M. Scharer, and Murray Moo-Young

Chemical Engineering Department, University of Waterloo
200 University Ave. West, Waterloo, Ontario, N2L 3G1, Canada
{knowruzi,aelkamel,jscharer,mooyoung}@uwaterloo.ca

Abstract. A systematic approach was developed to identify and optimize the essential amino acids in defined minimal medium for the production of recombinant human interleukin-3 (rHuIL-3) by *Streptomyces lividans*. Starvation trials were carried out initially to narrow down the number of probable essential amino acids from an initial number of twenty to eight. Then a screening mixture experiment was designed and performed with the eight identified amino acids and distance-based multivariate analysis was employed to rank the probable essential amino acids regarding both growth and product formation. Following this procedure, the search was narrowed to four amino acids (Asp, Leu, Met, and Phe). Finally, a mixture design experiment known as the simplex lattice design was carried out and the composition of the optimum minimal medium was found using both statistical and neural network models.

1 Introduction

Federal guidelines and regulations banning animal-derived medium components necessitate the refinement or re-development of industrial medium formulations [30]. Medium composition has been known to have a significant effect on product concentration, yield, and volumetric productivity. Therefore, the design of a fermentation medium is of critical importance during the development of an industrial fermentation. The development of defined medium without animal products is most desirable for the production of pure and safe biological products. The employment of defined medium often leads to better control strategies, higher growth rates, improved growth yields, and higher cell densities [28]. Several medium design strategies have been proposed [18]. Traditionally, medium design and optimization was done by varying one factor while keeping the other factors at a constant level. This one-factor-at-a-time technique [6, 22] is simple, but interactions between components are ignored. In addition, it involves a relatively large number of experiments and the optimal medium composition can be missed completely. Alternatively, statistically designed experiments allow

M. do Carmo Nicoletti, L.C. Jain (Eds.): Comp. Intel. Tech. for Bio. Mod., SCI 218, pp. 215–236.
springerlink.com

the evaluation of more than one factor at a time and the interactions between the factors [13, 16, 26, 27], but they require a large number of experiments and are not suited for minimal medium development. There have been also some attempts on the use of artificial neural networks and genetic algorithms have been used to model and optimize fermentation media [2, 25].

We present a unique combination of neural networks with statistical techniques as a a novel approach to identify the essential amino acids and determine their optimum concentrations for the growth of *Streptomyces lividans* and production of Recombinant Human Interleukin-3 (rHuIL-3). We introduce a set of mixture experiments and provide an overview of their structures in section 2. Section 3 is a brief introduction to distanced-based multivariate analysis for ranking amino acids regarding their importance for both biomass and rHuIL-3 production. In section 4 we present artificial neural networks and provide an overview of their structures, strengths, and limitations. Finally, section 5 is the case study employing the combination of these techniques to optimize a defined medium for rHuIL-3 production by *Streptomyces lividans.*

A schematic outline of the technique for design and optimization of medium is given in Fig.1. Initially starvation trials with growth limiting amino acid levels were performed to establish the baseline for nutritional requirement. Following these preliminary experiments, a screening mixture experiment was designed and distance based multivariate analysis was performed to establish the rank of the

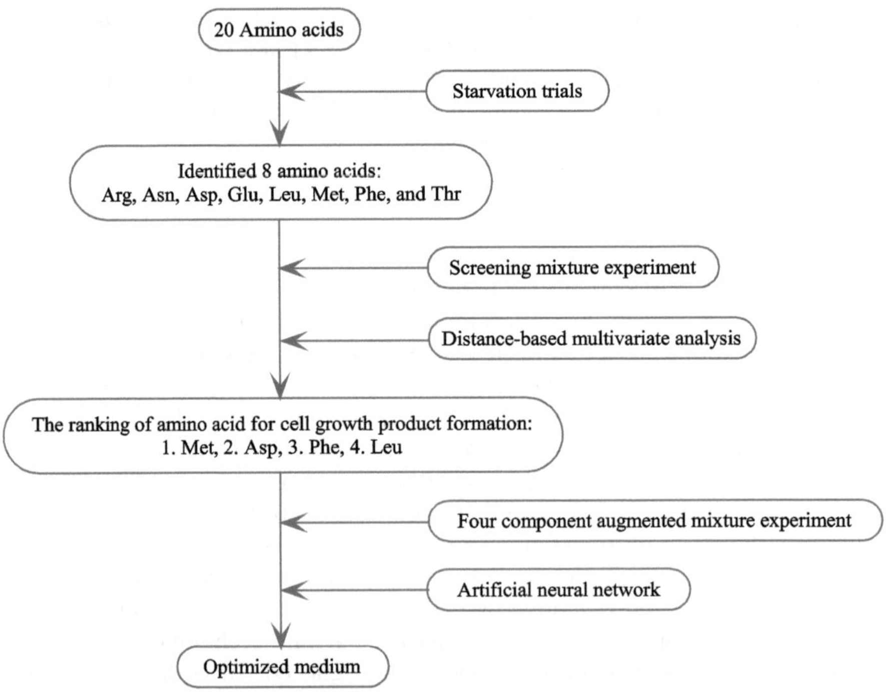

Fig. 1. Outline of the technique used for design and optimization of medium

amino acids. A mixture design known as a simplex lattice design [9, 24] was performed to obtain the optimum proportions of the essential amino acids. Finally, a neural network model was employed as an alternative to the statistical model.

2 Mixture Experiments

Statistical experimental design is a well-established concept for the planning and execution of the informative experiments [23]. The most common approach, factorial design, has been successfully used for the optimization of fermentation media. A two-level factorial design requires 2^n experiments if n factors have to be investigated. In our case, with twenty variables, this would lead to 1,048,576 experiments, which is a prohibitively large number. For this reason, an alternative statistical design, known as the mixture design was employed. This experimental design application concerns the preparation and modification of mixtures, in which the response is assumed to depend only on the relative proportions of the ingredients present in the mixture and not on the amount of the mixture [9]. In mixture experiments, the factors are the components of a mixture, and consequently their levels are not independent. Depending on the experimental objective, the response surface over the simplex region may be mapped by different mixture designs. A $\{q, m\}$ simplex lattice design for q components consists of points defined by the following coordinate settings: the proportions assumed by each component take the equally spaced values from 0 to 1,

$$x_i = 0, \frac{1}{m}, \frac{2}{m}, ..., 1; \qquad i = 1, 2, ..., q \tag{1}$$

and the $\{q, m\}$ simplex lattice consists of all possible combinations of the components where the proportions from Eq.1 are used. In general, the number of points in a $\{q, m\}$ simplex lattice design is

$$N = \frac{(q + m - 1)!}{m!(q - 1)!} \tag{2}$$

An alternative to the simplex lattice design is the simplex centroid design. In a $q-$component simplex centroid design, there are $2^q - 1$ points, corresponding to the q permutations of (1, 0, ..., 0) , the $\binom{q}{2}$ permutations of $(\frac{1}{2}, \frac{1}{2}, 0, ..., 0)$ the $\binom{q}{3}$ permutations of $(\frac{1}{3}, \frac{1}{3}, \frac{1}{3}, 0, ..., 0)$ and the overall centroid $(\frac{1}{q}, \frac{1}{q}, ..., \frac{1}{q})$. Fig.2 shows a simplex lattice and simplex centroid designs. For more information, the interested reader should consult Cornell [9] and Myers and Montgomery [24].

Mixture models differ from usual polynomials employed in response surface work because of the constraint $\sum_{i=0}^{q} x_i = 1$. The standard forms of mixture models that are in widespread use are:

Linear:

$$E(y) = \sum_{i=1}^{q} \beta_i x_i \tag{3}$$

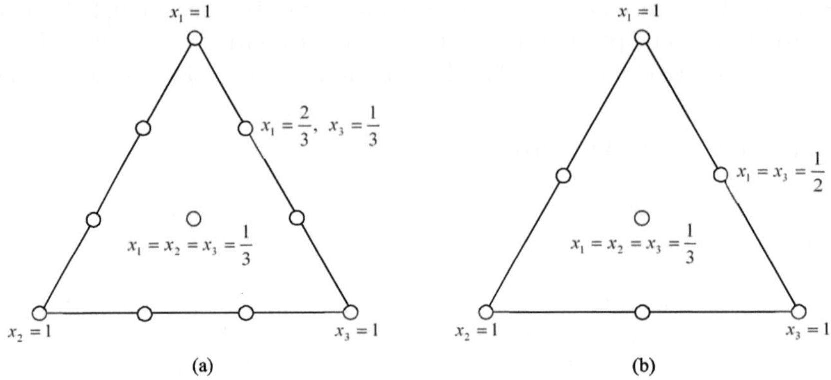

Fig. 2. (a) {3, 3} simplex lattice design, (b) simplex centroid design

Quadratic:

$$E(y) = \sum_{i=1}^{q} \beta_i x_i + \sum \sum_{i<j}^{q} \beta_{ij} x_i x_j \tag{4}$$

Full Cubic:

$$\begin{aligned} E(y) = \textstyle\sum_{i=1}^{q} \beta_i x_i + \sum \sum_{i<j}^{q} \beta_{ij} x_i x_j + \sum \sum_{i<j}^{q} \delta_{ij} x_i x_j (x_i - x_j) \\ + \textstyle\sum \sum \sum_{i,j,k}^{q} \beta_{ijk} x_i x_j x_k \end{aligned} \tag{5}$$

Special Cubic:

$$E(y) = \sum_{i=1}^{q} \beta_i x_i + \sum \sum_{i<j}^{q} \beta_{ij} x_i x_j + \sum \sum \sum_{i<j<k}^{q} \beta_{ijk} x_i x_j x_k \tag{6}$$

The terms in these models have relatively simple interpretations. In Eqs. 3 through 6, the parameter β_i represents the expected response to the pure blend $x_i = 1$ and $x_j = 0$ when $j \neq i$. The portion $\sum_{i=1}^{q} \beta_i x_i$ is called the linear blending portion. When there is a curvature arising from nonlinear blending between component pairs, the parameter β_{ij} represents either synergistic or antagonistic blending. Higher-order terms are frequently necessary in mixture models because of two reasons. First, the phenomena studied may be complex and secondly, the experimental region is frequently the entire operability region and is therefore large, requiring an elaborate model. When the experimental objective is screening, the design supporting a linear model is useful, but for optimization designs supporting quadratic or special cubic models are relevant [11]. Screening experiments are used in the beginning of the experimental work for investigating large number of components to screen out the unimportant components or single out the important ones. To do so, it is necessary to know how to measure the effect of individual components. According to Cornell [9] the effect of component i on the response is the change in the value of the response resulting from a change in the proportion of component i while holding the relative proportions of the other

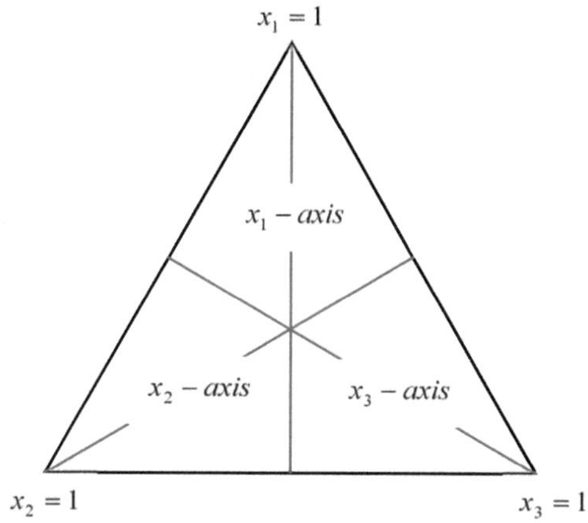

Fig. 3. The x_i−axes, i = 1, 2, and 3

components constant. The largest change that can be made in the proportion of x_i is one unit. If the change in x_i is made along the x_i−axis, which is the imaginary line extending from the base point $x_i = 0$, $x_j = \frac{1}{q-1}$ for all $j \neq i$, to the vertex where $x_i = 1$, $x_j = 0$ for $j \neq i$ all , the proportions x_j, $j \neq i$ of the other components will remain constant. In Fig.3 the three x_i−axis inside the three-component triangle are shown. The values of the proportions of each of the other $q-1$ components along the x_i−axis are $x_j = \frac{(1-x_i)}{(q-1)}$, for $j \neq i$. To measure the effect of all q components simultaneously, using Cox's model [9], the data should be collected at points located on all the axes to permit the estimation of all of the q parameters in the first-degree model.

As shown in Fig.1, the mixture experiment design concept has been employed twice during this study. As discussed in section 5.3, the objective in the first case was to identify the important amino acids while in section 5.4 an augmented simplex designed has been used for optimizing the medium.

3 Distance-Based Multivariate Analysis

Tests of interactions among factors form a very important component of multifactorial experiments. A significant interaction between two factors indicates that the effects of the one of the factors are not consistent across the levels of the other factor. The individual importance of single factors may be difficult or even impossible to isolate or interpret in a system where interactions are present [17]. Several nonparametric multivariate statistical methods are available [8, 20] which can be used in an analysis-of-variance approach, but these methods have the drawback of not generally allowing tests of multivariate interactions between factors in an ANOVA design. In particular, these methods are not capable of differentiating components of multivariate variation in structured models. The reason for this is twofold [19]. First, the distance used may not be metric, so a linear model cannot be applied directly. And second, the above methods rely on permutations based on multivariate normality. Distance-based redundancy analysis (db-RDA) which is described by Legendre and Anderson [19] is a comparatively new approach which has been developed for testing multispecies responses in multifactorial ecological experiments. This method is applicable to situations where the test is based on a distance measure of the researcher's choice. The special advantages of db-RDA are that (1) it can be used with distance measures that are non-Euclidean, (2) it can be used to test interaction terms, or any other term in a structured ANOVA model, and (3) it uses nonparametric permutation methods which do not rely on assumptions of multivariate normality. Steps in the procedure include: (1) calculating a matrix of distances among replicates using a distance measure of choice; (2) determining the principal coordinates, which preserve the distances; (3) creating a matrix of dummy variables corresponding to the design of the experiment; (4) analyzing the relationship between the principal coordinates and dummy variables using RDA; (5) implementing a test by permutation for particular statistics corresponding to the particular terms in the model. For a more detailed discussion one may consult Legendre and Anderson [19] and McArdle and Anderson [21]. A computer program called DISTLM *forward* [1, 21] that does a multivariate multiple regression on the basis of any distance measure and does a forward selection of the predictor variables with the tests by permutations was used for the analysis of our experimental data. The program uses the proportion of the total sum of squares that is explained by the individual variable as the criterion for the forward selection. Output from the program includes the results of the marginal tests individually, ignoring other variables followed by the results of forward selection procedure with the conditional tests. Also information on the correlation among all pairs of explanatory variables, which provides a further check on issues of multi-colinearity is included in the output file. Distance-based multivariate analysis has been employed is section 3 along with the screening mixture experiments to rank the importance of the amino acids based on both biomass concentrations and rHuIL-3 production.

4 Artificial Neural Networks

Artificial neural networks are so named because they are an attempt to simulate at least partially the behavior of brain and biological neurons and learn by trial and error. The goal of a neural network is to map a set of input patterns onto a corresponding set of output patterns. Networks are first subjected to sets of input and output correspondences for a given system. After a sufficient number of training iterations, the network learns the pattern in the data fed to it. The neural network then applies what it has learned to a new input pattern to predict the appropriate output. Neural networks are extensively interconnected parallel structures containing simple processing elements known as neurons or nodes shown in Fig.4. Neurons are arranged in parallel layers, with each neuron forming a weighted connection to all layers. There is an input layer where data are

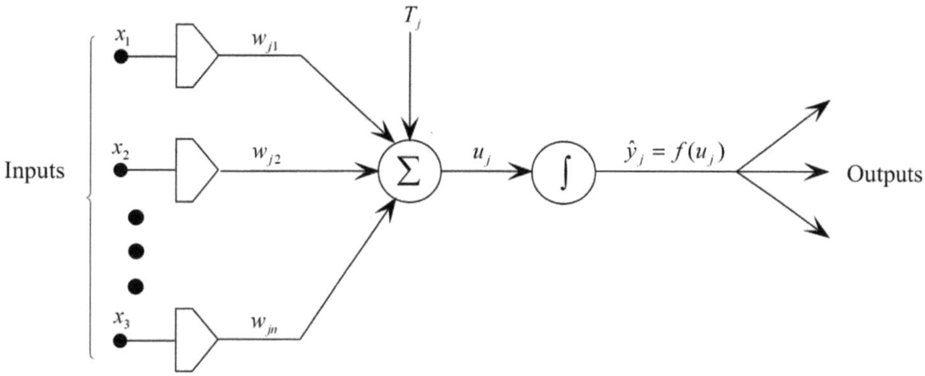

Fig. 4. A typical neuron j in a layer

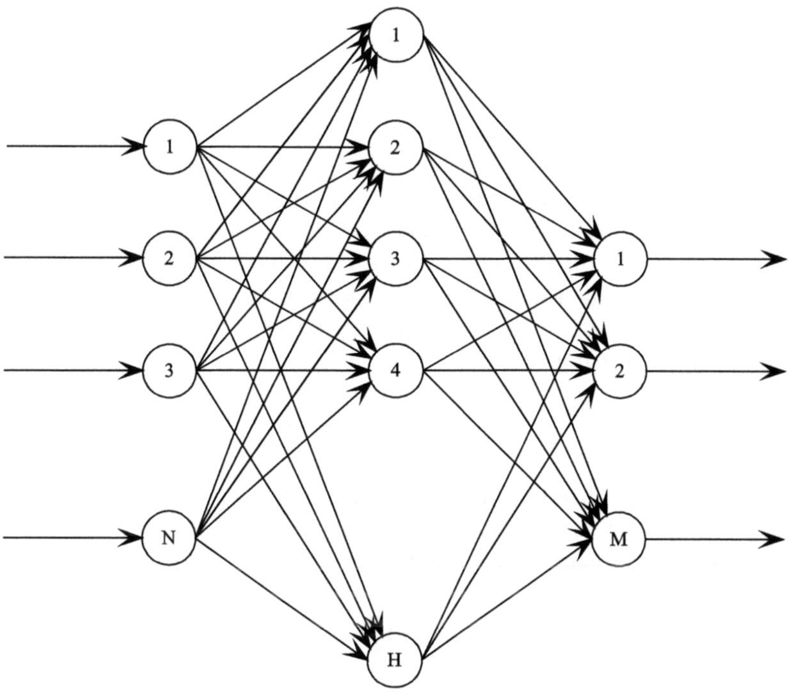

Fig. 5. A one hidden layer feed forward neural network

represented to the neural network, and an output layer that holds the response of the network to the input. In between lies at least one layer known as hidden layer, that enables neural networks to compute complicated associations between input and output patterns. According to universal approximation theory, a network with a single hidden layer with a sufficiently large number of neurons can map any input to any output, with an arbitrary degree of accuracy. Fig.5 shows a typical three-layered neural network with N input neurons, H hidden neurons and M output neurons. The number of input and output neurons is determined by the nature of the problem. The number of neurons in the hidden layer is often determined by the required accuracy, and is therefore a parameter in formulating a neural network model. Each neuron receives an input vector, $\mathbf{X}$, with components X_i (i = 1 to N). The output of each neuron is being determined by several factors other than the inputs. One of those factors is the weight factor, x_{ij} for the ith input, X_i, corresponding to the jth node. The other important factor determining the output of a neuron is the bias denoted by T_j for the jth neuron. The final factor governing the output of a neuron is the transfer function. The most commonly used transfer function is the S-shaped logistic sigmoid function, which is continuously differentiable, monotonic, symmetric and bounded between 0 and 1, i.e.,

$$f(x) = \frac{1}{1+e^{-x}} \tag{7}$$

Another useful transfer function is the hyperbolic tangent function, which is differentiable, continuous and bounded between -1 and 1:

$$f(x) = \tanh(x) = \frac{e^x - e^{-x}}{e^x + e^{-x}} \tag{8}$$

The hyperbolic tangent transfer function outperforms sigmoid transfer function in prediction networks, and this is because of its greater slope compared to sigmoid transfer function [3]. The input-output data should be normalized so that they are in the same range of the transfer function used. The normalization of the inputs also avoids overflows due to very large or very small weight factors. Data are normalized using:

$$\widehat{X} = \frac{X - X_{min}}{X_{max} - X_{min}} \tag{9}$$

where $\widehat{X}$ is the normalized value and X_{min} and X_{max} are the minimum and maximum values of $\mathbf{X}$, respectively. The output from neuron j, in the hidden and the output layers is given by Eq.10.

$$\widehat{y}_j = f(u_j) = f(\sum_{i=1}^{n}(w_{ji}\widehat{X}_i) + T_j) \tag{10}$$

Neural network training is an optimization process, in which an error function is minimized by adjusting the weight factors and the bias. The error function to be minimized is the individual sum-squared error between the actual output values and the predicted values, which is given by [10]:

$$E_i = \sum_{i=1}^{M}(\widehat{y}_{ji} - y_{ji})^2 \tag{11}$$

where $\widehat{y}_{ji}$ and y_{ji} are the ith measured and predicted network outputs that correspond to the ith input, respectively. Therefore, the training process requires a forward pass to calculate an output and a backward pass to update the weight factors and the bias. One pass through the set of training patterns along with updating the weight factors and the bias is called a cycle or epoch [15]. The details of the algorithm for adjusting the weight factors and bias to minimize the sum-squared error are described in Baughman and Liu [3]. The development of a neural network model consists of two steps. The first step is a training stage, where the network is subjected to a training set of input-output patterns. The second step is a testing stage, where the performance of the network is tested on patterns that have not been seen by the network during the training stage. The contribution of each predictor variable on the response could be determined by variable importance measure, VI, according to the procedure developed by Garson [12]. The procedure essentially involves partitioning of the weight factors of each hidden neuron into components associated with each input neuron. The equation proposed by Garson [12] for determining the variable importance measure of an input is given by:

$$VI_{X\,p} = \frac{\sum_{j=1}^{n_H}\left(\frac{|I|_{P_j}}{\sum_{k=1}^{n_P}|I|_{P_{j,k}}}|O|_j\right)}{\sum_{i=1}^{n_P}\left(\sum_{i=1}^{n_P}\frac{|I|_{P_{i,j}}}{\sum_{k=1}^{n_P}|I|_{P_{i,j,k}}}|O|_j\right)} \tag{12}$$

where n_P is the number of input variables, n_H the number of neurons in hidden layer,$|I|_{P_j}$ is the absolute value of the hidden layer weight factor corresponding to the Pth input variable in j^{th} neuron in hidden layer, and $|O|_j$ is the absolute value of the output layer weight factor corresponding to the jth neuron in hidden layer. For more details on the Garson's methodology consult Glorfeld [14]. Neural networks have been used in section 5.4.2 as an alternative to a regression model for optimizing the defined medium.

5 Case Study

The task at hand was to develop a systematic procedure for the design and optimization of a chemically defined medium. The study aimed at replacing casein peptone in conventional medium for *Streptomyces lividans* (strain 1326) with essential amino acids and determining the optimum proportion of the amino acids. The organism used secretes fully bioactive recombinant human interleukin-3 (rHuIL-3). The proposed methodology is believed to be robust and can be employed for the optimal medium design for any recombinant platform.

5.1 Materials and Methods

5.1.1 Microorganism and Medium

Recombinant *Streptomyces lividans* NCIMB 11416/IL3 p002, was kindly provided by Cangene Corporation, Mississuaga, Ontario, Canada. This strain was derived from *Streptomyces lividans 66* as designated by the John Innes Institute and contains multiple copies of plasmid encoding for the rHuIL-3 gene with a thiostrepton-resistance marker. The seed culture was prepared in thiostrepton containing complex medium as described previously [29]. The basal medium used for growth studies had the following composition (per liter of de-ionized water): 10.6g K_2HPO_4, 5.3g KH_2PO_4, 2.5g $(NH_4)_2SO_4$, 226μL polypropylene glycol (PPG) 2025, 100mL of 16% (w/v) glucose, 5mL of 20% (w/v)$MgSO_4$, and 1mL trace element solution. The pH was adjusted to 6.7 before autoclaving for 20min at 121°C. Glucose and magnesium sulfate solutions were autoclaved separately. Trace element solution consisted of (per liter of de-ionized water): 40g $Na_2EDTA.2H_2O$, 11g $CaCl_2.2H_2O$, 7g $FeSO_4.7H_2O$, 2g $MnCl_2.4H_2O$, 2g $ZnSO_4.7H_2O$, 0.4g $CuSO4.5H_2O$, 0.4g $CoCl_2.6H_2O$. Amino acid mixtures were added as described in the results section. The amino acid mixtures were dissolved in deionized water and the resulting solutions were sterilized by filtration through a 0.2μL filter (Millipore) at room temperature. *Streptomyces lividans* cultures were grown in 500mL baffled flasks at a volume ratio of 5:1 in medium containing 10$\mu g/L$ thiostrepton. The temperature was maintained at 32°C and agitation at 240rpm on an Innova 4300 incubation shaker (New Brunswick Scientific Co., USA).

5.1.2 Analytical Methods

For the estimation of the biomass, 10mL culture samples were centrifuged (5000g for 20min) in centrifuge tubes. The pellet was washed twice with distilled water and vacuum-filtered through a 0.45μL filter (Millipore). The biomass cake was dried to a constant weight in a convection oven at85°C. The first supernatant was kept for further analysis of the extracellular medium after filtration through 0.2μL filters (Millipore). For optical density (OD) measurements, approximately 5mL samples were homogenized in a 5mL cell grinder (Safe-Grind, Wheaton) by moving the pestle 20 times. The OD of the homogenized mixture was measured at 600nm in a DU 520UV/Vis spectrophotometer (Beckman Coulter). The cell grinder was rinsed with known volume of dilution water to minimize cell mass loss in the grinder. Since the optical density of the homogenized mixture must be less than 0.3 absorbance unit to remain in the linear range of the calibration curve, the ground samples were diluted by deionized water, accordingly. The supernatant fermentation broth served as the blank. The OD was linearly correlated with the gravimetric biomass measurement. Amino acid analysis of culture supernatants was performed by HPLC following pre-column derivatization with phenylisothiocyanate (PITC) as described by Bidlingmeyer et al. [4]. Total rHuIL-3 concentration was estimated using Direct Enzyme Linked Immunosorbent Assay (Direct ELISA). The primary and secondary antibodies for this assay were rabbit polyclonal Ab specific for IL3 (ab9620) (Abcam Inc.), and anti-rabbit IgG alkaline phosphate conjugate (Sigma, A3687), respectively. p-Nitrophenyl phosphate (Sigma, S0942) was used as the substrate for the secondary antibody. The absorbance of each well was measured at 405 nm in a DU 520UV/Vis spectrophotometer (Beckman Coulter). The amount of rHuIL-3 was estimated by comparing the absorbance of the unknown samples to the absorbance of rHuIL-3 standards obtained from Cangene Corp.

5.2 Starvation Trial

To establish growth limiting conditions, growth studies in equimolar amino acid solutions at suboptimal concentrations were carried out. For brevity, these tests are referred to as starvation trials. The results are shown in Fig.6. Growth was measured both in terms of optical density and biomass. The results in Fig.6 indicate that reducing the total concentration of amino acids below 0.25g/L limits growth noticeably. Thus, an initial amino acid concentration of 0.25g/L was employed as a minimum nutrient requirement for *Streptomyces lividans*. The supernatant samples corresponding to the stoppage point of growth were analyzed for amino acids. The results were overlaid on the initial amino acid profile to show the changes of amino acid concentration during the fermentation. A typical diagram of amino acid concentrations in fresh medium and the supernatant is shown in Fig.7. This figure corresponds to an initial concentration of 0.25 g/L of amino acids in the medium. Amino acid analysis of the fresh medium and the supernatant were carried out in triplicates and the reported data are the average values. The results of amino acid analysis for the rest of the fermentation with different amount of initial amino acids were the same as in Fig.7. It is evident that eight of the twenty amino acids were completely or almost completely consumed when growth ceased. These amino acids included: Arginine,. Asparagine, Aspartic acid, Glutamic acid, Leucine, Methionine, Phenylalanine, and Threonine. However this is not a definitive indication of absolute requirement of these amino acids by the bacterial culture. To clarify this ambiguity, fermentations were run with these identified eight amino acids only, the complementary twelve that were not exhausted, and all twenty amino acids in the medium. The results of these fermentations are given in Table 1. These results indicate that deleting the complementary twelve amino acids has no significant effect on growth.

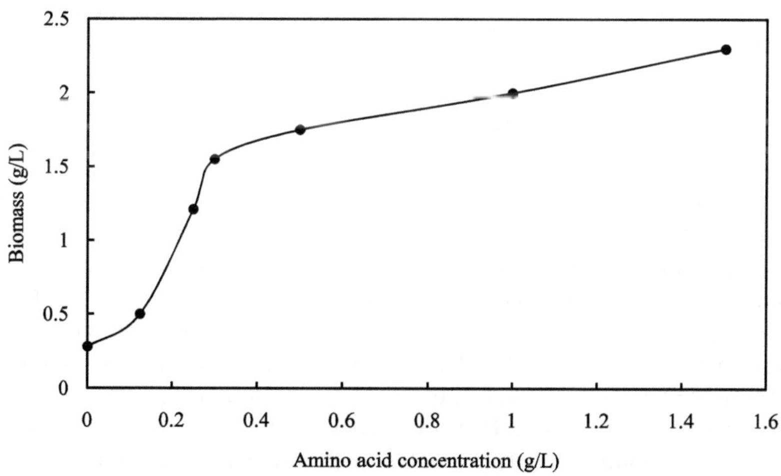

Fig. 6. The effect of total amino acid concentration on biomass production

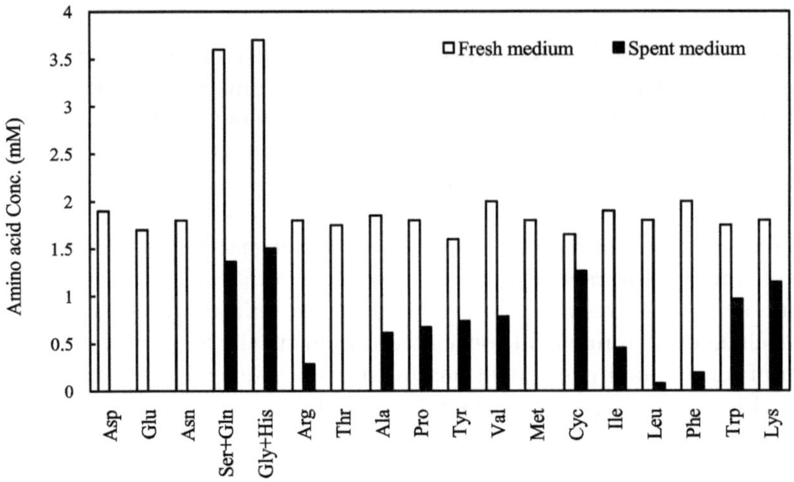

Fig. 7. Amino acid concentration in fresh medium and first supernatant

Table 1. Test of growth dependence on amino acid combinations

Number of Amino Acids	Total weight of Amino Acids (g/L)	Biomass (g/L)
Identified 8	0.630	1.85
Complementary 12	0.870	0.64
All 20	1.500	1.81
None	0.000	0.28

Accordingly, the identified eight amino acids were chosen for further study as potential essential amino acids and a screening mixture experiment was designed to rank the importance of the amino acids.

5.3 Screening Mixture Experiment

A screening mixture experiment was designed to rank the importance of the identified eight amino acids. The design was performed with the aid of Design Expert version 6.0.4 statistical software (Stat-Ease Inc., Minneapolis, MN). The corresponding design and results are shown in Table 2. A total amino acid concentration of 0.25 g/L was used in the experimental design. Then the DISTLM *forward* program [1] was employed for distance-based multivariate analysis using forward selection [21] to rank the identified amino acids. This program needs two input files in ASCII*.txt format. The first input file contains the data matrix of the response variables **Y** (rHuIL-3 concentration and biomass) normalized by dividing each column by the greatest value in the corresponding column, and the other input file is an X matrix containing 36 predictor variables. The predictor variables are 8 amino acid mole fractions in the experimental design

Table 2. Screening mixture experiment design and results

Run	Mole fraction								Biomass	rHuIL-3
	Arg	Asp	Asn	Glu	Leu	Met	Phe	Thr	(g/L)	(mg/L)
1	0.169	0.128	0.129	0.142	0.127	0.145	0.160	0.000	0.953	47.7
2	0.173	0.132	0.133	0.000	0.131	0.149	0.165	0.119	1.224	52.6
3	0.070	0.053	0.054	0.059	0.053	0.060	0.601	0.048	0.839	37.9
4	0.079	0.539	0.060	0.066	0.059	0.068	0.075	0.054	0.848	25.2
5	0.171	0.130	0.131	0.143	0.000	0.146	0.162	0.117	1.247	36.8
6	0.616	0.052	0.052	0.057	0.052	0.059	0.065	0.047	0.727	35.1
7	0.151	0.115	0.116	0.127	0.114	0.130	0.144	0.104	1.178	41.2
8	0.171	0.000	0.131	0.144	0.129	0.147	0.162	0.117	1.103	45.5
9	0.079	0.060	0.541	0.066	0.059	0.067	0.075	0.054	1.332	42.4
10	0.000	0.000	0.000	0.000	1.000	0.000	0.000	0.000	0.423	0.0
11	0.000	0.000	0.000	1.000	0.000	0.000	0.000	0.000	0.396	0.0
12	0.000	1.000	0.000	0.000	0.000	0.000	0.000	0.000	0.112	0.0
13	0.174	0.132	0.133	0.146	0.131	0.000	0.165	0.119	0.142	0.0
14	0.000	0.000	0.000	0.000	0.000	0.000	0.000	1.000	0.206	0.0
15	0.083	0.063	0.063	0.069	0.062	0.071	0.079	0.510	0.813	34.5
16	0.074	0.056	0.057	0.062	0.056	0.573	0.070	0.051	0.805	38.2
17	0.151	0.115	0.116	0.127	0.114	0.130	0.144	0.104	0.963	42.1
18	1.000	0.000	0.000	0.000	0.000	0.000	0.000	0.000	0.053	0.0
19	0.000	0.000	0.000	0.000	0.000	0.000	1.000	0.000	0.116	0.0
20	0.177	0.134	0.135	0.148	0.133	0.151	0.000	0.121	0.852	37.1
21	0.075	0.057	0.057	0.567	0.057	0.064	0.071	0.051	1.011	39.8
22	0.171	0.130	0.000	0.144	0.129	0.147	0.162	0.117	0.772	24.6
23	0.151	0.115	0.116	0.127	0.114	0.130	0.144	0.104	0.866	49.0
24	0.000	0.135	0.136	0.150	0.134	0.153	0.169	0.122	0.915	32.7
25	0.000	0.000	1.000	0.000	0.000	0.000	0.000	0.000	0.246	4.9
26	0.000	0.000	0.000	0.000	0.000	1.000	0.000	0.000	0.495	25.4
27	0.079	0.060	0.061	0.066	0.537	0.068	0.075	0.054	1.050	31.0
28	0.151	0.115	0.116	0.127	0.114	0.130	0.144	0.104	1.005	40.0
29	0.151	0.115	0.116	0.127	0.114	0.130	0.144	0.104	0.985	40.7

(primary predictor variables) and the remaining 28 predictor variables are the binary interaction of those 8 primary predictor variables. Euclidean distance was the choice of the distance measure. After running the program 18 out of 36 variables explained hundred percent of the variability in the data which meant that the degrees of freedom was eighteen.

A summary of the computer output is shown in Fig.8. This figure presents the percent variability of the response matrix (normalized rHuIL-3 and biomass concentration) as explained by the individual amino acids and their binary combinations. Using this information, the probable amino acid requirement (starting with the highest ranked single amino acid, Methionine) was calculated by summing the corresponding probability values.

The predicted rank is shown in Table 3. Verification of the rank predicted by the computer program was done by another series of fermentations, in which

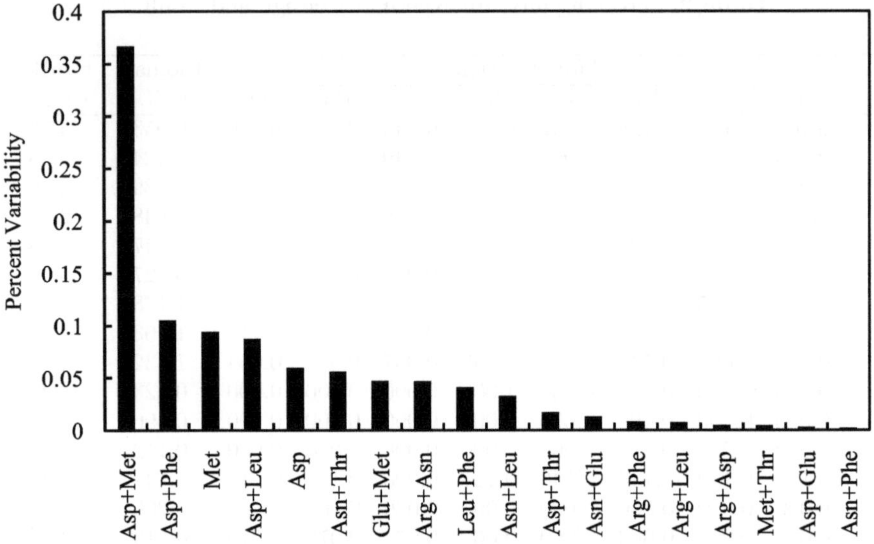

Fig. 8. Percent variability of distance-based multivariate response matrix

the total amino acid concentration was 0.25 g/L. The first run contained only Methionine (Met), the second Met and Aspartic acid (Asp) with equimolar concentration, the third one Met, Asp, and Phenylalanine (Phe) with equimolar concentration, and so on. The results of these fermentation runs are shown in Table 3. It is obvious that after Leucine (Leu), addition of the rest of the identified amino acids decreases the growth. This is due to dilution of the essential amino acids. As mentioned previously, the total amount of the amino acids concentration was fixed, therefore, by adding more amino acids to the solution the essential amino acids the concentrations decreased. Based on these findings, Met, Asp, Phe and Leu were singled out as the essential amino acid candidates.

Table 3. Predicted rank of amino acid combinations and corresponding biomass concentration (ranking obtained by DISTLM *forward* computer program)

Variable	Probability	Biomass (g/L)
Met	0.09	0.52
Met+Asp	0.47	1.23
Met+Asp+Phe	0.47	1.84
Met+Asp+Phe+Leu	0.53	1.92
Met+Asp+Phe+Leu+Glu	0.67	1.49
Met+Asp+Phe+Leu+Glu+Asn	0.82	1.55
Met+Asp+Phe+Leu+Glu+Asn+Arg	0.97	-
Met+Asp+Phe+Leu+Glu+Asn+Arg+Thr	1.00	-

5.4 Defined Medium Optimization by Mixture Design Method

The mixture experiment method was rearranged to obtain the best possible minimal medium composition for *Streptomyces lividans*. The standard simplex design is a boundary point design. With the exception of the overall centroid, all design points are on the boundaries of the simplex. To make predictions about properties of complete mixtures more runs are required in the interior of the simplex. Hence, the usual simplex design was augmented with four axial runs and five design points were done in replicate to check the accuracy of the model and the lack of fit. Therefore, the total number of experiments supporting a quadratic model with four components would be 24. The experimental design and the results for both biomass and rHuIL-3 are summarized in Table 4. The total amount of amino acids used in these experiments was 0.25 g/L.

5.4.1 Multiple Regression Analysis

Based on the experimental values, statistical testing using Fisher's statistical test was carried out. The experimental design and statistical analysis of the

Table 4. Experimental design and results for four-component mixture experiment

Run	Mole fraction				rHuIL-3 (mg/L)		
	Asp	Leu	Met	Phe	Experimental	Regression	Neural network
1	0.000	0.000	0.525	0.475	28.13	27.13	28.29
2	0.000	0.000	1.000	0.000	17.48	17.78	15.70
3	0.475	0.000	0.525	0.000	55.31	59.20	53.64
4	0.000	0.475	0.525	0.000	15.34	19.75	17.80
5	0.594	0.119	0.169	0.119	40.35	54.77	43.78
6	0.000	0.950	0.050	0.000	20.43	21.71	19.74
7	0.000	0.950	0.050	0.000	20.42	21.71	19.74
8	0.119	0.119	0.644	0.119	35.72	34.52	33.32
9	0.000	0.475	0.050	0.475	20.91	29.10	18.62
10	0.475	0.000	0.050	0.475	78.01	68.79	76.00
11	0.950	0.000	0.050	0.000	45.30	44.03	43.83
12	0.119	0.119	0.169	0.594	49.49	43.93	49.25
13	0.000	0.000	0.050	0.950	32.96	36.49	34.16
14	0.317	0.317	0.050	0.317	62.74	46.75	61.56
15	0.000	0.000	1.00	0.000	15.06	17.78	15.70
16	0.119	0.595	0.169	0.119	26.78	29.41	29.53
17	0.238	0.238	0.288	0.238	50.91	44.21	51.44
18	0.000	0.000	0.050	0.950	35.36	36.49	34.16
19	0.317	0.000	0.367	0.317	72.95	58.02	73.85
20	0.000	0.317	0.367	0.317	28.52	25.33	29.65
21	0.317	0.317	0.367	0.000	39.35	40.41	39.04
22	0.475	0.475	0.050	0.000	40.36	32.87	40.67
23	0.950	0.000	0.050	0.000	44.10	44.03	43.83
24	0.475	0.000	0.050	0.475	74.02	68.79	76.00

Table 5. ANOVA table fort he improved model

Source of variation	SS	df	MS	F	$P > F$
Model	2.130	5	0.430	11.51	< 0.0001
Linear mixture	1.480	3	0.490	13.27	< 0.0001
Asp Met	0.280	1	0.280	7.45	0.0137
Asp Phe	0.400	1	0.400	10.88	0.0040
Residual	0.670	18	0.037		
Lack of Fit	0.540	13	0.041	1.56	0.3275
Pure error	0.130	5	0.026		
Total	2.800	23			

results was performed with the aid of Design Expert version 6.0.4 statistical software (Stat-Ease Inc., Minneapolis, MN). The values of regression coefficients were calculated and the following equation which includes all terms regardless of their significance was derived for rHuIL-3:

$$\begin{aligned} rHuIL-3 &= 33.7 \times Asp + 18.8 \times Leu + 15.9 \times Met + 26.6 \times Phe \\ &+6.1 \times Asp \times Leu + 97.7 \times Asp \times Met + 151.4 \times Asp \times Phe \\ &+3.1 \times Leu \times Met + 24.7 \times Leu \times Phe + 59.8 \times Met \times Phe \end{aligned} \tag{13}$$

The model F-value was 4.04 implying that the model is significant. The P-values for and terms were less than 0.05 indicating that these terms are significant at a probability level of 95% and the remaining interaction terms were insignificant at that level. The insignificant terms of the model of Eq.13 were eliminated to give the following improved model:

$$\begin{aligned} rHuIL-3 &= 44.0 \times Asp + 21.7 \times Leu + 17.8 \times Met + 36.5 \times Phe \\ &+113.2 \times Asp \times Met + 114.1 \times Asp \times Phe \end{aligned} \tag{14}$$

The analysis of variance for the improved model is summarized in Table 5. An F-value of the improved model of 12.76, and a P-value of less than 0.0001 indicate that there is less than 0.05% chance that a model F-value this large could occur due to noise. P-values less that 0.05 indicate that the model terms are significant at probability level of greater than 95%. In this case, linear component interactions, and are significant model terms. The of the regression model was 0.78. Because the coefficient of Asp in the model was greater than the coefficients of Phe, Leu, and Met, one would conclude that Asp has higher impact on rHuIL-3 production by *Streptomyces lividans* than Phe, Leu, and Met. Furthermore, since the interaction between Asp and Met, and Asp and Phe has a positive coefficient and these coefficients are greater than the coefficients of the individual amino acids, the interaction terms have higher impact on product formation than the individual amino acids.

5.4.2 Artificial Neural Network

The architecture of the neural network developed in this work was an input layer of 4 neurons and an output layer of one neuron only corresponding to rHuIL-3

Table 6. Verification of regression and neural network models

Run	Mole fraction				rHuIL-3 (mg/L)		
	Asp	Leu	Met	Phe	Experimental	Regression	Neural network
1	0.250	0.250	0.250	0.250	39.98	44.21	37.13
2	0.500	0.000	0.500	0.000	54.16	59.20	49.18
3	0.000	0.500	0.500	0.000	22.69	19.75	17.89
4	0.000	0.000	0.500	0.500	23.18	27.14	21.73
5	0.170	0.160	0.500	0.170	45.31	38.97	47.60
6	0.500	0.160	0.170	0.170	50.77	54.04	48.66
7	0.170	0.500	0.170	0.160	29.62	33.58	30.96
8	0.170	0.170	0.160	0.500	47.23	45.04	47.12

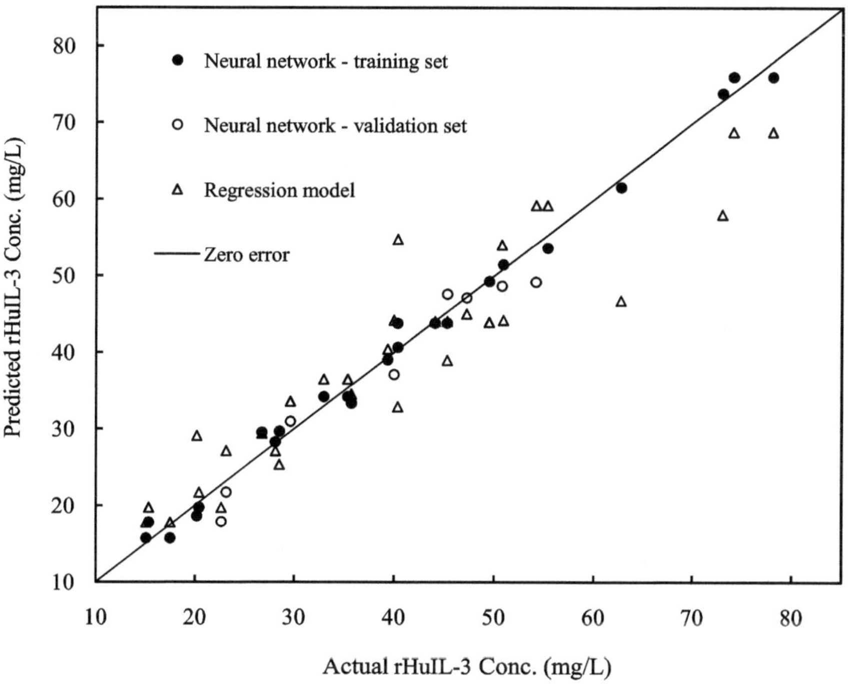

Fig. 9. Predicted vs. actual rHuIL-3 concentration

concentration. Only one hidden layer was used. Different numbers of neurons (4, 5, and 6 neurons) were tried in the hidden layer, checking each time if the developed neural network succeeded in reading an error target. It was found that a network with one hidden layer of four neurons gave sufficiently accurate predictions. Both sigmoid and tangent hyperbolic functions were tried as transfer functions and it was found that because of its greater slope compared to sigmoid function, tangent hyperbolic function produced better results. The data set obtained from mixture experiments comprising 24 experimental runs was used as the training set to optimize the weights of the neural network. The R^2 of the neural network for the training data set was found to be 0.9987. The trained network was used to furnish predictions of rHuIL3 concentrations for the data points that were not used in the training network. This validation step is important to check generalization characteristics of the developed neural network. Another set of 8 experimental runs were carried out by choosing combinations of mole fractions of amino acids which were not taken earlier to validate the neural network model and verify the final regression model. The results of the verification experiments are given in Table 6. The R^2 of the neural network for the testing data was 0.9559. Fig.9 shows the network-predicted rHuIL3 concentrations for training and validation data sets plotted against the corresponding experimental data. The solid circles represent the network-trained outputs while the open circles denote the network-predicted outputs for input variables belonging to the validation set. The neural network model not only fits the training data very well but also provides predictions of the validation data that very close to the experimental measurements. The rHuIL-3 concentrations calculated from the multiple regression equation (Eq.14) is also shown in Fig.9 for comparison. It is obvious that the neural network predictions are much closer to the line of perfect prediction than those of regression polynomial equation, confirming the usefulness of the neural networks as empirical models in response surface analysis.

5.4.3 Optimization of Defined Medium

Contour plot obtained from the regression model over the simplex region (Fig.10) showed the existence of a local optimum within the experimentally explored range. The location of the optimum could be obtained by solving the optimization problem using sequential quadratic programming algorithm. The response surface and neural network models for rHuIL-3 production in terms of the amino acid composition were used to find the optimum level of amino acid by solving the following nonlinear optimization problem:

$$\begin{aligned}
&Max[rHuIL-3 \quad production] \\
&s.t. \\
&Asp + Leu + Met + Phe = 1 \\
&0 \leq Asp \leq 0.95 \\
&0 \leq Leu \leq 0.95 \\
&0.05 \leq Asp \leq 1 \\
&0 \leq Phe \leq 0.95
\end{aligned} \tag{15}$$

This problem was solved using MATLAB (The MathWorks, 2007). MATLAB includes an optimization toolbox that implements various numerical optimization routines, including sequential quadratic programming algorithm to solve for the constrained optima. The MATLAB function fmincon solves the constrained problems and was employed in this research to find the optimal composition of amino acids in the medium. The maximum value of the objective function (rHuIl-3 production) using multiple regression model was found to be 67.95 mg/L, and according to the neural network model, the maximum achievable

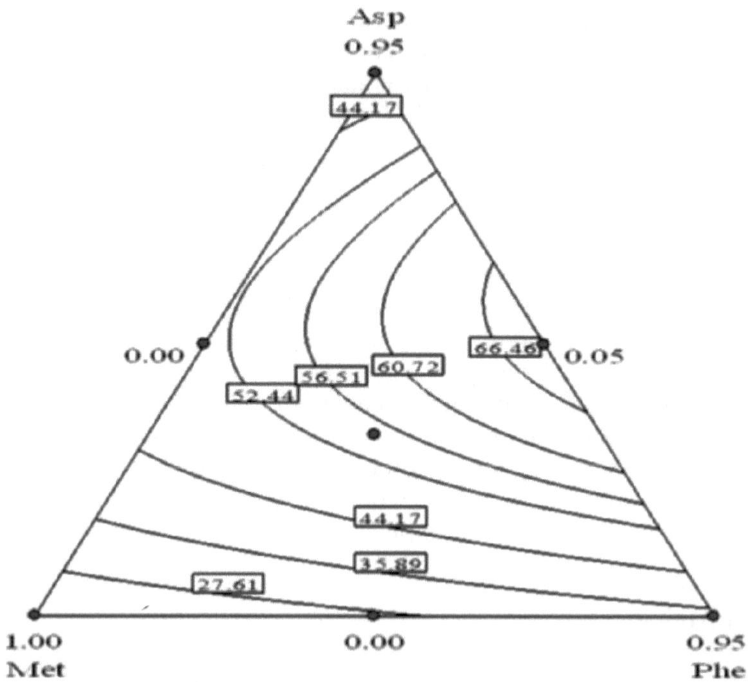

Fig. 10. Contour plot of rHuIL-3 production (mg/L) over simplex region

Table 7. Maximum rHuIL-3 concentration and the optimum input sets

Model	rHuIL-3	Mole fraction			
	(mg/L)	Asp	Leu	Met	Phe
Multiple regression	67.95	0.53	0.00	0.05	0.42
Neural network	76.00	0.56	0.00	0.05	0.39

rHuIL-3 concentration for this fermentation was 76.00 mg/L. The combinations of mole fractions of the four amino acids giving the maximum concentration according to both models are listed in Table 7. To verify the results of the optimization, 4 shake flask fermentations were carried out using the optimal amino acid composition of the minimal medium and rHuIL-3 concentration was found to be 80.03 $\pm$ 8.24 mg/L. The maximum identified by the neural network model is 10.6% higher than the maximum identified by the multiple regression analysis. This difference indicates that solution obtained by multiple regression analysis with poor modeling capacity is not guaranteed to be optimum. Therefore, the ability of the model to approximate the true response surface with a high degree of accuracy is of the key importance in optimization step.

The recombinant host strain employed in this study has an absolute requirement for methionine. In cells methionine has multiple functions. Besides being an amino acid residue in proteins, it acts as an initiator of protein synthesis and is a methyl donor in several biochemical reactions. The genes leading to methinonine synthesis in *Streptomyces lividans* are not very well known, but the disruption of the methylenehydrofolate gene (metF) has been reported to lead to methionine dependence [5]. The beneficial effect of aspartic acid in the medium on the growth of Streptomyces has been observed previously [7]. The overproduction of recombinant protein is known to impose a metabolic burden on the host. It appears that aspartic acid provides a convenient entry into the Krebs cycle and acts as a supplementary energy source. The role of aspartic acid as an energy supplement is supported by our findings that the optimal level of aspartic acid is replaceable with a glutamic acid/aspartic acid mixture with nearly identical results. Phenylalanine provides a ready-made aromatic ring and is a known precursor for the synthesis of tyrosine. The important aspect of this study is not with regard to the specific acid requirements of *Streptomyces lividans* for the optimal production of a recombinant protein. Rather, this paper presents a novel, statistics-based methodology for the rapid identification of the key amino acids and the determination of their optimal composition for producing a recombinant product. The methodology can be applied for any other organism.

6 Conclusion

The hybrid neural network-sequential quadratic programming algorithm approach is described in this work. This novel methodology serves as a viable alternative to the traditional multiple regression approach for modeling and optimization of fermentation media. The input space of a neural network model can be optimized using a sequential quadratic programming algorithm. It appears that neural networks provide a better fit to experimental data than multiple regression models. Starvation trials used in the first step were efficient to provide initial screening for those amino acids which affect rHuIL-3 production and cell growth. Mixture experiments and distance-based multivariate analysis used in the second step to rank the essential amino acids and finally another mixture experiment combined with neural network models was used to find the optimum composition of the medium.

References

[1] Anderson, M.J.: DSTLM forward: a FORTRAN computer program to calculate a distance-based multivariate analysis for a linear model using forward selection (2003), `http://www.stat.auckland.ac.nz/~mja/Programs.htm`

[2] Baishan, F., Hongwen, C., Xiaolan, X., Ning, W., Zongding, H.: Using genetic algorithms coupling neural networks in a study of xylitol production: medium optimisation. Process Biochem. 38, 979–985 (2003)

[3] Baughman, D.R., Liu, Y.A.: Neural networks in bioprocessing and chemical engineering. Academic Press Limited, California (1995)

[4] Bidlingmeyer, B.A., Cohen, S.A., Tarvin, T.L.: Rapid analysis of amino acids using pre-column derivatization. J. Chromatogr. 336, 93–104 (1984)

[5] Blanco, J., Coque, J.J., Martin, J.F.: The folate branch of the methionine biosynthesis pathway in Streptomyces lividans: Disruption of the 5,10-methylenetetrahydrofolate reductase gene leads to methionine auxotrophy. J. Bacteriol. 180, 1586–1591 (1998)
[6] Chary, C.V.K., Rambhav, S., Venkateswerlu, G., Ramachandran, L.K.: Possible precursor for thiostrepton in *Streptomyces azureus* - culture medium production optimization. Indian J. of Microbiol. 15, 414–417 (1989)
[7] Cheng, Y.R., Fang, A., Demain, A.L.: Effect of amino acids on rapamycin biosynthesis by Streptomyces hygroscopicus. Appl. Microbiol. Biotechnol. 43, 1096–1098 (1995)
[8] Clarke, K.R.: Nonparametric multivariate analyses of changes in community structure. Aus. J. Ecol. 18, 117–143 (1993)
[9] Cornell, J.A.: Experiments with Mixtures: Designs, Models, and the Analysis of Mixture Data. John Wiley & Sons, Inc., New York (2002)
[10] Elkamel, A., Abdul Wahab, S., Bouhamra, W., Alper, E.: Measurment and prediction of ozone levels around a heavily industrialized area: a neural network approach. Adv. Enviro. Res. 5, 47–59 (2001)
[11] Eriksson, L., Johansson, E., Wikström, C.: Mixture design design generation, PLS analysis, and model usage. Chemometr. Intell. Lab. 43, 1–24 (1998)
[12] Garson, G.D.: Interpreting neural-network connection wights. In: AI Expert, pp. 47–51 (1991)
[13] Gheshlaghi, R., Scharer, J.M., Moo Young, M., Douglas, P.L.: Medium optimization for hen egg white lysozyme production by recombinant Aspergilus niger using statistical methods. Biotechnol. Bioeng. 90, 754–760 (2005)
[14] Glorfeld, L.W.: A methodology for simplification and interpretation of backpropagation-based neural network models. Expert Syst. Appl. 10, 37–54 (1996)
[15] Goh, A.T.C.: Back-propagation neural networks for modeling complex systems. Artif. Intell. Eng. 9, 143–151 (1995)
[16] Gouveia, E.R., Baptista Neto, A., Badino Jr., A.C., Hokka, C.O.: Optimization of medium composition for clavulanic acid production by Streptomyces clavuligerus. Biotechnol. Lett. 23, 157–161 (2001)
[17] Hilborn, R., Stearns, S.C.: On inference in ecology and evolutionary biology: the problem of multiple causes. Acta Biotheror. 31, 145–164 (1982)
[18] Kennedy, M., Krouse, D.: Strategies for improving fermentation medium performance: a review. J. Ind. Microbiol. Biotech. 23, 456–475 (1999)
[19] Legendre, P., Anderson, M.J.: Distance-based multivariate analysis: Testing multispecies responses in multifactorial ecological experiments. Ecol. Monogr. 69, 1–24 (1999)
[20] Mantel, N., Valand, R.S.: A technique of nonparametric multivariate analysis. Biometrics 26, 547–558 (1970)
[21] McArdle, B.H., Anderson, M.J.: Fitting multivariate models to community data: A comment on distance-based redundancy analysis. Ecology 82, 290–297 (2001)
[22] Monot, F., Martin, J.R., Petitdemange, H., Gay, R.: Aceton and butanol production by clostridium acetobutylicum in synthetic medium. Appl. Environ. Microbiol. 44, 1318–1324 (1982)
[23] Montgomery, D.C., Runger, G.C.: Applied statistics and probability for engineers. John Wiley & Sons, Inc, New York (2006)
[24] Myers, R.H., Montgomery, D.C.: Response surface methodology: process and product optimization using designed experiments. John Wiley & Sons, Inc, New York (2003)

[25] Nagata, Y., Chu, K.H.: Optimization of a fermentation medium using neural networks and genetic algorithms. Biotechnol. Lett. 25, 1837–1842 (2003)

[26] Ooijkaas, L.P., Wilkinson, E.C., Tramper, J., Buitelaar, R.M.: Medium optimization for spore production of coniothyrium minitans using statistically-based experimental designs. Biotechnol. Bioeng. 64, 92–100 (1999)

[27] Techapun, C., Sinsuwongwat, S., Watanabe, M., Sasaki, K., Poosaran, N.: Production of cellulase-free xylanase by a thermotolerant streptomyces sp. grown on agricultural waste and media optimization using mixture design and Plackett & Burman experimental design methods. Biotechnol. Lett. 24, 1437–1442 (2002)

[28] Yee, L., Blanch, H.W.: Defined medium optimization for growth of recombinant Escherichia coli X90. Biotechnol. Bioeng. 41, 221–230 (1992)

[29] Yun, S., Yahya, A.R.M., Cossar, D., Anderson, W.A., Scharer, J.M., Moo Young, M.: Temperature downshift increases recombinant cytokine titer in Streptomyces lividans fermentation. Biotechnol. Lett. 23, 1903–1905 (2001)

[30] Zhang, J., Greasham, R.: Chemically defined media for commercial fermentations. Appl. Microb. Biotechnol. 51, 407–421 (1999)

9

Bioprocess Modelling for Learning Model Predictive Control (L-MPC)

María Antonieta Alvarez[1,*], Stuart M. Stocks[2], and S. Bay Jørgensen[1,**]

[1] CAPEC, Department of Chemical and Biochemical Engineering, Technical University of Denmark,
Søltofts Plads, 2800 Kgs. Lyngby, Denmark
sbj@kt.dtu.dk
[2] Novozymes, Bagsværd, Denmark

Abstract. Batch and Fed-Batch cultivation processes are used extensively in many industries where a major issue today is to reduce the production losses due to sensitivity to disturbances occurring between batches and within batches. In order to ensure consistent product quality by eliminating the influence of process disturbances it is very important to consider implementation of monitoring and control and thereby significantly improve the economic impact for these industries. A data driven modeling methodology is described for batch and fed batch processes which is based upon data obtained from operating processes. The chapter illustrates how additional production experiments may be designed to improve model quality for control. The chapter also describes how the developed models may be used for process monitoring, for ensuring process reproducibility through control and for optimizing process performance by enforcing learning from previous batch runs through Learning Model Predictive Control (L-MPC).

1 Introduction

Fermentation processes are spreading into traditional chemical industries as more environmentally friendly production methods gain importance. Most often fermentations are carried out as batch or fed-batch cultivations which have been used extensively in bio-technical and pharmaceutical industries. However microbial cultivations exhibit significant sensitivity to disturbances occurring between batches and within batches which as evidenced by a relatively large batch to batch variability often experienced in production. Therefore there is significant interest in development of methods which may ensure more consistent product quality by eliminating or counteracting the influence of process disturbances. In order to ensure more consistent product quality it is very important to consider implementation of monitoring and control, and thereby significantly improve the economic impact for these industries.

* Present address: Facultad de Ingeniera en Electricidad y Computación. Escuela Superior Politécnica del Litoral. Guayaquil. Ecuador.

** Corresponding author.

M. do Carmo Nicoletti, L.C. Jain (Eds.): Comp. Intel. Tech. for Bio. Mod., SCI 218, pp. 237–280.
springerlink.com

Given the main goal of reducing batch to batch variability the question then is which possibilities there are or may be developed for satisfying this purpose. When viewed from the properties of microorganisms then these cultivations are influenced by many variables which render the control problem multivariate from the outset. Furthermore a control design which enables tracking the desired batch trajectory within some margin is highly desirable in order to avoid that the cultivation triggers a one of the many stress responses possible and thereby deviate significantly from the desired behavior. Hence a multivariate control design which enables control actions relatively often throughout the fed-batch production phase would be desirable. However multivariate control calls inevitably for a model based control design. Thus the question is how to develop a model which can be suitable for such a control design?. Given the present knowledge of microorganisms and especially the limited understanding of the behavior of their regulatory network precludes a first principles modeling approach. Hence a data driven approach seems to be the only viable alternative. Therefore such an approach is investigated in this chapter.

The batch processing types covered in this chapter includes Batch, Fed-batch and periodic operation which all have the common traits of a repeated operation which start from nearly the same initial conditions. Thus the time within the batch and the batch number are the two characteristic independent variables. Batch processing is subject to variations in raw material properties, in start-up initialization and other disturbances during execution. These different disturbances introduce often significant variations in the final product quality. Compensating for these disturbances have been difficult in the past due to the nonlinear and time-varying behavior of batch processing and to the fact that reliable on- or in-line sensors for monitoring final product quality rarely are available. Consequently development of a systematic methodology which can ensure reliable reproducible operation may provide significant benefits for batch processing.

Each batch operation may be defined as a series of operational tasks, i.e. mixing, reaction and separation. Within each task a set of subtasks, e.g. heating/cooling, (dis-)charging is handled. There may be more than one feasible set of operational tasks that can produce the specified product(-s). Consequently an optimal sequence of tasks and subtasks with respect to a defined objective needs to be identified. This set of operational tasks is labeled the *optimal batch operations model.* Thus the Batch Operations Model combines the batch processing tasks normally specified in a generic recipe with the batch equipment under availability and other resource constraints.

The question of which methodology to select is based modeling using general empirical models combined with the possibility for executing control throughout the fed-batch phase. A method based on prediction of end of batch properties [4] considers midterm adjustment, which may be too late for many industrial cultivations. Therefore experimental adaptation (optimization) of the Batch Operations Model, e.g. the method based on "the solution model" through tracking constraints such as the Necessary Conditions of Optimality (NCO) [10,11] or determination of the presently (most) constrained variable [12] may be feasible. However for the present study the Grid of Linear Model (GoLM) method was selected since that method actually is quite general in its approach.

This chapter presents methodologies based upon the previous ideas for batch or periodic operation to provide a predictive modeling which can enable reliable reproducible operation using control and which may enable optimizing operation. The contribution comprises data driven time series modeling of batch processes based upon a large set of locally linear models and a learning model predictive control methodology. The modeling methodology produces both a Linear Time-Invariant (LTI) state space model representation for inter-batch prediction and a Linear Time-Varying (LTV) state space model representation for intra-batch prediction. The modeling approximates the non-stationary and nonlinear behavior of batch processes with a set of local but interdependent linear regression models parameterized as AutoRegressive Moving Average models with eXogenous (ARMAX) inputs. Tikhonov Regularization (TR) is applied to reliably estimate the many parameters of this model set. However since the model is intended for control it is important to ensure that the model is suitable for this purpose. Therefore design of identification experiments with input perturbations is proposed in this chapter and investigated both in simulation and on an industrial cultivation. The resulting model may be applied for Learning Model Predictive Control (L-MPC). This control methodology is presented for control of repeated operation of stochastic Linear Time-Varying systems with finite time horizons together with tuning requirements for ensuring guaranteed convergence and hence closed-loop stability. The methodologies have been implemented as a Matlab toolbox named Grid of Linear Model (GoLM). The application of these methods is illustrated through both the simulation study and the experimental pilot plant investigation.

The chapter is structured such that the methods are briefly introduced and subsequently they are illustrated on a simulation benchmark example. Here the experimental design, the model estimation and the L-MPC design and performance. Subsequently an industrial cultivation is modeled with the aim to implement an L-MPC control design. Here both the model estimation and implementation of a Kalman filter for estimating unmeasured variables, especially enzyme activity are illustrated. The implementation of a L-MPC control design is briefly discussed, before the conclusions are drawn.

2 Methods

This section briefly presents the different methods used in this chapter. For modeling estimation of a fermentation process, the methodology applied is Grid of Linear Model (GoLM), as explained in subsection 2.1. The Learning Model Predictive Control (L-MPC) methodology used for the batch fermentation process is presented in subsection 2.2. To develop GoLM models from a data set consisting of a relatively low number of batches it is desirable to design experiments with perturbations in the actuator variables. The experimental design method applied is presented in subsection 2.3.

2.1 Modeling for Control

Batch processes are modeled with the toolbox GoLM as a sets of N LTI models. Such a set of LTI models could also be referred to as one LTV batch model. These LTI models can be parameterized in a number of ways, but in the present contribution an ARMAX parameterizations was chosen. This choice of parametrization offers a simple multivariable system description with a moderate number of model parameters. The objective of the model set is to quantify the causal correlations between the process outputs $y_{k,i} \in \mathbb{R}^{n_y}$, inputs $u_{k,i-1} \in \mathbb{R}^{n_u}$, and disturbances $v_i \in \mathbb{R}^{n_y}$, for $i = 1, \ldots, t$, at times $t = 1, \ldots, N$ in batch k. To simplify notation, define the input u_k, output y_k, shifted output y_k^0, and disturbance v_k profiles in batch k as

$$\begin{aligned} u_k &= \left[u'_{k,0}\ u'_{k,1} \cdots u'_{k,N-1}\right]' \\ y_k &= \left[y'_{k,1}\ y'_{k,2} \cdots y'_{k,N}\right]' \\ y_k^0 &= \left[y'_{k,0}\ y'_{k,1} \cdots y'_{k,N-1}\right]' \\ v_k &= \left[v'_{k,1}\ v'_{k,2} \cdots v'_{k,N}\right]' \end{aligned} \tag{1}$$

The GoLM toolbox models the differences between two successive batches with an ARMAX model

$$\Delta y_k = y_k - y_{k-1} = -A\Delta y_k^0 + B\Delta u_k - Cv_k \tag{2}$$

where $\Delta u_k = u_k - u_{k-1}$ and $\Delta y_k^0 = y_k^0 - y_{k-1}^0$. The batch ARMAX model (2) may be converted into different representations dependent on the particular application task. If the task at hand is to predict (or simulate) the behavior of a batch before it is started, the Eq. 3 is convenient

$$\Delta y_k = H\Delta y_{k,0} - G\Delta u_k + Fv_k \tag{3}$$

where $\Delta y_{k,0} = y_{k,0} - y_{k-1,0}$. The change in the initial conditions $\Delta y_{k,0}$ can be considered as either an input/control variable or a disturbance. Eq. (3) is also convenient for the task of classification (e.g. normal or not) of a batch after it has been completed. Furthermore, Eq. (3) can be used to determine open-loop optimal recipes in the sense of optimizing an objective for the batch. If the objective is to minimize the deviations $e_k = \left[e'_{k,1}, e'_{k,2}, \ldots, e'_{k,N}\right]'$, $e_{k,t} \in \mathbb{R}^{n_y}$, from a desired Batch Operations Model $\bar{y}$, then Eq. (3) can be modified into

$$e_k = \bar{y} - y_k = e_{k-1} - H\Delta y_{k,0} + G\Delta u_k - Fv_k \tag{4}$$

The two Eqs. (3) and (4) of the batch ARMAX model above are applicable to off-line or inter-batch type applications. For on-line estimation, monitoring, feedback control, and optimization however, it is convenient to use a state space realization of the batch ARMAX model.

In an observer canonical form, which is structurally a minimal realization, the state space realization is given as

$$\begin{aligned} x_{k,t} &= A_t x_{k,t-1} + B_t \Delta u_{k,t-1} + E_t v_{k,t} \\ \Delta y_{k,t} &= y_{k,t} - y_{k-1,t} = Cx_{k,t} \end{aligned} \tag{5}$$

with $\Delta u_{k,t-1} = u_{k,t-1} - u_{k-1,t-1}$ and the state space model dimension $n_x = n_y \max(n_i(t) \mid 1 \leq t \leq N, i = A, B, C)$ and the initial condition $x_{k,0} = C'\Delta y_{k,0}$. The state space model matrices $A_t \in \mathbb{R}^{n_x \times n_x}$, $B_t \in \mathbb{R}^{n_x \times n_u(t)}$, $E_t \in \mathbb{R}^{n_x \times n_y}$, and $C \in \mathbb{R}^{n_y \times n_x}$ contain the corresponding block columns in the batch ARMAX model (A, B, C). To exemplify, assume that $n_A(t) = n_B(t) = n_C(t) + 1 = 3$ and the state space model matrices will be given as

$$A_t = \begin{bmatrix} -a_{t,t-1} & I & 0 \\ -a_{t+1,t-1} & 0 & I \\ -a_{t+2,t-1} & 0 & 0 \end{bmatrix}, \quad E_t = \begin{bmatrix} -I \\ -c_{t+1,t} \\ -c_{t+2,t} \end{bmatrix}$$

$$B_t = \begin{bmatrix} b_{t,t-1} \\ b_{t+1,t-1} \\ b_{t+2,t-1} \end{bmatrix}, \quad C = \begin{bmatrix} I & 0 & 0 \end{bmatrix}$$

Just as Eq. (3), the state space model form (5) is convenient for prediction, monitoring, and optimization type applications, and it facilitates on-line implementation of such applications. Furthermore, the state space model form (5) is particularly well suited for closed-loop or feedback control applications. For tracking control applications the state space model form (5) can be modified to

$$\begin{aligned} x_{k,t} &= A_t x_{k,t-1} + B_t \Delta u_{k,t-1} + E_t v_{k,t} \\ e_{k,t} &= \bar{y}_t - y_{k,t} = e_{k-1,t} - C x_{k,t} \end{aligned} \tag{6}$$

In order to use Eq. (6) for tracking control, it is necessary to estimate the states $(\hat{x})$ based on noisy observations of the outputs. Assume that during a batch (k), observations $z_{k,t}$ of the outputs $y_{k,t}$ are collected at times $t = 0, 1, \ldots, N$ and let the optimal estimate of the state x_{k,t_1} in batch k at time t_1 given data up to and including time t_2 be given as the conditional mean (e.g. obtainable with a Kalman Filter) $\hat{x}_{k,t_1|t_2} = E\{x_{k,t_1} \mid I_{k,t_2}\}$ where the information $I_{k,t}$:

$$I_{k,t} = \{z_{k,t}, \Delta u_{k,t}, I_{k,t-1}\}, \quad I_{k,-1} = I_{k-1,N}, \quad I_{0,-1} = \{y_{-1}, u_{-1}, z_{-1}\} \tag{7}$$

Then the tracking error e_{k,t_1} in batch k at time t_1 given data up to and including time t_2 is estimated as $\hat{e}_{k,t_1|t_2} = \hat{e}_{k-1,t_1|N} - C\hat{x}_{k,t_1|t_2}$ where the smoothed estimate of the error profile in batch $k-1$ is given as

$$\hat{e}_{k-1|k-1} = \begin{bmatrix} \hat{e}'_{k-1,1|N} & \hat{e}'_{k-1,2|N} & \cdots & \hat{e}'_{k-1,N|N} \end{bmatrix}' = \bar{y} - \hat{y}_{k-1|k-1} \tag{8}$$

and $\hat{y}_{k-1|k-1}$ is the smoothed output profile estimate from batch $k-1$ (e.g. obtainable with a Kernel Smoother). Note that the superscript signifies an estimate.

2.2 Learning – Model Predictive Control

Given the previous developed models for batch operation then it is possible to develop a control paradigm which handles both intra batch and inter batch disturbances. This is achieved by combining Model Predictive Control, which handle intra batch disturbances, with a Learning Control, which handle batch to batch disturbances. To utilize available information during a batch to obtain the best possible tracking performance, the following Learning Model Predictive Control formulation

$$\begin{aligned} \{\Delta u_{k,l,t}\}_{l=t}^{N-1} = \arg\min_{\{\Delta u_{k,i}\}_{i=t}^{N-1}} & \left[\sum_{i=t+1}^{N} \hat{e}'_{k,i|t} Q_i \hat{e}_{k,i|t} + \Delta u'_{k,i-1} R_i \Delta u_{k,i-1}\right] \\ s.t. \quad & \hat{x}_{k,i|t} = A_i \hat{x}_{k,i-1|t} + B_i \Delta u_{k,i-1} \\ & \hat{e}_{k,i|t} = \hat{e}_{k-1,i|N} - C\hat{x}_{k,i|t} \\ & u_{\min i-1} \leq \Delta u_{k,i-1} + u_{k-1,i-1} \leq u_{\max i-1} \\ & y_{\min i} \leq \bar{y}_i - \hat{e}_{k,i|t} \leq y_{\max i} \end{aligned} \tag{9}$$

is solved at times $t = 0, 1, \ldots, N-1$ in batch k and the control sequence

$$\Delta u_k^\star = \left[\Delta u'_{k,0,0}\ \Delta u'_{k,1,1} \ldots \Delta u'_{k,N-1,N-1}\right]' \tag{10}$$

then approximates a closed-loop optimal control sequence. I.e., at time t in batch k, Eq. (9) is solved based on updated state estimates, and the input $u_{k,t} = \Delta u_{k,t} + u_{k-1,t}$ is implemented on the process.

2.3 Experimental Design for Control

In order to efficiently reject disturbances in bioprocesses it is important to be able to act as soon as deviations from a desired performance is detected, otherwise the bioprocess may develop into operating regions from which it is not possible to recover. One such situation may occur if the microorganisms experience a lack of oxygen, which may lead to excessive cell death. Therefore it is essential to be able to apply control throughout the batch operation to enable tracking of the desired operations model. However this desire implies that the data driven modeling must be based upon data where the actuator variables have been moved during the process to provide sufficient information for model parameter estimation. Thus experimental design for identification becomes very important for providing such informative data.

The subject of experimental design is well known and studied for continuous processes, e.g. [7], however, for batch processes additional aspects come into play, just as for model estimation as treated briefly above. These aspects were investigated by [8]. The key point is that by introducing perturbations around the nominal Batch Operations Model it is possible to obtain informative data for identification of batch models. These include a.o.:

1. The batch operation should be perturbed around the nominal batch operations model for the potential control actuator variables.
2. The magnitude of the perturbations should ensure a balance between proximity to the nominal batch and sufficient perturbation to obtain informative data, as judged, e.g. by the Fisher information matrix.
3. For noise-corrupted measurement data, the noise content can be reduced by smoothing, e.g. Kernel Smoothing.

The detailed design of the perturbations is treated in connection with the benchmark case and the industrial pilot plant case, described in Section 3 and Section 4, respectively.

3 Benchmark Example

The GoLM methodology is used to develop a discrete-time state space model representing a batch process based on historical operating data. In subsection 3.1, a simple Benchmark case is used to represent a fermentation process (Penicillin Fermentation). From the simulation of the benchmark model a set of batches is obtained, with this data set a state space model is estimated using the GoLM methodology (subsection 3.2). In subsection 3.3 the L-MPC methodology is explained for control design. Based upon the obtained model a Learning Model Predictive Control (L-MPC) controller is designed (subsection 3.4) and demonstrated through simulation under two scenarios: one with inter batch disturbances and another with intra batch disturbances. This simple benchmark case is used to illustrate the methodology behind GoLM and to investigate model development for prediction of process behavior using a data set with a limited number of batches for modeling. In subsection 3.5 the conclusions of the benchmark case are given.

3.1 Benchmark Simulation Model

The example fed batch process is simulated penicillin fermentation. The model is a modified version of the model in [2] which describes the growth of biomass and formation of the single product from a pure substrate simulated with the parameters from [5], these values are shown in Table 1. The model equations are:

$$\frac{dX}{dt} = \mu X - \frac{FX}{V} \tag{11}$$

$$\frac{dS}{dt} = -\frac{\mu X}{Y_X} - \frac{\theta X}{Y_p} - M_X X + \frac{F(S_f - S)}{V} \tag{12}$$

$$\frac{dP}{dt} = \theta X - KP - \frac{FP}{V} \tag{13}$$

$$\frac{dV}{dt} = F \tag{14}$$

for $t \in [t_0, t_f]$, where X (g/L) is the biomass concentration, S (g/L) is the substrate concentration, P (g/L) is the product concentration, V (L) is the reactor volume, F (L/h) is the feed flow rate, Y_X and Y_P are yield coefficients and S_f (g/L) is the substrate feed concentration. M_X represents a constant specific maintenance demand of the cells and K represents a constant first-order decay rate for the product. For the kinetic equations, specific growth rate μ uses

Table 1. Simulation model parameters

Y_X	0.47	K_1	0.006
Y_P	1.2	K_{21}	0.0001 $\frac{g}{l}$
S_f	400 $\frac{g}{l}$	K_{22}	10 $\frac{l}{g}$
M_X	0.029 h^-1	X_0	1 $\frac{g}{l}$
K	0.01 h^-1	S_0	0.5 $\frac{g}{l}$
μ_{max}	0.11 h^-1	P_0	0 $\frac{g}{l}$
θ_{max}	0.004 h^-1	V_0	250 l

Michaelis Menten kinetics while the product formation rate θ were modeled with Contois kinetics [1]:

$$\mu = \mu_{max} \frac{S}{S + K_1 X} \tag{15}$$

$$\theta = \theta_{max} \frac{S}{K_{22} S^2 + S + K_{21}} \tag{16}$$

This simple model (physical model Eqs. 11 - 14) provides the example process used to generate a data set for model identification using the GoLM methodology (black box model). After the black box model is development, the same physical model is used to simulated the "real process" under control.

The penicillin simulation case represents a cultivation with two input variables (substrate feed concentration S_f and feed flow rate F) and four measured output variables (biomass X, substrate S, product P (penicillin) and reactor volume V). The simulation generates a data set with 7 batches where every batch runs for 150 hours with measurements sampled once per hour (150 samples). The number of batches was selected as a realistic case to test the estimation capability of the GoLM methodology to a realistic case, where perturbations are used to ensure the information content in data. The Penicillin simulation starts with a batch phase in which substrate, nutrients and inoculum have been loaded into the sterilized reactor where only oxygen is supplied during cultivation. Initially in the batch phase the nominal value of substrate concentration is high (80 g/L) while the biomass formation rate reaches its highest value after an almost exponential growth. The initial substrate composition and amount varies with a standard deviation of 20% between the seven batches. When the substrate concentration falls below 5 g/L the fed-batch phase is set to start, the feed flow rate is 2 L/h and the substrate feed concentration is 400 g/L, the initial reactor volume is 250 L.

Independent Pseudo Random Binary Signal (PRBS) sequences are used with a clock period of 3 hours to perturb each input variable. The perturbation amplitude for the feed flow rate is 1 L/h, whereas, the perturbation amplitude for the substrate concentration in the feed is 200 g/L. The amplitude of both signals is chosen sufficiently large to generate measurable process variations which are sufficient to obtain reliable information on the process behavior but not so large that the process behavior drifts away from its normal behavior. Figure 1 shows the input sequences to the process with perturbation for the seven batches simulated.

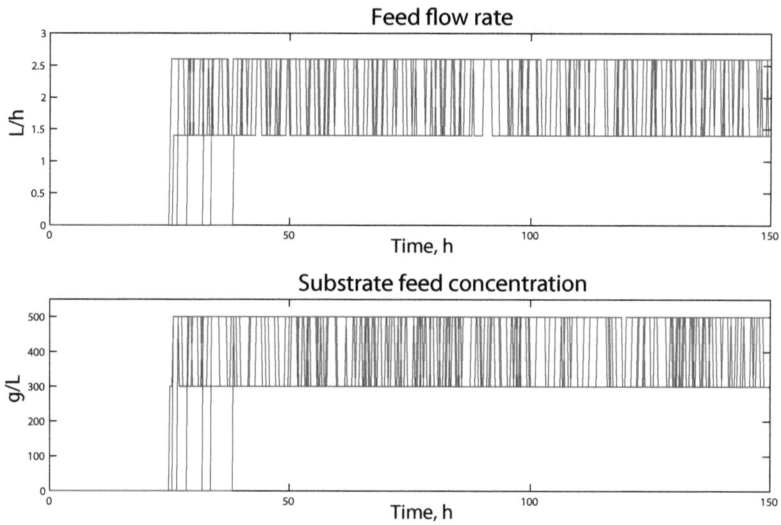

Fig. 1. Feed flow rate and Substrate concentration signals with disturbance. Top: Feed flow rate signal vs time; bottom: Substrate concentration in the feed flow vs time.

The outputs of the process for theses inputs are depicted in Figure 2. It is clearly possible to see the effect which the perturbations generate in the process variables.

To investigate the influence of the perturbation amplitude for model estimation, three data sets again each with seven batches have been simulated with different perturbation amplitudes. The perturbations that characterize each dataset are described in Table 2:

Table 2. Characterization of the perturbations for each dataset in the seven simulated benchmark experiments

Data	Amplitude perturbed		
	% of the Signal	Feed flow rate (L/h)	Substrate feed concentration (g/L)
a	± 10	0.4	80
b	± 25	1	200
c	± 37.5	1.5	300

3.2 Benchmark Model Estimation and Validation

The following sections are based on [3]. A model is estimated using a novel interpretation of Tikhonov Regularization, the estimated model is validated on independently perturbed simulations. Due to the repetitive nature and the finite horizon, the measurements are collected from a grid of sample points in time. With these samples points it is possible to model the evolution between two consecutive sample points in a batch with local LTI models, which are labeled *grid-point* models (Figure 3). Estimation of the parameters in the grid point models

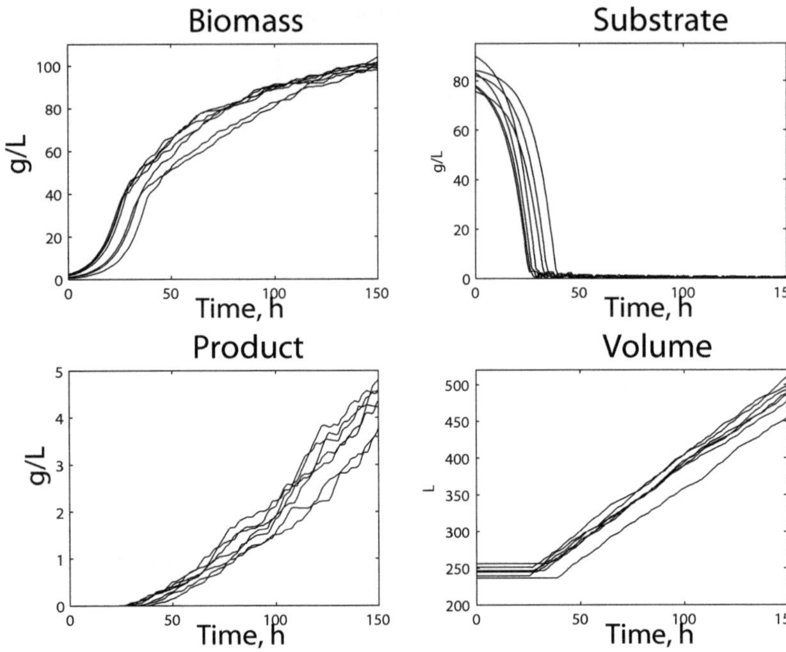

Fig. 2. Results of the simulation. Top left: Biomass conc. vs time; top right: substrate conc. vs time; bottom lef: product conc. vs time; bottom right: volume vs time.

is possible because at each sample time the measurements are repeated as many times as there are batches. Clearly however the Linear Time-Invariant (LTI) model which represent the behavior from batch to batch is rather large since the number of parameters is proportional to the product of the number of measurement intervals multiplied with the number measured process variables plus the number of input variables. Therefore it is essential to use a sound regularization method which ensures that only parameters which extract information from the measured data are actually estimated. This is achieved by using regularization. Note that when considering the behavior within a batch then the batch process may be represented as a Linear Time-Varying (LTV) batch model.

Input-Output Modeling: The model parametrization may be carried out in two ways, using either AutoRegressive models with eXogenous (ARX) inputs or AutoRegressive Moving Average models with eXogenous (ARMAX) inputs. The software GoLM gives the option to use different parameter estimation methods, i.e. as Least Squares (LS), Ridge Regression (RR) and Tikhonov Regularization (TR). Here TR is preferred, since it provides a stable and efficient method for determining the model parameter estimates.

To illustrate the modeling concept the different formulations of the ARX and ARMAX models used for GoLM modeling is presented next.

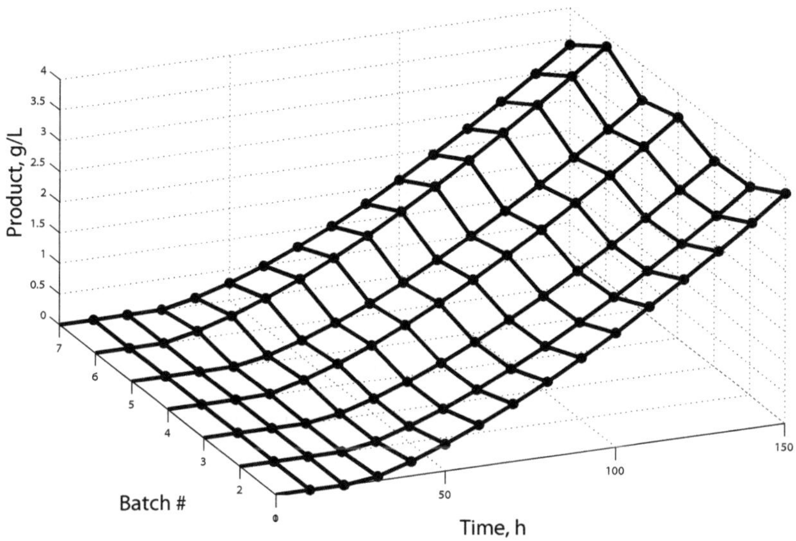

Fig. 3. The three dimensions of batch data for one particular variable, e.g. biomass concentration. The filled circles represent the sample points taken every 10 hours. The smooth connecting lines symbolize the development of the particular variable through the batch. The connecting lines between batches illustrate the development form batch to batch. Note the finite duration of the batch.

The inputs and outputs of the batch operation are defined as deviations from a reference trajectory for the batch. The reference trajectory is indicated with an overbar. Thus the model at each discrete time instant t is given as:

- Input variable $u_t \in \mathbb{R}^{n_u(t)}$ with reference $\bar{u}_t \in \mathbb{R}^{n_u(t)}$
- Output variable $y_t \in \mathbb{R}^{n_y(t)}$ with reference $\bar{y}_t \in \mathbb{R}^{n_y(t)}$
- Disturbance variable $w_t \in \mathbb{R}^{n_y(t)}$

Using an ARX model parametrization, the output deviation from the reference $\bar{y}_t - y_t$ at time t may be given in the following locally linear equation

$$\begin{aligned}\bar{y}_t - y_t = &-a_{t,t-1}(\bar{y}_{t-1} - y_{t-1}) - \ldots - a_{t,t-n_A(t)}(\bar{y}_{t-n_A(t)} - y_{t-n_A(t)}) \\ &+b_{t,t-1}(\bar{u}_{t-1} - u_{t-1}) + \ldots + b_{t,t-n_B(t)}(\bar{u}_{t-n_B(t)} - u_{t-n_B(t)}) \\ &+w_t\end{aligned} \tag{17}$$

where $n_A(t)$, $n_B(t) \in [1,\ldots,t]$ are the structured grid-point ARX model orders and $a_{i,j} \in \mathbb{R}^{n_y(i) \mathrm{x} n_y(j)}$ and $b_{i,j} \in \mathbb{R}R^{n_y(i) \mathrm{x} n_u(j)}$ are the structure grid point ARX model parameter matrices, w_t is the disturbance. The model type (ARX or ARMAX), depends on how the disturbance is modeled. For the entire batch the input-output model is expressed as

$$\bar{y} - y = -A(\bar{y}^0 - y_0) + B(\bar{u} - u) + w \tag{18}$$

where the input u, output y, shifted output y^0, and the disturbance profile w are

$$\begin{aligned} u &= [u_0' \quad u_1' \quad \dots \quad u_{N_1}']' \qquad & y &= [y_0' \quad y_1' \quad \dots \quad y_N']' \\ y^0 &= [y_0' \quad y_1' \quad \dots \quad y_{N_1}']' \qquad & w &= [w_0' \quad w_1' \quad \dots \quad w_N']' \end{aligned} \tag{19}$$

The process disturbances are caused by several influences which include bias in the reference input profile $\bar{u}$, the effect of process upsets, modeling errors from linear approximations and errors due to bias in transition times between sets of active constraints. The disturbance profile w contains contributions from both batch wise persistent disturbances and random disturbances with no batch wise correlation. Hence, the disturbance is modeled as a random walk model with respect to the batch index k

$$w_k = w_{k-1} + \Delta w_k \tag{20}$$

The difference between two successive batches is given by

$$\begin{aligned} \Delta y_k &= y_k - y_{k-1} \\ &= -A(y_k^0 - y_{k-1}^0) + B(u_k - u_{k-1}) - \Delta w_k \end{aligned} \tag{21}$$

the disturbance profile increment Δw_k is modeled with a Moving Average (MA) model with respect to time

$$\Delta w_{k,t} = v_{k,t} + c_{t,t-1} v_{k,t-t} + \cdots + c_{t,t-n_C(t)} v_{k,t-n_C(t)} \tag{22}$$

where the model order $n_C(t) \in [0, \dots, t-1]$. In matrix form the disturbance model is:

$$\Delta w_k = C v_k \tag{23}$$

Then the following ARMAX model can describe the difference profile between two successive batches

$$\Delta y_k = -A(y_k^0 - y_{k-1}^0) + B(u_k - u_{k-1}) - C v_k \tag{24}$$

A key difference between ARMAX and ARX models is that an ARMAX model includes the model of the disturbance. Depending on the particular application purpose the ARMAX model (Eq. 24) can be converted into different representations. The following Eq. (25) is convenient if the task is to simulate the behavior of a batch or for classification (e.g. normal or not) of a batch after it has been completed, an other task can be to determine open-loop optimal recipes in the sense of optimizing an objective for the batch.

$$\Delta y_k = H \Delta y_{k,0} - G \Delta u_k + F v_k \tag{25}$$

The form of the batch ARMAX model (Eq. 25) is used for off-line or inter-batch type applications, while for on-line application such as on-line estimation, monitoring, feedback control and optimization; it is convenient to use a Space State (SS) realization of the batch ARMAX model:

$$\begin{aligned} x_{k,t} &= A_t x_{k,t-1} + B_t \Delta u_{k,t-1} + E_t v_{k,t} \\ \Delta y_{k,t} &= C x_{k,t} \end{aligned} \tag{26}$$

with the initial condition

$$x_{k,0} = C' \Delta y_{k,0} \tag{27}$$

This Space State model is precisely the form given in Eq. 5 after which the matrices are defined. Like Eq. 25, the Space State model form (Eq. 26) is convenient for prediction, monitoring and optimization type applications. In addition the Space State model form (Eq. 26) is well suited for close-loop control applications.

Parameter Estimation: Tikhonov Regularization (TR) provides an efficient method to obtain a trade off between bias and variance of the model parameter estimates. It is a coefficient shrinkage based parameter estimation method that can incorporate model properties into weighted Least Squares (wLS) estimates. The formulation for the TR estimation problem used in GoLM is

$$\begin{aligned} \hat{\theta}_{TR}(\hat{V}, W, \Lambda) &= \arg\min_\theta \left[\mathbf{J}_{wLS}^{TR}\right] \\ s.t. \quad \mathbf{J}_{wLS}^{TR} &= \|Y - X\theta\|_W^2 + \|\mathbf{L}\theta\|_{\Lambda^2}^2 \end{aligned} \tag{28}$$

where $\mathbf{L}$ is a structured penalty matrix which maps the parameter vector θ into the desired parameter differences, $\mathbf{\Lambda}$ is a diagonal weighting matrix that weights the parameter differences. The penalty matrix $\mathbf{L}$ consists of five sub-matrices

$$\mathbf{L} = [\mathbf{L}_1' \quad \mathbf{L}_2' \quad \mathbf{L}_3' \quad \mathbf{L}_4' \quad \mathbf{L}_5'] \tag{29}$$

the five sub-matrices are weighted individually by block diagonal weighting matrices

$$\mathbf{\Lambda} = \begin{pmatrix} \mathbf{\Lambda}_1 & & \\ & \ddots & \\ & & \mathbf{\Lambda}_5 \end{pmatrix} \tag{30}$$

Each of the five sub-matrices $\mathbf{L_i}$ for $i = 1, ..., 5$ imposes a penalty on certain properties of the parameters as briefly described in the following.

$\mathbf{L_1}$: Penalizes the model parameter time evolution by penalizing the approximate first order time derivative of the parameters θ. This means that the quantities (using $\mathbf{A}$ matrix as an example)

$$\left(\hat{a}_{t,t-l}(i,j) - \hat{a}_{t+1,t-l+1}(i,j)\right)^2 \tag{31}$$

for $i, j = 1, ..., n_y(t), t = 1, ..., N-1$ and $l = 1, ..., n_A(i,j,t)$ are penalized according to the corresponding scalar weight $\lambda_{1,i,j}^a$.

$\mathbf{L_2}$: Penalizes non-smoothness of the model parameter time evolution by penalizing the approximate second order time derivative of the parameters θ. This means that the quantities (using $\mathbf{A}$ matrix as an example)

$$\left(\hat{a}_{t-1,t-l-1}(i,j) - 2\hat{a}_{t,t-l}(i,j) + \hat{a}_{t+1,t-l+1}(i,j)\right)^2 \tag{32}$$

for $i, j = 1, ..., n_y(t), t = 2, ..., N-1$ and $l = 1, ..., n_A(i,j,t) - 1$ are penalized according to the corresponding scalar weight $\lambda^a_{2,i,j}$.

L₃: Penalizes the time evolution of the impulse responses by penalizing the approximate first order time derivative of the impulse response of the local models θ_t. This means that the quantities (using **B** matrix as an example)

$$\left(\hat{b}_{t+l-1,t-1}(i,j) - \hat{b}_{t+l,t-1}(i,j)\right)^2 \tag{33}$$

for $i = 1, ..., n_y(t), j = 1, ..., n_u(t), t = 1, ..., N-1$ and $l \in \{\, l \mid n_B(i,j,t+l) > l, l \in [1, ..., N-t]\}$ are penalized according to the corresponding scalar weight $\lambda^b_{3,i,j}$.

L₄: Penalizes the non-smoothness of the impulse responses by penalizing the approximate second order time derivative of the impulse response of the local models θ_t. This means that the quantities (using **B** matrix as an example)

$$\left(\hat{b}_{t+l-1,t-1}(i,j) - 2\hat{b}_{t+l,t-1}(i,j) + \hat{b}_{t+l+1,t-1}(i,j)\right)^2 \tag{34}$$

for $i = 1, ..., n_y(t), j = 1, ..., n_u(t), t = 1, ..., N-2$ and $l \in \{\, l \mid n_B(i,j,t+l) > l, l \in [1, ..., N-t]\}$ are penalized according to the corresponding scalar weight $\lambda^b_{4,i,j}$.

L₅: Penalizes the variance of the model parameter estimates $\hat{\theta}$ ($\mathbf{L_5} = \mathbf{I}$). This means that the quantities (using **C** matrix as an example)

$$\hat{c}_{t,t-l}(i,j)^2 \tag{35}$$

for $i, j = 1, ..., n_y(t), t = 2, ..., N$ and $l = 1, ..., n_C(i,j,t)$ are penalized according to the corresponding scalar weight $\lambda^c_{5,i,j}$. It is noted that if $\mathbf{L} = \mathbf{L_5}$, the TR problem (Eq. 28) is reduced to a standard form TR problem or a RR problem.

As for other regularization methods such as Ridge Regression, this method introduces bias to the model parameter estimates. The trade off between variance and bias of the parameter estimates determines the predictive capabilities of the model. The weighting matrix $\mathbf{\Lambda}$ determines the coefficient shrinkage. Selection of which elements to include and to what extend constitutes important parameters for the design of an optimal model for a given purpose.

Model Selection: The quality of the model is evaluated by how well the model generalizes to an independent data set. The method used is cross-validation, which estimates the generalization error G when a model $\hat{\theta}$ is applied to an independent data set [3].

$$G(\hat{\theta}) = \frac{1}{N_B^{test}} \sum_{k=1}^{N_B^{test}} \left\| \Delta\hat{y}_k - \Delta\hat{y}_k(\hat{\theta}) \right\|^2_{\bar{W}} \tag{36}$$

where $\Delta\hat{y}_k$ is the model prediction of the regressors in batch k of the independent validation data set and $\bar{W}$ is a block diagonal weighting matrix with symmetrical, positive definite, block elements $\bar{W}_t \in \mathbb{R}^{n_y(t),n_y(t)}$ and the weighting matrix $\bar{W}$ should be chosen as

$$\bar{W}_t(i,i) = ((N-1)n_y max\{|\Delta\hat{y}_{k,j}(i)|\, k=1,...,N_B; j=1,...,N\})^{-1} \quad (37)$$

for i=1,...,n_y and t=1,...,N. The size of the validation data set N_B^{val} is typically 50% of the size of the data set used for model parameter estimation N_B^{est}.

In practice the validation is performed through a combination of visual inspection of pure simulation, i.e. prediction of the entire batch. This is supported with comparison of model performance using a scalar measure of the FIT, or rather the misfit, which also is used by GoLM for quantitative analysis of the quality of the model on validation data. Thereafter the model to be used for the specific modelling purpose is selected.

Estimated Benchmark Model: As mentioned, there are three data sets (a, b and c - see Table 1)each containing seven batches to be modeled; the objective of the modeling is to determine the influence of the input perturbation amplitude on model identification when only a few batches are available. In GoLM the user selects the model type (ARMAX or ARX), the maximum model order and the estimation method (weighted Least Squares (wLS), Ridge Regression (RR) or Tikhonov Regularization (TR)).

To model each of the three data sets, GoLM splits up the data into an estimation set, a model validation set and a test set. Batches 1-3 have been used for identification, batches 4-5 have been used for validation and batches 6-7 have been used for model test. The model type used is ARX with a maximum model order of two and the estimation method used is TR. GoLM can estimate the model using Pure Simulation (PS) and One Step Ahead (OSA) prediction error profiles: PS estimates the outputs based on initial conditions at the start of the cultivation, this means that an early error will remain throughout the batch. Whereas in OSA estimation, the output estimated is based on using the measured output as initial condition. To provide a compromise between these two types of prediction, then another user parameter is introduced *Model Purpose* (ρ) which is a number between 0 and 1, where 0 estimates the model using One Step Ahead Prediction while 1 is for Pure Simulation Prediction. Thus the *Model Purpose* (ρ) weights between prediction of a one-step ahead error and a pure simulation - where the whole batch is predicted - error. The larger the ρ value the closer the estimate is to a pure simulation prediction. The default setting in GoLM Software is the golden cut $\rho = 0.618$.

The models estimated will be compared using a Pure Simulation Prediction generalization error (Eq. 36), as calculated by the GoLM Software. This is complemented with visual inspection of the model parameters and predictions. This FIT (or rather misfit) measure is between 0 and 1, where near 0 means that the model predicts the process trajectory very well, whereas a FIT near 1 implies a poor prediction of the process behavior. The parameters for the estimation setting in GoLM are shown in Table 3 for the three data sets (a, b and c - see Table 2).

Table 3. Parameters for GoLM estimation for the three data set (a, b and c).

Model Structure	Max. model order	Model Purpose	Regression Type
ARX	2	0.618	Tikhonov Regularization

Table 4. FIT for Pure Simulation Prediction for ARX models

Data set	FIT Biomass	FIT Substrate	FIT Product	FIT Volume	FIT Total
model a	0.7375	0.3951	0.9665	1.0834	0.8378
model b	0.1709	0.0315	0.2819	0.0458	0.1672
model c	0.1952	0.1411	0.1740	0.0478	0.15049

With the FIT of the three models (see Table 4), it is possible to conclude that when there are few batches available to be used for modeling, it is extremely important that there is sufficient information in the data for identification. Thus, it is essential to realize designed experiments for model identification. With perturbation in the inputs, the outputs contain information about the dynamics from these inputs and such data is sufficiently good to be used for modeling for control purposes.

The criteria used to choose the model is the FIT obtained by each model as listed in Table 4, and visual inspection of Figure 4. For the first data set (*model a*) (Figure 4(a)), the perturbed amplitude is 10% of nominal value of the signal. The model obtained does not simulate the real process well, hence the perturbed inputs do not generate sufficient information in the outputs. Whereas for data set *model b* (Figure 4(b)), the perturbed amplitude is 25% of the nominal value and with this perturbation in the inputs, it is possible to obtain a good model of the process. The data set *model c* -with a perturbed amplitude of 37% of the nominal value- in general, only provides a little improvement in the total FIT. But Figure 4(c) shows that the model estimated for substrate develops negative values during the batch phase. Thus, perturbation amplitude for identification is 25% of the nominal value (data set *b*) provides the most satisfactory data for model identification of this benchmark case.

The graphics that presents GoLM Software for the estimation results show the comparison between measurement, Pure Simulation Prediction, One Step Ahead Prediction and the previous batch. Note that a naive prediction represents the previous batch. PS Prediction represents the most demanding application of a batch process model. Hence, if a model quality based on PS model prediction is reasonable in terms of the estimated generalization error, then the model will have credibility in any similar application. There is a tendency that model selection based on OSA Prediction under-smoothes, while PS Prediction error has a tendency to over-smoothing. As Figure 4 show, PS for *model b* provides clearly the most reasonable prediction. Obviously OSA provides a seemingly better FIT, however for such a model the prediction horizon is basically limited to just one sample.

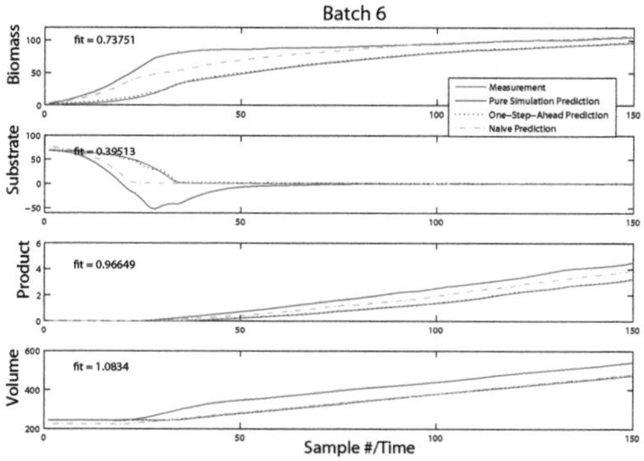

(a) *model a*

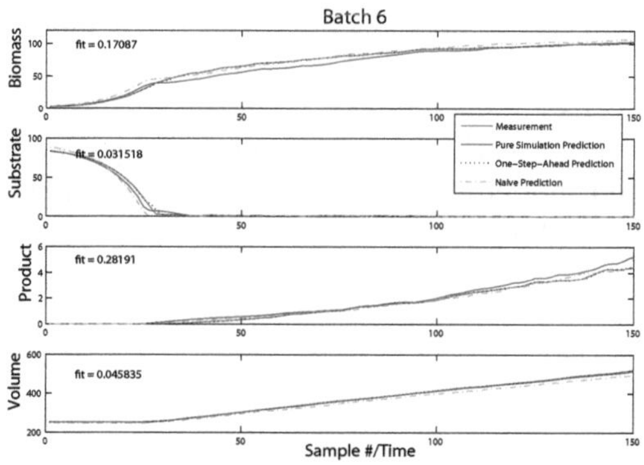

(b) *model b*

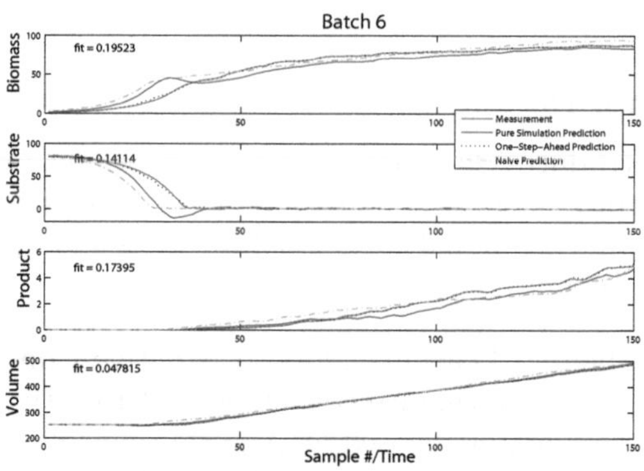

(c) *model c*

Fig. 4. (a): Simulation of batch 6 using a model obtained from the model estimation using the data set *a*; (b): Simulation of batch 6 using a model obtained from the model estimation using the data set *b*; (c): Simulation of batch 6 using a model obtained from the model estimation using the data set *c*

For the previous model estimation (model estimated *a, b and c*). The analysis was done keeping the $\rho = 0.618$ (see Table 3) and varying the input perturbation amplitude for the three data sets (*a, b and c* - see Table 2). The model obtained with data set *b* (*model b*) is attempted further improved using the decreasing weight of Pure Simulation as shown in Table 5.

Table 5. Parameters for GoLM model estimation for data set *b* for estimating three models with different values of Model Purpose

Model name	Model Structure	Max. Model order	Model Purpose	Regression Type
model 1	ARX	2	0.854	Tikhonov Regularization
model 2	ARX	2	0.618	Tikhonov Regularization
model 3	ARX	2	0.382	Tikhonov Regularization

Figure 5 visualizes the obtained FIT for the three models (*model 1, 2 and 3*), while Table 6 presents the FITs for each variable in each model. Clearly *model 1* in Figure 5 (a) shows a significant improvement of the model FIT. This investigation illustrates that the Model Purpose (ρ) parameter can be used to improve the estimated model according to the model application which here is to predict the entire batch duration.

Table 6. FIT for Pure Simulation Prediction of batches 6-7 for ARX model using data set *b* with different value of ρ

Model name	FIT Biomass	FIT Substrate	FIT Product	FIT Volume	FIT Total
model 1	0.072	0.0515	0.0353	0.0217	0.0489
model 2	0.1709	0.0315	0.2819	0.0458	0.1672
model 3	0.1065	0.2002	0.8725	0.0833	0.4526

3.3 Batch Control Design

This subsection introduces the use of Learning Model Predictive Control (L-MPC) in batch processes, first (subsection 3.3) introduces of L-MPC applying GoLM modeling. Then, subsection 3.4 presents an implementation of the controller on the Benchmark example.

L-MPC: The idea behind Learning Model Predictive Control (L-MPC) is to combine the asymptotic performance convergence of Iterative Learning Control (ILC) with the approximately closed-loop performance of Model Predictive Control (MPC). [3] proposes a Learning Model Predictive Control of batch process which is described by stochastic Linear Time-Varying (LTV) system directly obtained from Eq. (26):

$$\begin{aligned} x_{k,t} &= A_t x_{k,t-1} + B_t \Delta u_{k,t-1} + E_t v_{k,t} \\ e_{k,t} &= \bar{y}_t - y_{k,t} \\ &= e_{k-1} - C x_{k,t} \end{aligned} \tag{38}$$

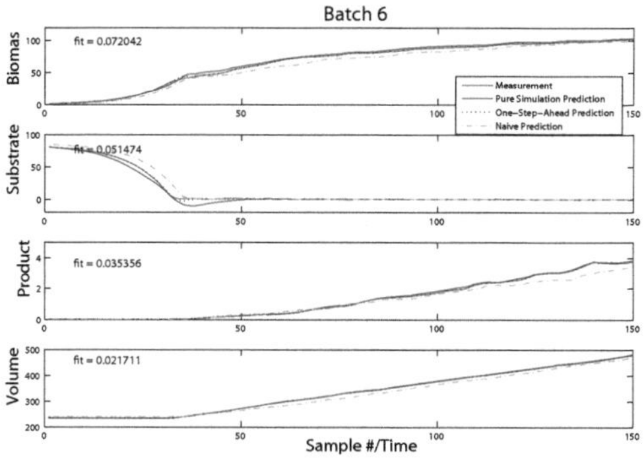

(a) *model 1*

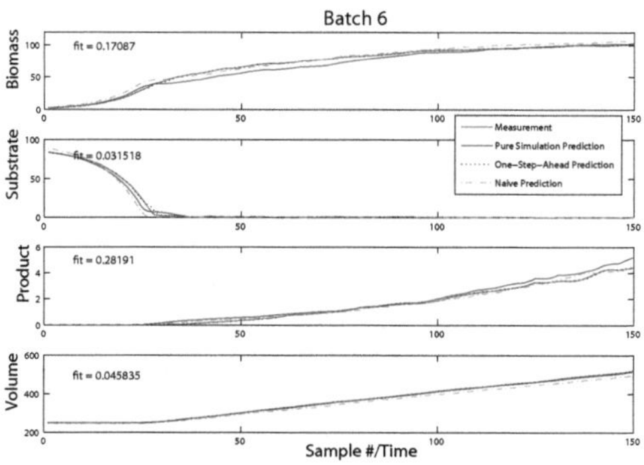

(b) *model 2*

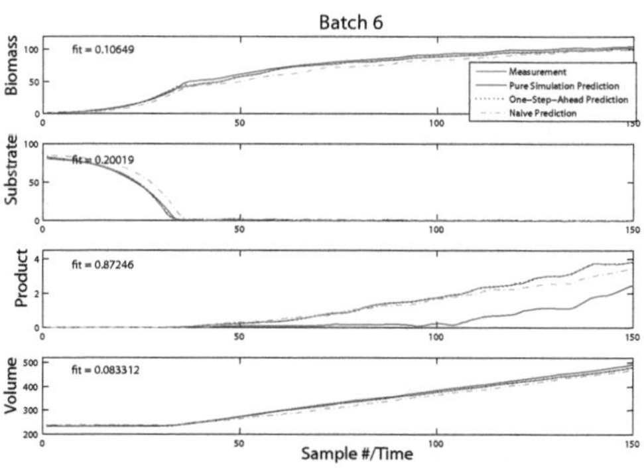

(c) *model 3*

Fig. 5. (a): Simulation of batch 6 using *model 1*; (b): Simulation of batch 6 using *model 2*; (c): Simulation of batch 6 using *model 3* for $t=1,\ldots,N$ with initial condition

$$x_{k,0} = -Cv_{k,0} \tag{39}$$

This means that the state $x_{k,t} \in \mathbb{R}^{n_x}$ of the system (38) at time t in batch k, is given by linear mappings of the state $x_{k,t-1}$ at time $t-1$ in batch k; the control correction $\Delta u_{k,t-1} = u_{k,t-1} - u_{k-1,t-1} \in \mathbb{R}^{n_u(t-1)}$ and a zero-mean Gaussian disturbance $v_{k,t} \in \mathbb{R}^{n_y}$. The tracking error $e_{k,t} \in \mathbb{R}^{n_y}$ of the system (38) is given as the difference between the system output $y_{k,t} \in \mathbb{R}^{n_y}$ and the desired output reference $y_{\bar{k},t} \in \mathbb{R}^{n_y}$.

The formulation for the L-MPC is

$$\begin{array}{rl} \{\Delta u_{k,l,-1}\}_{l=0}^{N-1} = \arg\min_{\{\Delta u_{k,i}\}_{i=0}^{N-1}} & \left[\sum_{i=1}^{N} \hat{e}'_{k,i|t} Q_{k,i} \hat{e}_{k,i|t} + \Delta u'_{k,i-1} R_{k,i} \Delta u_{k,i-1}\right] \\ s.t. & \hat{x}_{k,i|-1} = A_i \hat{x}_{k,i-1|-1} + B_i \Delta u_{k,i-1} \\ & \hat{e}_{k,i|-1} = \hat{e}_{k-1,i|N} - C\hat{x}_{k,i|-1} \\ & u_{\mathtt{min}\, i-1} \leq \Delta u_{k,i-1} + u_{k-1,i-1} \leq u_{\mathtt{max}\, i-1} \\ & y_{\mathtt{min}\, i} \leq \bar{y}_i - \hat{e}_{k,i|-1} \leq y_{\mathtt{max}\, i} \end{array} \tag{40}$$

if the weighting matrices Q_k and R_k are block diagonal

$$\begin{aligned} Q_k &= \mathtt{diag}(Q_{k,t}) \\ R_k &= \mathtt{diag}(R_{k,t}) \end{aligned} \tag{41}$$

and the weighting matrices $Q_{k,t}$ and $R_{k,t}$ are all symmetric and positive definite. [3] proves that the optimal solution $\Delta u_{k,-1} = \{\Delta u_{k,l,-1}\}_{l=0}^{N-1}$ to the L-MPC satisfies the stochastic ILC design requirement and guarantees convergence (see [3] for details).

Thus, the cost function J defined is (Eq. 42). Where $t = -1, 0, \ldots, N$ and $-1 \leq j < N$.

$$J_k(\hat{e}_{k,t}(\Delta u_{k,j}), \Delta u_{k,j}) = \|\hat{e}_{k,t}(\Delta u_{k,j})\|_{Q_k}^2 + \|\Delta u_{k,j}\|_{R_k}^2 \tag{42}$$

State and Output Estimation: For state and output estimations, GoLM implements the Kalman Filter recursive equations

$$\begin{aligned} \hat{x}_{k,t|t} &= \hat{x}_{k,t|t-1} + K_{k,t}(z_{k,t} - C\hat{x}_{k,t|t-1} - \hat{y}_{k-1,t|N}) \\ \hat{x}_{k,t|t-1} &= A_t^S \hat{x}_{k,t-1|t-1} + B_t \Delta u_{k,t-1} + E_t S_t R_{\epsilon_t}^{-1}(z_{k,t-1} - \hat{y}_{k-1,t-1|N}) \end{aligned} \tag{43}$$

for $t = 0, 1, \ldots, N$, with initial condition

$$\hat{x}_{k,0|-1} = 0 \tag{44}$$

The Kalman Filter gain matrix $K_{k,t}$ and state estimate covariance matrix $P_{k,t|t}$ are propagated by the recursions

$$\begin{aligned} P_{k,t|t} &= P_{k,t|t-1}(I - C'K'_{k,t}) \\ K_{k,t} &= P_{k,t|t-1}C'(CP_{k,t|t-1}C' + R_{\epsilon_t} + \bar{P}_{k-1,t|N})^{-1} \\ P_{k,t|t-1} &= A_t^S P_{k,t-1|t-1} A_t^{S'} + E_t \Sigma_t^S E_t' \end{aligned} \tag{45}$$

with the initial condition

$$P_{k,o|-1} = C'\Sigma_0 C \tag{46}$$

In most of bio-chemical batch processes there are no observations of the initial conditions, thus

$$\begin{aligned} \hat{x}_{k,0|0} &= 0 \\ P_{k,0|0} &= C'\Sigma_0 C \end{aligned} \tag{47}$$

However, if these are available they can be specified. The Kalman Filter recursion is used for estimation of outputs or states when the measurement is not available on-line.

3.4 Benchmark Control Design

To implement the controller to the benchmark model a Learning Model Predictive Control is used which was developed in the GoLM methodology. In the subsection 3.3 was presented the controller. In this subsection the L-MPC is implemented to the benchmark example (Penicillin Production). The control objective for this case study is batch reproducibility. The model used for the controller is the second best model estimated by GoLM in the previous section 3.2, the objective is to test the robustness of the controller instead the model used is different from the plant.

The duration of the batch is 150 hours, the samples measurement (N) for the outputs variables ($y_{k,t}$) and inputs variables ($u_{k,t}$) are each hour ($N = 150$). The process starts with a batch phase where no feed is added, the initial substrate added is sufficient to ensure biomass growth during this phase. When the substrate concentration is less than 5 g/L the feed starts. The initial conditions are given as

$$y_{k,0} = \begin{bmatrix} 0 \\ 80 \\ 0 \\ 250 \end{bmatrix} + \begin{bmatrix} 1.5 & & & \\ & 5 & & \\ & & 0 & \\ & & & 10 \end{bmatrix} p_k \tag{48}$$

where p_k is a random variable sampled from a zero mean uniform distribution with variance 1. The formulation of the control objective used is

$$\begin{aligned} \{\Delta u_{k,l,t}\}_{l=t}^{N-1} &= \arg\min_{\{\Delta u_{k,i}\}_{i=t}^{N-1}} [J_{k,i,t}] \\ s.t. \quad J_{k,i,t} &= \textstyle\sum_{i=t+1}^{N} (u_{k,i-1} - u_{k,i-2})' T_i (u_{k,i-1} - u_{k,i-2}) + \hat{e}'_{k,i|t} Q_i \hat{e}_{k,i|t} \\ &\quad + \Delta u'_{k,i-1} R \Delta u_{k,i-1} \\ \hat{x}_{k,i|t} &= A_i \hat{x}_{k,i-1|t} + B_i \Delta u_{k,i-1} \\ \hat{e}_{k,i|t} &= \hat{e}_{k-1,i|N} - C\hat{x}_{k,i|t} \\ u_{k,i-1} &= \Delta u_{k,i-1} + u_{k-1,i-1} \\ 0 \begin{bmatrix} u(1)_{k,i-1} \\ u(2)_{k,i-1} \end{bmatrix} &\leq \begin{bmatrix} 5 \\ 600 \end{bmatrix} \\ y_{\mathtt{min}_i} \leq \hat{y}_i &- \hat{e}_{k,i|t} \leq u_{\mathtt{max}_i} \end{aligned} \tag{49}$$

with $u_{k,-1} = u_{k-1,0}$. Where $u(1)$ is the feed flow rate and $u(2)$ is the substrate concentration in the feed. The output constraints for the process are

$$y_{\mathtt{min}i} = \begin{bmatrix} 0 \\ 0 \\ 0 \\ 250 \end{bmatrix} \qquad y_{\mathtt{max}i} = \begin{bmatrix} 120 \\ 90 \\ 10 \\ 600 \end{bmatrix} \tag{50}$$

for the samples $i = 1, \ldots, N$. The measurements are: $y(1)$: biomass concentration (g/L), $y(2)$: substrate concentration (g/L), $y(3)$: product concentration (g/L) and $y(4)$: volume of the reactor (L). According to the control objective the weighting matrices were for this preliminary design selected as

$$Q_t = \begin{bmatrix} 10 & 0 & 0 & 0 \\ 0 & 1 & 0 & 0 \\ 0 & 0 & 10 & 0 \\ 0 & 0 & 0 & 1 \end{bmatrix} \qquad R_t = \begin{bmatrix} 1 & 0 \\ 0 & 1 \end{bmatrix} \qquad T_t = 1 \tag{51}$$

The control objective is batch product reproducibility using a previous batch as reference. For cultivation industries this type of control is interesting for reducing the influence of variation of initial conditions or substrate composition on the batch product concentration. The above preliminary control design is tested introducing variations on initial conditions (as specified in Eq. 48). Figure 6 presents the open loop performance for simulation of 20 batches (Penicillin model), where the inputs vary within 15% of the nominal value. The mean

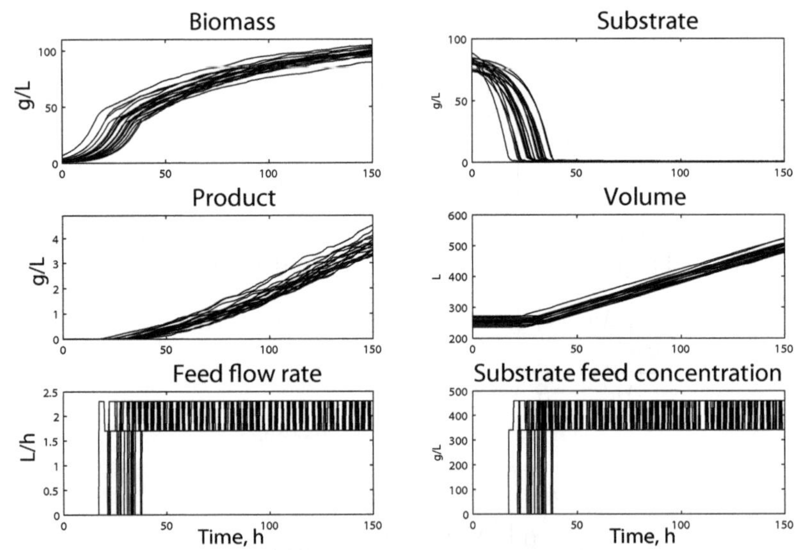

Fig. 6. Open-loop performance simulation of 20 batches. The inputs are perturbed with an amplitude of 15% of the nominal value. Stochastic initial conditions.

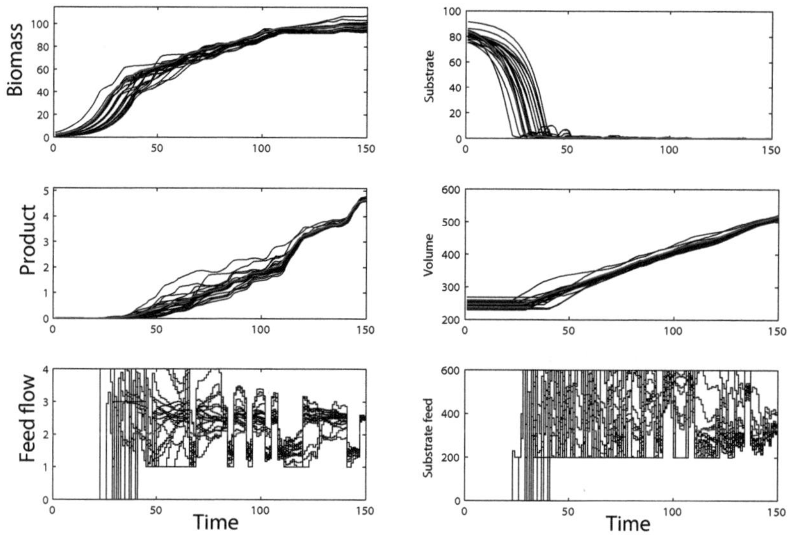

Fig. 7. Closed-loop performance of L-MPC control with stochastic initial conditions. The control objective is to ensure product reproducibility.

Table 7. FIT for Pure Simulation Prediction of batches 6-7 for ARX model

Operation	Mean Product Conc.	Productivity
Disturbance on initial conditions		
Open-loop	3.7	100%
Ref. Batch	4.7	127%
Closed-loop	4.7	127%
At 50 hours μ_{max} decreases 30%		
Open-loop	3.3	100%
Ref. Batch	4.5	136%
Closed-loop	4.1	124%

product concentration of the 20 batches is 3.7 g/L. The batch with the highest productivity has a final product concentration of 4.5 g/L. This batch is chosen as reference batch for the control design.

The closed-loop performance simulation for 20 batches, the learning MPC controller efficiently rejects the perturbation produced by variation on the initial conditions (see Figure 7). Note especially the behavior towards the end of the batches for product concentration. Table 7 shows the mean value of the product concentration without control and with control.

Another scenario for control testing is generating a disturbance during the batch, e.g. by decreasing the maximal specific growth rate (μ_{max}) with 30% of the original value after 50 hours. This disturbance simulates a change in the metabolism of the microorganism due to an inhibitory effect. Figure 8(a)

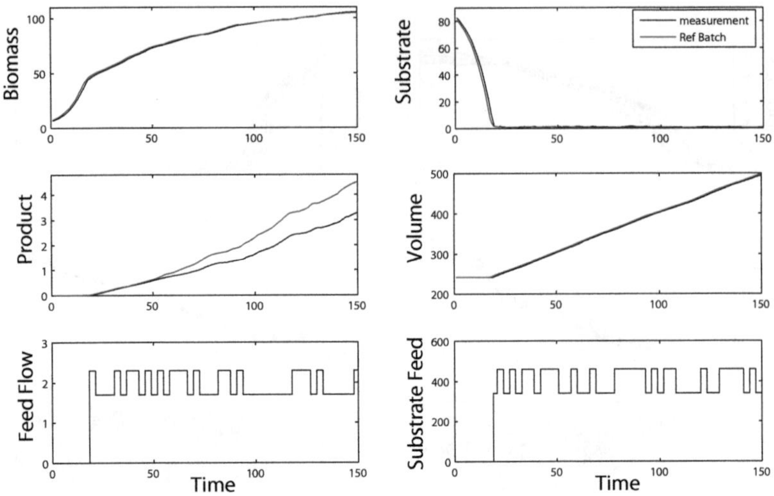

(a) Open-loop of the system with disturbance

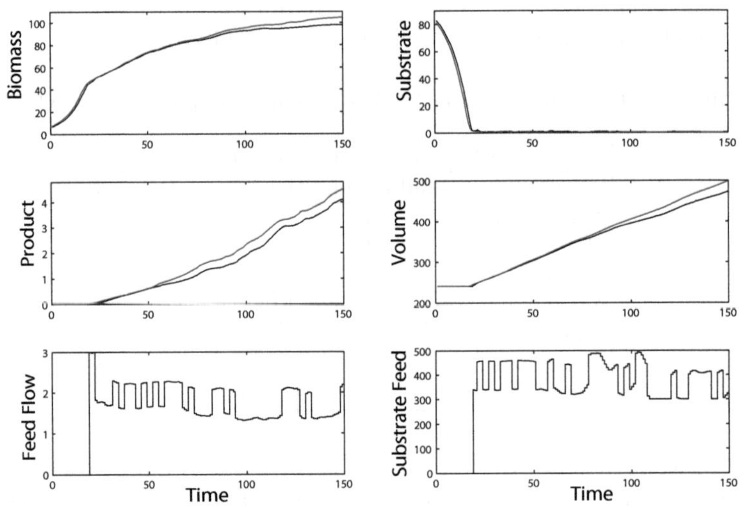

(b) Closed-loop disturbance rejection

Fig. 8. (a):Open-loop with disturbance. The 30 % reduction of μ_{max} is applied at time 50 hr. (b): Close-loop performance of a learning MPC control for disturbance rejection intra-batch.

illustrates the open-loop performance with the disturbance in μ_{max} at 50 hours. Figure 8(b) depicts the closed-loop performance of L-MPC control, and shows that even though the plant changes its behavior, the controller attempts to minimize the error by reducing the feed rate while increasing the feed concentration. Table 7 shows that this is achieved quite well.

3.5 Conclusions on Benchmark

On the benchmark simulation case it is demonstrated that it is possible to develop a data driven model from a limited number of batches using designed batch runs with perturbations in the input variables which are going to be used as actuator variables in the control design. There are parameters in GoLM that can be used for model improvement such as, changing the number of batches for identification, validation and testing (selecting the batches with input perturbation for identification); the weighting matrix (λ) for Tikhonov Regularization; the model type (ARX or ARMAX); maximum model order; the Model Purpose (ρ) (a value between 0 to 1). The Benchmark also illustrates that it is possible to develop a soft sensor for the enzyme activity, which otherwise normally is not available during the cultivation. Finally a preliminary L-MPC control design is demonstrated to provide reasonable control both in case of intra-batch and inter-batch disturbances. As expected, even thought the plant changes its behavior, the controller tries to reject the disturbance in spite of the model used for the controller is different from the plant. This allows to conclude that the L-MPC seems to be a robust controller. Hence a suitable background has been established for investigating the application of the GoLM methodology on an industrial case.

4 Industrial Case

Based upon the above benchmark investigation it was decided to investigate application of the GoLM methodology on a challenging fungal cultivation. The main challenge stems from the fact that the theology of the cultivation is complicated by the growth of filaments wherein the main activity during growth and production actually resides.

4.1 Introduction to Pilot Plant

The Industrial Case Study is the production of an α-amylase called Fungamyl using the microorganism *Aspergillus oryzae*. The fermentation process is operated following a standard recipe at the Novozymes Pilot Plant. The available on-line and off-line variables are presented in Table 8.

The fermenter feed flow design has the physical connections illustrated in Figure 9. Two feed lines are available for the experimental cultivation to excite the inlet total feed flow rate and the substrate feed concentration to the reactor. As Figure 9 shows the inlet feed flow rate is the total feed ($F_T = F_d + F_w$). The inlet substrate concentration (S_f) is total amount of substrate per mole of total feed.

The dosing tank is prepared before the main tank is inoculated according to the recipe; this tank is connected to the reactor to transfer the substrate, with the flow rate F_d which is manipulated using control valve V_2. The seed tank is filled with water after the inoculation has been transfer to the main tank; the water flow is manipulated with a control valve by a previously designed signal. These two flows are connected before the fermenter, and the substrate is

Table 8. Variables available from Pilot Plant

On-line Variables	**Unit**
Output variables	
Fermentation time	Hours
NH_3flow rate	g NH_3/h
Accumulated NH_3	g
Dissolved O_2 tension (DOT)	% of saturation
pH	No unit
Back pressure	Bar gauge
O_2 uptake rate (OUR)	moles/h
Accumulated O_2	moles
CO_2 evolution rate (CER)	moles/h
Accumulated CO_2 evolution	moles
Weight	kg
Viscosity	cP
Refractive index (RI)	No unit
Respiratory quotient (RQ)	No unit
Temperature	C
Air flow rate	NL/min
Input variables	
Feed dosing flow rate ($Feed_d$), Measured	L/h
Accumulated dosing $Feed_d$ flow	L
Dosing $Feed_d$ flow rate, Set Point	L/h
Water Feed flow rate ($Feed_w$), Measured	L/h
Accumulated water $Feed_w$ flow	L
$Feed_w$ flow rate, Set Point	L/h
Off-line Variables	**Unit**
Enzyme Activity	FAU/g
Biomass, CTS	g/L
Ammonia concentration	mg/L
Laboratory pH	No unit

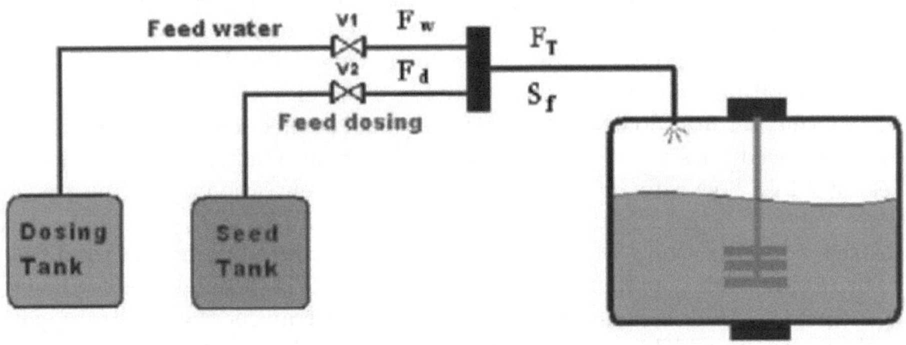

Fig. 9. Flowsheet for Pilot Plant feed line connections

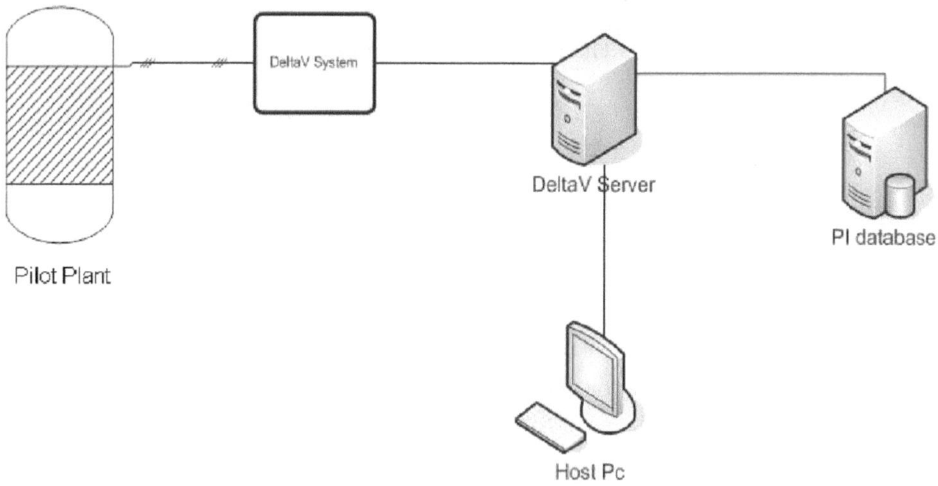

Fig. 10. Connections between Pilot Plant computers

mixed with water and diluted to generate the perturbation in the inlet substrate concentration and in the total flow rate.

The physical connections of the computer system are shown Figure 10. All sensor signals are connected to the DeltaV System that communicated with the DeltaV Server by an Ethernet. The PI Database Server stores all the data acquired from the DeltaV Server using a "swinging door" algorithm. The host PC directly collects data from the DeltaV Server every 10 seconds and this data set is further prepared for analysis and model identification.

Given the above flow sheet (Figure 9), it is important to carefully design of the perturbation signals in order to ensure that substrate concentration and total flow rate indeed are independently perturbed. Obviously this would not be the case if such signals were applied directly to the valves V_1 and V_2. Therefore special consideration is given to the design of these two perturbation sequences.

4.2 Input Perturbation Design

The input variables F_T and S_f are, as explained previously, closely related due to the physical connections (see Figure 9). This fact is also evident from the mass balances:

$$F_T = F_d \left(1 + \frac{F_w}{F_d}\right) \tag{52}$$

$$S_f = S_d \frac{1}{1 + \frac{F_w}{F_d}} \tag{53}$$

where S_d is the Dosing Substrate. Therefore, the aim is to design F_d and F_w, which both are physically correlated, such that F_T and S_f will be perturbed nearly independently. The strategy is to linearize the equations (52) and (53), and then decouple the equations. Defining the steady state values as $ps = [F_d^s, F_w^s]$, and $a = \frac{S_d}{(F_d^s+F_w^s)^2}$ then the linearized model becomes:

$$\begin{pmatrix} F_T \\ S_f \end{pmatrix} = \begin{pmatrix} F_T^s \\ S_f^s \end{pmatrix} + \begin{pmatrix} 1 & 1 \\ -aF_d^s & aF_w^s \end{pmatrix} \begin{pmatrix} \Delta F_w \\ \Delta F_d \end{pmatrix} \tag{54}$$

Define the covariance matrix for $\begin{bmatrix} F_T \\ S_f \end{bmatrix}$. such that their cross-covariance will be zero:

$$P = \begin{pmatrix} F_T \\ S_f \end{pmatrix} \begin{pmatrix} F_T & S_f \end{pmatrix} \tag{55}$$

The Lyapunov equation for the covariance of $[F_T \; S_f]$ is (assuming no process noise):

$$P = ARA^T \tag{56}$$

$$P = \begin{pmatrix} \sigma_1^2 & 0 \\ 0 & \sigma_2^2 \end{pmatrix} \tag{57}$$

where the A matrix is defined by equation 54:

$$A = \begin{pmatrix} 1 & 1 \\ -aF_d^s & aF_w^s \end{pmatrix}$$

and σ_1^2 and σ_2^2 are the variance of F_w and F_d respectively. An R matrix is defined as:

$$R = HH^T \tag{58}$$

The aim is to determine R such as the cross-covariance between F_T and S_f is zero. Given the square root H matrix it is possible to determine ΔF_d and ΔF_w:

$$\begin{pmatrix} \Delta F_w \\ \Delta F_d \end{pmatrix} = H \begin{pmatrix} e_1 \\ e_2 \end{pmatrix} \tag{59}$$

where e_1 and e_2 are the perturbed signals. In this case these signals are uncorrelated PRBS signal with period of 4 hours for e_1 signal and 8 hours for e_2 signal, and hence F_w and F_d become:

$$\begin{pmatrix} F_w \\ F_d \end{pmatrix} = \begin{pmatrix} F_w^s \\ F_d^s \end{pmatrix} + H \begin{pmatrix} e_1 \\ e_2 \end{pmatrix} \tag{60}$$

F_w and F_d were designed following the above analysis. The reference signals for F_T and S_f are ramp signals, both signals start at the beginning of the fed-batch phase. From these signals F_w and F_d are designed according to Eq. 58. Figure 11 shows the resulting signals.

A correlation analysis for F_T and S_f shows there is a cross-correlation of 0.0646, which indicates that both signals are almost independent. Thus with the above linear based design it is possible to satisfactorily decouple the signals in spite of the physical coupling.

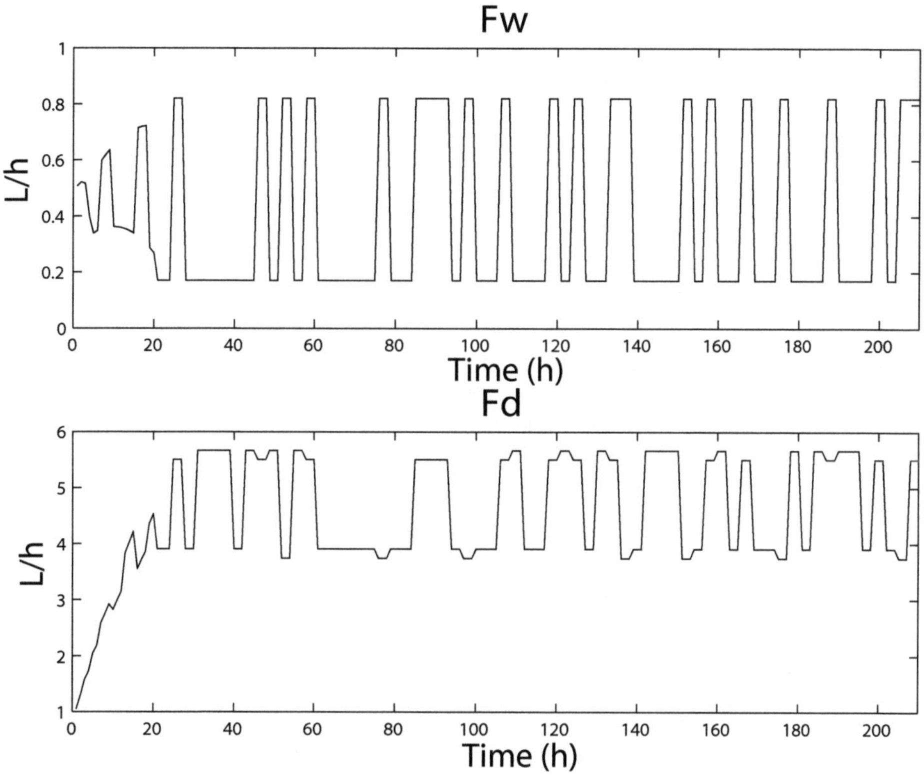

Fig. 11. Top: Feed water flow rate signal; bottom: Feed dosing flow rate signal

4.3 Model Estimation

For most microbial batch processes it is difficult to develop a model using first principle modeling, thus, there is an interest to use black-box models. The aim in this case is to illustrate estimation and validation of a model using the GoLM methodology for an α-amylase cultivation of the Novozymes Pilot Plant. First, it is presented the data of the batches available for identification. Then, it is shown the estimation of the model using GoLM, the validation of the model.

Data Selection: The eleven batches presented below are from production of α-amylase in *Aspergillus oryzae* cultivations from the Novozymes Pilot Plant. The available results in Table 9 are produced in the same reactor.

Classifying the batch behaviors revealed that five of these batches fall out of the normal operation pattern of the process [8]. From the available data set, only batches 1082, 1098, 1099, 1108, 1110 and 1111 are selected to be potentially useful for model identification. After selection of the potentially useful batches it is necessary to pre-process the data as explained next.

Variable Selection and Data Pretreatment for Modeling: It is important to select variables for model estimation that contain sufficient information. The

Table 9. Variables available for each batch

Variable	Type	1082	1098	1099	1100	1101	1102	1103	1108	1109	1110	1111
Input perturbation (F_T-S_f)	online	-	-	-	-	-	-	-	-	-	X	X
RI	online	-	X	X	X	X	X	X	X	X	X	X
Viscosity	online	X	X	-	-	X	X	X	X	X	X	X
NH_3 flow rate	online	-	-	X	X	X	X	X	X	X	X	X
Biomass	offline	X	-	-	-	X	X	X	X	X	X	X
Enzyme Activity	offline	X	X	X	X	X	X	X	X	X	-	X
Feed-back control		-	-	-	X	X	X	X	-	X	-	-

Table 10. Carbon concentration in dosing flow. The concentration is constant during each batch process, except for batches 1110 and 1111 where the carbon concentration is perturbed during the batch as explained in section 4.2.

Batch	Calc. Carbon concent. (C moles/L)
1082	12.48
1098	11.08
1099	15.68
1108	14.20

pH, back pressure and temperature are controlled, for that reason they are not used for modeling but their control actuator signals may be used. The air flow rate is kept constant during the cultivation and depends on the capacity of the central compressor. The remaining variables do contain information about the process and are used to estimate a model with the GoLM methodology.

The input variables to the process are: Feed flow rate and Substrate feed concentration. The decoupling design illustrated in Figure 11 for the two input variables was used in batch 1111, while correlated perturbations were introduced in batch 1110. It is necessary to calculate the substrate concentration in the feed. Table 10 summaries the carbon concentration in the dosing for the batches used for model estimation. Figure 12 depicts the total feed flow rate F_T and substrate feed concentration S_f for all barches.

The data preparation is carried out for the variables than can be used for model estimation, first outliers is removed from the data. That is, data that is inconsistent with the normal operation by visual inspection is removed. At this point the dead time produced by delay response of sensors is removed. Then a kernel smoother is used for re-sampling data to reduce the effect of noise and of slow sampling interval for a few of the off-line sampled variables. Thereby high frequency measurement noise is removed and missing data reconstructed. Care must be taken not to remove system dynamics while filtering the measurement noise. Not all the measurements are available for all the batches (see Table 9) and depending of the variable different methods were used as explained next. Figure 13 shows the filtered variables for all the batches available for model identification.

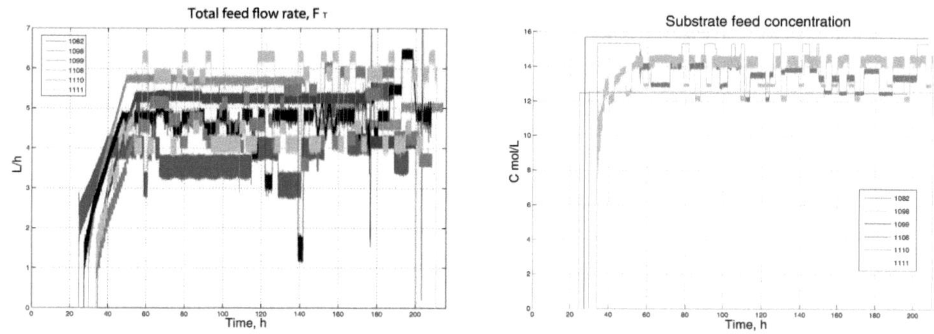

(a) Total flow rate (b) Substrate feed concentration

Fig. 12. Left: Total feed flow rate for all the batches; Right: Substrate concentration in the feed for all the batches. Batches 1110 and 1111 were design with perturbations in total flow rate and substrate feed. Only batch 1111 used the designed uncorrelated perturbations.

The measurement of ammonia flow rate is not available for the batches 1082 and 1098. To estimate the ammonia flow rate, the accumulated ammonia that is measured for each batch was used. The accumulated NH_3 is a measure of the total amount of NH_3 added to the fermentation tank and it is the integrated NH_3 flow rate. Therefore, the differentiation of the accumulated ammonia provides the ammonia flow rate. This is relevant for batches 1082 1nd 1098.

The variables used for model estimation are:

- **Inputs:** Total feed flow rate and Substrate feed concentration
- **Outputs:** NH_3 flow rate, CER, DOT, RI, Viscosity and Enzyme activity

Model Estimation: Several different GoLM models have been estimated with different combinations of outputs variables (the same inputs is using for estimation of all the models [F_T and S_f]), different sets of batches and different values of Model Purpose (ρ). For the purpose of obtaining a model with reasonable prediction ability to be used for the implementation of the control on the Pilot Plant. The following mostly sequential methodology was applied:

1. Selection of output to be used for modeling.
2. Selection of batches used for estimation, validation and an independent set for testing.
3. Selection of model type.
4. Selection of maximum local model order.
5. Set the scalar for Model Purpose (ρ).

This section only presents the combination used to estimate the best model obtained for the Fungal Process. After many estimation attempts it was possible to conclude that batch 1082 is not a good batch for identification. Table 11 shows a comparison between two estimated models, one using batch 1082 and

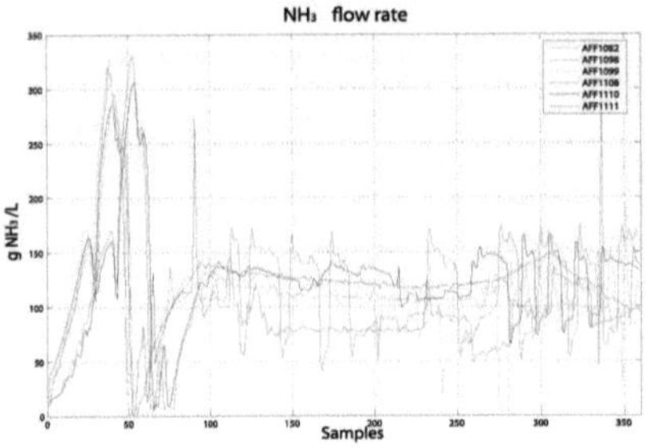

(a) NH_3 flow rate

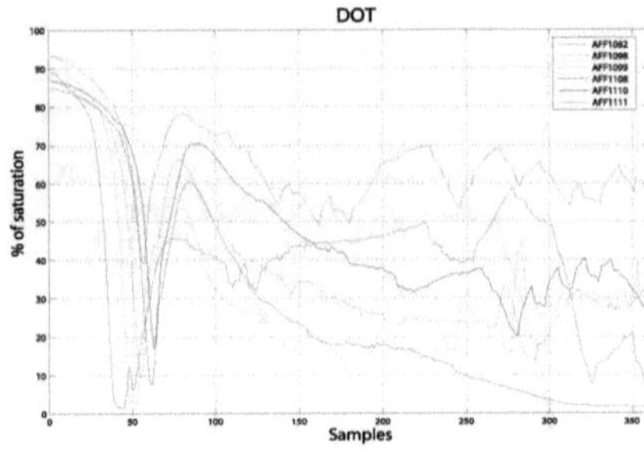

(b) DOT

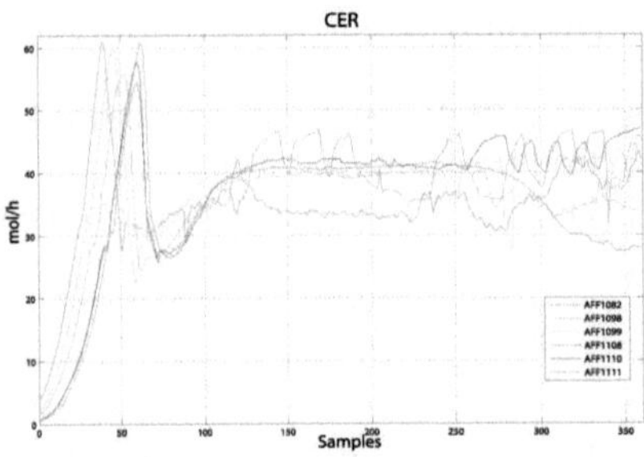

(c) CER

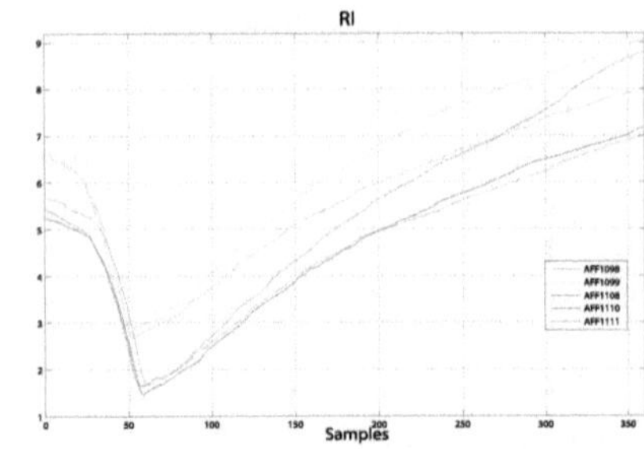

(d) RI

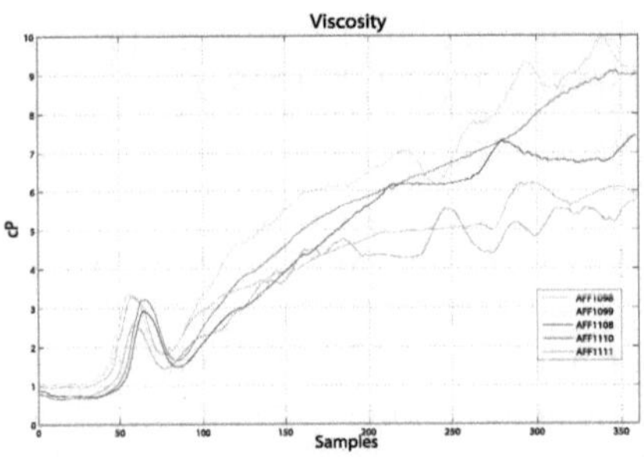

(e) Viscosity

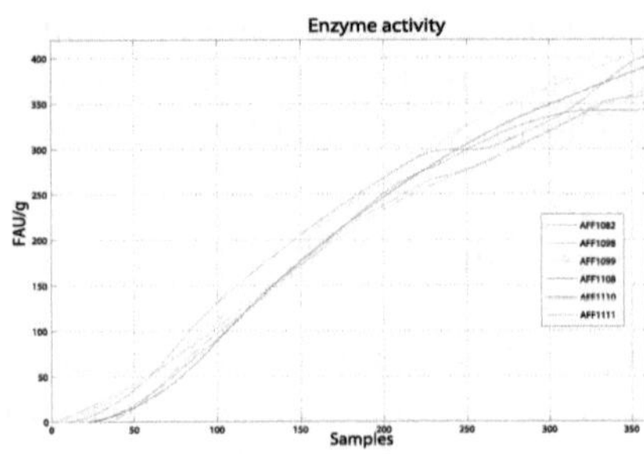

(f) Enzyme Activity

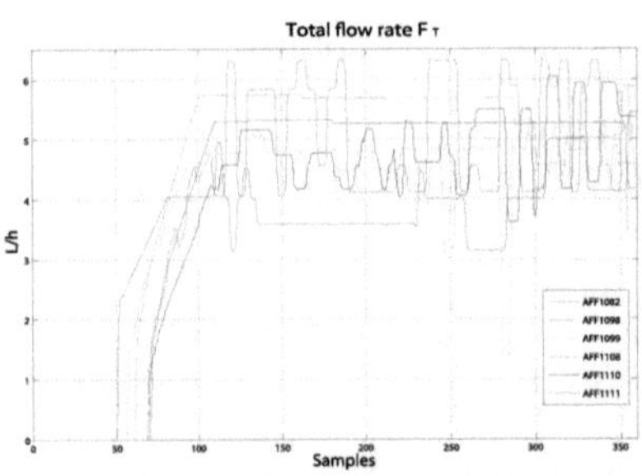

(g) Total flow rate

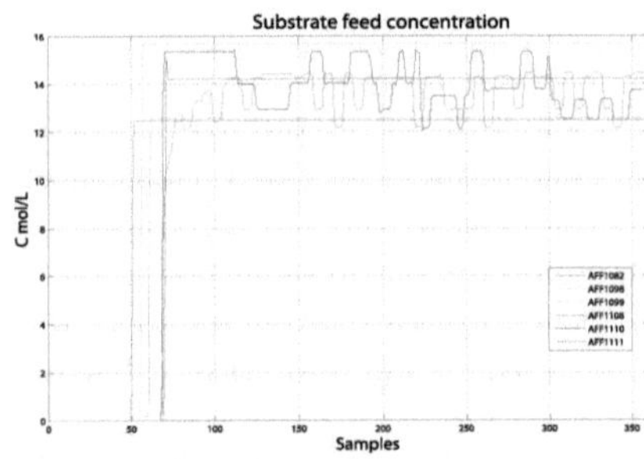

(h) Substrate feed concentration

Fig. 13. Pilot plant data available for data driven model estimation

Table 11. Comparison between two models, one with batch 1082 and the other without this batch

Model	Inputs	Ouputs	ARX	Batches order	Fit				Total fit
					NH_3	DOT	CER	Enz.	
fung12	F_t S_f	NH_3 DOT CER Enz.	2	N_{est}:1110,1082,1099,1111 N_{val}:1098,1108 N_{test}:1098,1108	0.0756	0.1149	0.0868	0.1448	0.1089
model1	F_t S_f	NH_3 DOT CER Enz.	2	N_{est}:1110,1111,1099 N_{val}:1098,1108 N_{test}:1098,1108	0.0837	0.1012	0.074	0.1158	0.095

the other without this batch. It is interesting to note that without batch 1082, the estimated model (*model1*) improves its prediction. This indicates that batch 1082 does not follow the real behavior of the process and its data does not provide reliable information. Therefore, care should be taken when data is chosen for model estimation. As only there are available six batches for model estimation, the batches 1110,1082,1099,1111 are used for estimation, batches 1098,1108 are used validation and testing. Figure 13 shows the variables used for identification including data from batch 1082.

According to the above results, it was decided to not include batch 1082 for model estimation. When the estimation of a model is using a black-box identification as the case of GoLM, it is necessary to use data that contains a lot of information of the process. For that reason, a model was estimated with two more measured outputs (RI (Refractive Index) and Viscosity), as soon as these measured variables became available. Both new measurements provide information about biomass and product concentration during the cultivation. Table 12 shows the FITs of four models estimated.

Since only five batches are available for the estimation, these are divided into three batches for estimation and two for validation. Therefore, there is not an independent set of batches for testing hence the batch set for validation is also used for testing. Calculation of the generalization error is, as for GoLM, used to continue the iteration until convergence where the Estimated Generalization Errors (G_ρ) is less than the user specified value. This is supplemented with visual validation.

$$\mathcal{G}(\hat{\theta}^*) = \frac{1}{N_B^{test}} \sum_{k=1}^{N_B^{test}} \left\| \Delta\hat{y}_k - \Delta\hat{y}_k(\hat{\theta}^*) \right\|_{\bar{W}}^2 \tag{61}$$

where $\hat{\theta}^*$ is the selected near optimal model

$$\hat{\theta}^* = \hat{\theta}_{TR}(\hat{V}, W, \Lambda^*, n_A^*, n_B^*, n_C^*) \tag{62}$$

The model *model2* in Table 12 which includes RI (Refractive index) is not as good as *model1*, while *model3* which includes the viscosity measurement clearly

Table 12. Model estimated with different combinations of outputs

Model	Inputs	Ouputs	ARX	Batches	Fit						Total fit
					NH_3	DOT	CER	Enz.	RI	Visc.	
model1	F_t S_f	NH_3 DOT CER Enz.	2	N_{est}:1110,1111,1099 N_{val}:1098,1108 N_{test}:1098,1108	0.0837	0.1012	0.074	0.1158	-	-	0.095
model2	F_t S_f	NH_3 DOT CER Enz. RI	2	N_{est}:1110,1111,1099 N_{val}:1098,1108 N_{test}:1098,1108	0.0896	0.1078	0.085	0.1382	0.107	-	0.1052
model3	F_t S_f	NH_3 DOT CER Enz. Visc.	2	N_{est}:1110,1111,1099 N_{val}:1098,1108 N_{test}:1098,1108	0.0754	0.0841	0.0788	0.0625	-	0.1083	0.0832
visc2	F_t S_f	NH_3 DOT CER Enz. RI Visc.	2	N_{est}:1110,1111,1099 N_{val}:1098,1108 N_{test}:1098,1108	0.0755	0.0873	0.0666	0.0873	0.0409	0.0838	0.0718

is better than the other two. However including both RI and the viscosity measurement *visc2* improves the fit significantly, which indicates that these two variables contain complementary information probably on substrate breakdown and on the influence of the microorganism filaments on the rheology.

Figure 14 shows the validation of the best model *visc2* on batch 1098. In the lower part of the figure the batch increment deviation between the model *visc2* and batches 1098 and 1108 clearly illustrates that even though the fit is quite reasonable for *visc2* there are deviations in the estimated enzyme activity. Note however that there only are four measurements during a batch of this variable. The other variables actually fit quite well, except during the shift from batch to fed-batch. The estimated model *visc2* is being used in the following for Kalman Filter and control design.

To verify the influence of the perturbation on the substrate feed concentration using water to dilute the dosing two models were estimated. The models were estimated using two inputs (F_T and S_f) and four outputs (NH_3, DOT, CER and Enz.), the model structure used is ARX with a maximum local model order of two and with a value of $\rho = 0.618$. The first model was estimated using batches 1099 and 1098 for identification and batches 1108, 1110 and 1111 for validation, for testing of the model was used the same batch set as for validation. Table 13 shows the results of the fits for each model. The second model estimated was used batches 1110 and 1111 for identification and batches 1099, 1098 and 1108 for validation and testing. Figure 15 presents the validation of both models. The most pronounced effect is seen on the DOT (and the ammonia) fit which is poor when there are no deliberate perturbations (model *test1*) on the input variables. Also the overall fit is significantly better with perturbations for (model *test2*) which has perturbations on the inlet flow variables.

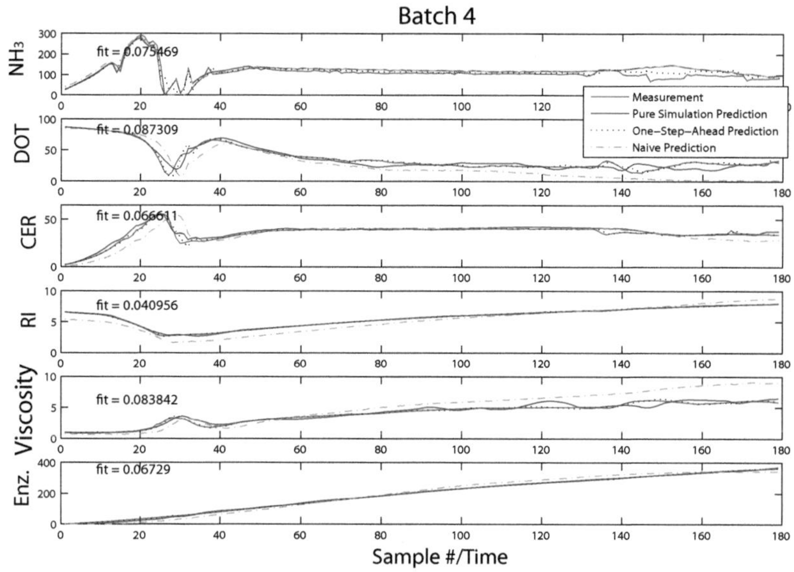

(a) Batch 1098

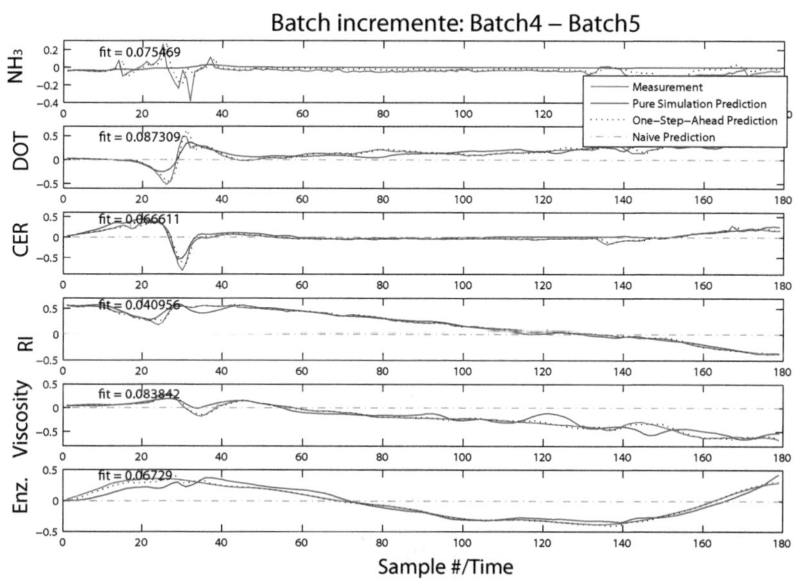

(b) Batch increment: Batch 1098 - 1108

Fig. 14. (a): Validation of the model *visc2* using batch 1098. (b): Batch increment between batches 1098 and 1108.

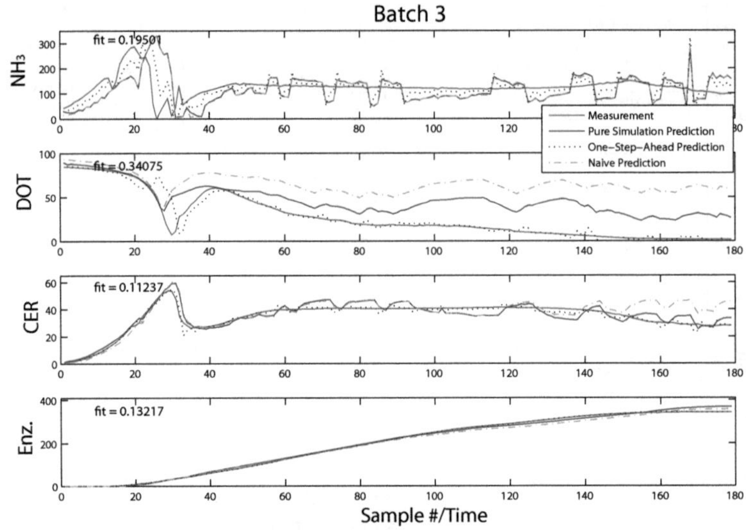

(a) Batch 1108 - model *test1*

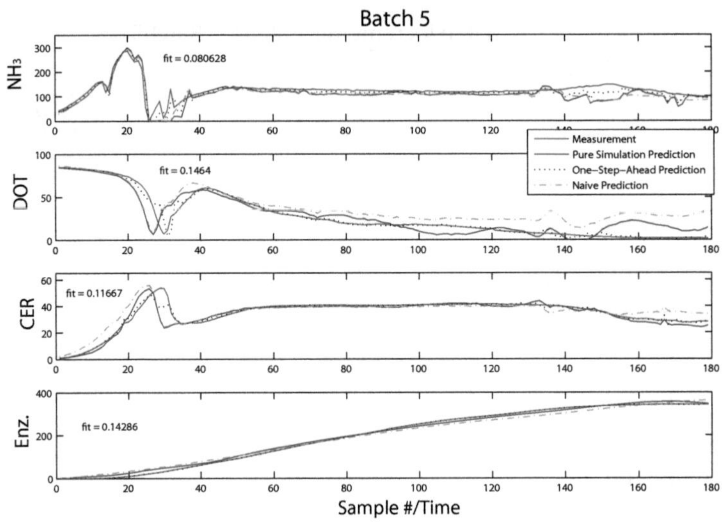

(b) Batch 1108 - model *test2*

Fig. 15. (a): Validation of the model *test1* to simulate batch 1108 using as previous batch 1098. (b): Validation of the model *test2* to simulate batch 1108 using as previous batch 1110.

Table 13. Model estimated with different combinations of outputs

Model	Inputs Perturb	Ouputs	ARX	Batches	Fit				Total fit
					NH_3	DOT	CER	Enz.	
test1	F_t	NH_3 DOT CER Enz.	2	N_{est}:1099,1098 N_{val}:1108,1110,1111 N_{test}:1108,1110,1111	0.195	0.3407	0.1124	0.1322	0.1947
test2	F_t S_f	NH_3 DOT CER Enz.	2	N_{est}:1110,1111 N_{val}:1098,1098,1108 N_{test}:1098,1098,1108	0.0806	0.1464	0.1168	0.1429	0.1367

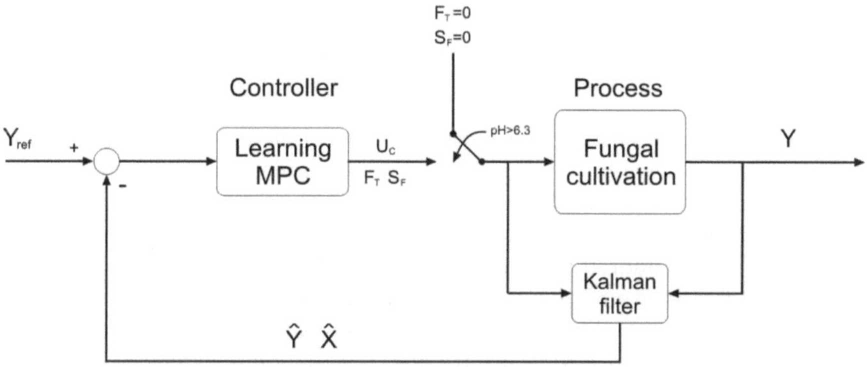

Fig. 16. Closed-loop Scheme for the Fungal Cultivation. It is identifying the controller block, process block and Kalman Filter block conections.

Having developed a model with reasonable predictive power, then the following section details the implementation of a soft sensor for enzyme activity on a simulation and of the design and closed-loop simulation of L-MPC control of the Fungal Cultivation (Figure 16).

4.4 Pilot Plant Control Simulation

The model obtained (*visc2*) is intended for the designing a L-MPC controller for implementation on the Pilot Plant. However first the designed controller is tested in simulation where the same estimated model, as used for control design, is used to simulate the process. However since the enzyme activity is an off-line (seldom) measured variable it is necessary to estimate it. For that purpose a Kalman Filter is designed as a soft sensor to estimate the values of these outputs. The basis for the filter design is also the estimated model. Below the open-loop model is simulated to investigate the tuning of the Kalman Filter. Once the Kalman Filter is working, the closed loop process is simulated to investigate the controller tuning.

The control objective is batch reproducibility and the control design methodology is as specified in section 3.4. The following formulation is used:

$$
\begin{aligned}
\{\Delta u_{k,l,t}\}_{l=t}^{N-1} &= \arg\min_{\{\Delta u_{k,i}\}_{i=t}^{N-1}} [J_{k,i,t}] \\
s.t. \quad J_{k,i,t} &= \textstyle\sum_{i=t+1}^{N} (u_{k,i-1} - u_{k,i-2})' \breve{T}_i (u_{k,i-1} - u_{k,i-2}) + \hat{e}'_{k,i|t} \breve{Q}_i \hat{e}_{k,i|t} \\
&\quad + \Delta u'_{k,i-1} \breve{R} \Delta u_{k,i-1} \\
\hat{x}_{k,i|t} &= A_i \hat{x}_{k,i-1|t} + B_i \Delta u_{k,i-1} \\
\hat{e}_{k,i|t} &= \hat{e}_{k-1,i|N} - C \hat{x}_{k,i|t} \\
u_{k,i-1} &= \Delta u_{k,i-1} + u_{k-1,i-1} \\
0 \leq &\begin{bmatrix} u(1)_{k,i-1} \\ u(2)_{k,i-1} \end{bmatrix} \leq \begin{bmatrix} 8 \\ 15 \end{bmatrix} \\
y_{\mathtt{min}_i} \leq &\ \hat{y}_i - \hat{e}_{k,i|t} \leq u_{\mathtt{max}_i}
\end{aligned} \tag{63}
$$

Soft sensor Design: During the Fungal Cultivation the measurement of enzyme activity is not available on-line, as this variable is an off-line measurement where it takes days to obtain the biochemical analysis result. Therefore, a Kalman Filter is used to estimate the enzyme activity. The Kalman Filter is designed using the estimated fermentation model. The process variable measurements are assumed containing a normally distributed measurement noise with mean 0 and variance 0.1. The Kalman Filter tuning used is: The diagonal elements of P (Sate Covariance Matrix) are one, the diagonal elements of R (Noise Covariance) are 0.01. The Disturbance Covariance matrix is estimated during model estimation by the GoLM Software. Note that this disturbance covariance matrix estimate is most useful for obtaining a suitable tuning of the Filter. Figure 16 illustrates the performance of the Kalman Filter on batch 1098 with noise on the output variables. Note that even though the enzyme activity is an off-line measurement, it is indeed possible to estimate the variable on-line using the Kalman Filter.

Closed-loop Simulation: A closed-loop simulation of the Fungal cultivation is carried out using the model estimated and a L-MPC control design as specified below and detailed in section 4.3. The control objective is batch reproducibility and the batch used as control reference is batch 1111. This batch was chosen because the DOT is kept high with a low viscosity, this enables increasing the inputs without the problem of lack of oxygen, and furthermore, the enzyme activity is higher. The output constraints for the process are

$$
y_{\mathtt{min}_i} = \begin{bmatrix} 0 \\ 0 \\ 0 \\ 0 \\ 0 \\ 0 \end{bmatrix} \qquad y_{\mathtt{max}_i} = \begin{bmatrix} 400 \\ 100 \\ 70 \\ 15 \\ 20 \\ 700 \end{bmatrix} \tag{64}
$$

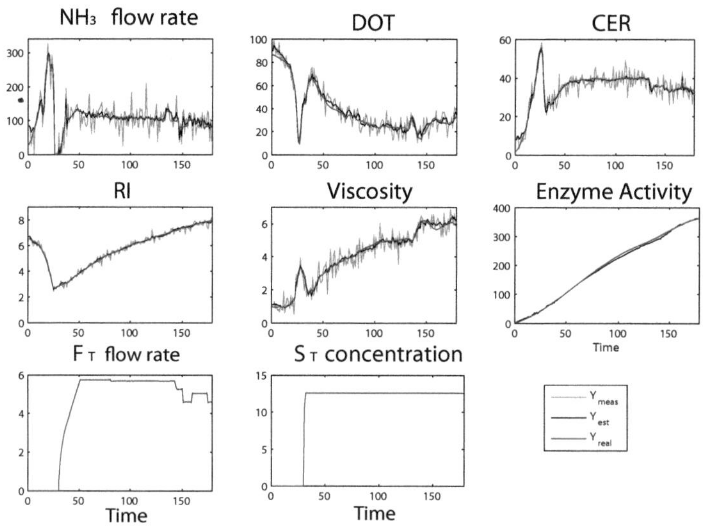

Fig. 17. Green: Process Measurement with noise; black: The Kalman Filter estimated output; red: The real output measurement of batch 1098.

for $i = 1, \ldots, N$ and $y(1)$ is NH_3 flow rate (g/h), $y(2)$ is DOT (% of saturation), $y(3)$ is CER (mole/h) $y(4)$ RI, $y(5)$ is the viscosity (cP) and $y(6)$ is the enzyme activity (FAU/g). Following the control objective the weighting matrices were designed as

$$Q_t = \begin{bmatrix} 100 & & & & & \\ & 100 & & & & \\ & & 10 & & & \\ & & & 10 & & \\ & & & & 10 & \\ & & & & & 100 \end{bmatrix} \qquad R_t = \begin{bmatrix} 0.1 & \\ & 0.1 \end{bmatrix} \qquad T_t = 1 \tag{65}$$

The results are shown in Figure 18. GoLM estimates the model for the difference between two batches, as is $y_k = \Delta\hat{y}_k + y_{k-1}$. Therefore, to simulate the plant using a GoLM model, it is necessary to use a previous batch, in this cases batch 1098 was used. The control objective is product reproducibility. During the batch phase the controller is open since no feed is added, the criterion to start the feed is when the pH is above 6.3. For the simulation, there is no measurement of pH hence time is used to start the feed because all the batches start more and less at the same time. Thus at time 35 the control loop is closed and feed starts. Figure 18 shows that the enzyme activity is tracked rather well, even though the fed batch is started a little later than for the reference batch. However this promising controller needs a more careful investigation of the tuning before being implemented in practice.

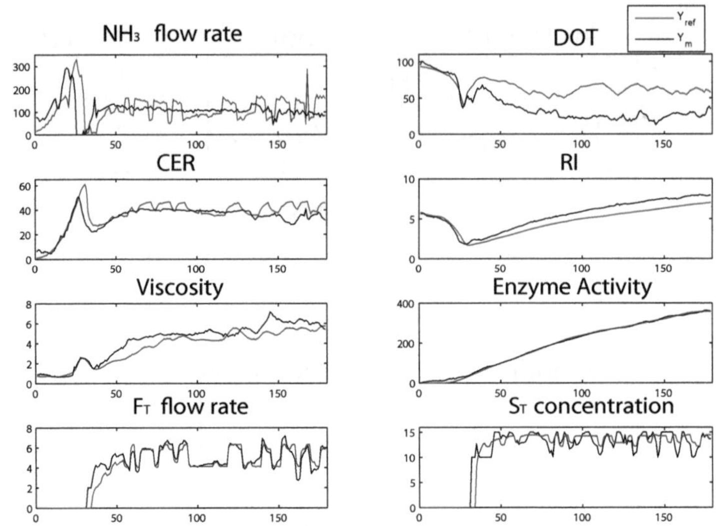

Fig. 18. Closed-loop simulation. The plant is simulated using the model *visc2* with batch 1098 as previous batch while batch 1111 is the reference batch for the control trajectory.

5 Discussion and Conclusions

5.1 Discussion

Computational intelligence is a branch of science dealing with problems that cannot be solved using only effective computational algorithms. Computational intelligence combines elements of learning, adaptation, and modelling to develop solutions that are, in some sense, intelligent. The modelling methodology presented in this chapter may be viewed as intelligent control ([6]) in the sense of qualitative control where a qualitative model is established based upon operation experience in the form of knowledge about the batch plant purpose combined with modelling the behaviour between samples. The plant purpose knowledge is essential for for structuring the more detailed data driven modelling of the behaviour between samples. The type of models selected for modelling the behavior between sample points could be, e.g. based on neural nets. However It is important to stress that even though the behaviour between measurement points is approximated with linear models then the behaviour is indeed timevarying since the parameters may vary between measurement points. Thus the selected linear models seem justified provided the measurement points are sufficiently close. Hence the methodology presented and applied in this chapter, is intelligent in the sense that intelligent modeling, which could have used fuzzy logic or neural networks instead of ARX/ARMAX model structure, is applied to model the batch behavior based upon our knowledge of batch structure.

Due to the repetitive nature and the finite horizon of the process, the measurements are collected from a grid of sample points in time. This means that a sample of all possible measured variables is collected as often as possible, i.e. at each sampling period. With these sampled data, it is possible to model the evolution between two consecutive sample points in a batch with local LTI models, which are labeled grid-point models. This modeling is made possible since at each sample time there are several measurements available due to the repetition of batches. The GoLM Methodology estimates a LTI model which contains a large set of smaller LTI models describing the process behaviour from batch to batch, while the model describing the behaviour within the batch is LTV.

The methodological choice in this paper is dictated by our knowledge of three functional requirements or facts. The first of these requirements is the desire to be able to reject inter and intra-batch production process disturbances. The second is the fact of limited available knowledge of the microbial physiology where the regulatory networks still are largely unknown. Consequently a data driven modeling methodology has to be selected. The third requirement originates from considering the control methodology where it is considered mandatory to be able to execute control actions fairly often during cultivation to prevent the cultivation from entering into operating regions where irrecoverable microbial stress responses may be triggered which might lead to opening of new pathways where substrate may be lost to other purposes than production of the desired product. This third overriding consideration lead to the choice of a discrete time model. The discrete time model is developed using the GoLM approach from a combination of existing operating data and designed fed batch runs with input perturbations in order to obtain informative data for the modeling for control. Furthermore the importance of using measurement variables that contain most relevant information about the process is demonstrated.

The modeling results of this paper demonstrates that it is actually feasible to obtain a reasonable cultivation model from a few designed production experiments. In fact it turned out when the experienced personnel saw the responses from the input perturbations they were enthusiastic and wanted to employ these for other purposes as well.

The obtained discrete time model is demonstrated to be directly applicable for providing a soft sensor signal of the measured variables, such as enzyme activity in this particular case. The tuning of this filter is greatly facilitated by utilizing covariance estimates from the GoLM estimation.

The control contribution presents how asymptotic convergence of Iterative Learning Control may be combined with the closed-loop performance of Model Predictive Control to form an asymptotically stable optimal controller labeled Learning Model Predictive Control for ensuring reliable and reproducible operation of batch and periodic processes. Consequently this controller type may also be used for optimizing control, [9]. The data driven approach obviously restricts the usability of such a model to similar cultivations of the same microorganism, however if the product is a bulk product which is produced relatively often, then many batches will be able to benefit from a soft sensor and control design, such that there is a sound economical basis for such an investment. The benefit is obviously even larger if a high value product is produced in large amount.

5.2 Conclusions

The GoLM Methodology provides a methodology for black-box model estimation complemented with the L-MPC for handling control both within the batch and between batches. The GoLM methodology has in this chapter been applied to an Aspergillus oryzae cultivation, where the produced enzyme is used to speed up the degradation of starch to glucose. Since the microorganism is a filamentous fungus this production process is very challenging. However it is demonstrated that the GoLM methodology can be used for model estimation when only few batches are available. However, it is important that the batches have been designed for identification, this means, that the data set recorded has relevant information on actuator movement. The importance of the information content is demonstrated on the benchmark case study through applying different perturbation amplitude on the inputs.

With the obtained model a L-MPC was developed and implemented. The controller was tested under two different scenarios; the first scenario simulates disturbances in the initial condition, which is a typical disturbance in Batch Fermentations. The control objective is batch reproducibility, the control rejects this disturbances and a model obtained with the same productivity as specified by the reference batch. Such a reproducibility is a significant advantage when the cost of loss for the variations on the final product is significant. In the second scenario an intra-batch disturbance occurred. In this case the robustness of the controller was tested during the simulation where the maximal growth rate was changed, this. The controller effectively minimized the error, even though the model used for the controller did not represent the plant due for the parameter change.

The experimental design for the industrial cultivation illustrates the importance of perturbing the process input. An amplitude of 25% of the nominal value yielded informative measurement data, as realized during two cultivations. Concatenating these batches to previous unperturbed batches it was possible to have sufficient data for GoLM model identification. From the previous batches it was most important to eliminate batches with a qualitatively different behaviour mainly using classification.

Based upon the discrete time GoLM model the development of a soft sensor for the measured variables including enzyme activity is demonstrated. This sensor represents perhaps the first direct application of the model. The model may also be used to optimize the Batch Operations Model for the cultivation. And furthermore the model is directly applicable for design of a L-MPC controller for handling intra- and inter-batch disturbances as illustrated in simulation. It is important to emphasize that even though the model of the process used for the control may not be perfect, then the controller can iteratively minimize this error to reject intra and inter batch disturbances and ensure stable learning by suitable design of the control weighting.

Acronyms

ARMAX AutoRegressive Moving Average models with eXogenous
ARX AutoRegressive models with eXogenous
G_ρ Estimated Generalization Errors
GoLM Grid of Linear Model
ILC Iterative Learning Control
L-MPC Learning Model Predictive Control
LS Least Squares
LTI Linear Time-Invariant
LTV Linear Time-Varying
MA Moving Average
ρ Model Purpose
MPC Model Predictive Control
NCO Necessary Conditions of Optimality
OSA One Step Ahead
PRBS Pseudo Random Binary Signal
PS Pure Simulation
RR Ridge Regression
SS Space State
TR Tikhonov Regularization
wLS weighted Least Squares

References

1. Bajpai, R.K., Reuss, M.: A mechanistic model for penicillin production. J. Chem. Technol. Biotechnol. 30, 332–344 (1980)
2. Bajpai, R.K., Reuss, M.: Evaluation of Feeding Strategies in Carbon-Regulated Secondary Metabolite Production through Mathematical Modeling. Biotechnology and Bioengineering 23, 717–738 (1981)
3. Bonné, D.: Optimal and Reproducible Operation of Batch Processes. Technical University of Denmark, Department of Chemical Engineering. Ph. D. Thesis (2005)
4. Flores-Cerrillo, J., MacGregor, J.F.: Control of Batch Product Quality by Trajectory Manipulation Using Latent Variable Models. Journal of Process Control 14, 539–553 (2004)
5. Kristensen, N.R.: Fed-Batch Process Modelling for State Estimation and Optimal Control. Technical University of Denmark, Department of Chemical Engineering (2003)
6. Lind, M.: Status and Challenges of Intelligent Plant Control. A. Rev. Control 20, 23–41 (1996)
7. Ljung, L.: System Identification: Theory for the User. Prentice-Hall, Englewood Cliffs (1999)
8. Petersen, N.: Multivariable Modeling for Control of Industrial Fed Batch Cultivations. Department of Chemical Éngineering, Technical University of Denmark, M.Sc. Thesis (2006)
9. Rawlings, J.B., Bonné, D., Jørgensen, J.B., Venkat, A.N., Jørgensen, S.B.: Unreachable Setpoints in Model Preditive Control. Transactions of Automatic Control (2008)

10. Srinivasan, B., Bonvin, D.: Dynamic Optimization under Uncertainty Via NCO Tracking: A Solution Model Approach. In: BatchPro - Symposium on Knowledge Driven Batch Processes, Kipparisidis, C, pp. 17–35 (2004)
11. Srinivasan, B., Biegler, L., Bonvin, D.: Tracking the Necessary Conditions of Optimality with Changing Set of Active Constraints Using a Barrier-Penalty Function. Comput. Chem. Engng. 32, 572–579 (2008)
12. Åkesson, M., Hagander, P., Axelsson, J.P.: Probing Control of Fed-Batch Cultures: Analysis and Tuning. Control engng. practice 9, 709–723 (2001)

10

Performance Monitoring and Batch to Batch Control of Biotechnological Processes

Julian Morris and Jie Zhang

Centre for Process Analytics and Control Technology
School of Chemical Engineering and Advanced Materials
Newcastle University, Newcastle upon Tyne, NE1 7RU, UK
julian.morris@newcastle.ac.uk, jie.zhang@newcastle.ac.uk

Abstract. This Chapter describes two approaches to ensuring the production quality of batch biotechnological processes. The first makes use of the multivariate statistical data analysis and multivariate statistical process control (MSPC) or better termed multivariate statistical process performance monitoring (MSPM). An industrial application is described to the interrogation of data from a reaction vessel producing an active pharmaceutical ingredient (API) which enabled the realization of a better understanding of the factors causing the onset of an impurity formation to be obtained as well demonstrating the power of multivariate statistical data analysis techniques to provide an enhanced understanding of the process. In the second application, a simulation study of batch-to-batch iterative learning control strategy is presented where the batch control actions for the next batch are adjusted using the information obtained from current and previous batches. The control policy updating is calculated using a model linearized around a reference batch. In order to cope with process variations and disturbances, the reference batch can be taken as the immediate previous batch. After each batch, the newly obtained process operation data is added to the historical data base and an updated linearized model is re-identified. Since the control actions during different stages of a batch are usually correlated, it is proposed here that the linearized model can be identified from partial least square regression.

1 Introduction

The advancement of bioprocess performance monitoring and control will play a key role to meet the future requirements of bioprocessing and pharmaceutical development and production technologies. Major challenges relate to the faster product and process development to reduce the time to the market and to ensure full adoption of the guidelines of the recent FDA Process Analytical Technology initiative[1]. Due to the complexity of biological systems and the lack of appropriate sensor and analyzer systems biological production processes are still poorly understood and it becomes necessary to develop comprehensive process monitoring and closed loop process

[1] "PAT - A Framework for Innovative Pharmaceutical Development, Manufacturing, and Quality Assurance", FDA, 2004.

M. do Carmo Nicoletti, L.C. Jain (Eds.): Comp. Intel. Tech. for Bio. Mod., SCI 218, pp. 281–310.
springerlink.com

control concepts by taking full account of the multidimensional – multivariate nature of the biological process and its environment involving the most significant process variables and their interactions. "Although research for the development of novel sensor systems is progressing their applicability in bioprocessing is very limited with respect to on-line and in-situ measurement due to specific requirements of aseptic conditions, high number of analytes, drift, and often rather low physiological relevance. Modeling enables the continuous prediction of otherwise discrete off-line data whereby critical process states can be more easily detected" (from [1]).

Batch processes are of considerable importance in many sectors of industry including pharmaceuticals, food, fine chemicals and crystallizations, for the production of high-value added goods where the aim is to simultaneously achieve consistent quality and high yield. Batch processing is synonymous with flexible manufacturing and has the advantage over continuous production in that the system can be more easily modified to take account of changes in product specifications resulting from changes in or evolving customer requirements. Financial savings can be made in low volume manufacturing by using one set of vessels to produce a variety of products.

Traditionally, control of fed-batch processes has been based on highly scheduled systems where addition sequences and operational adjustments to the process environment are clearly defined and rigorously followed. In particular the sequencing of a fermentation is frequently determined from a basic understanding of the process and any adjustments to the process, such as the modification of process conditions and addition times by operators and process engineers are performed as the batch evolves. However, the actual biological and dynamic interactions of particular processes are often little understood due to the complexity of the reaction kinetics and the difficulties in sampling and analyzing the broth.

With increasing global competition, it is of critical importance that each fermentation produces high quality product at maximum yield. This has resulted in process performance monitoring becoming an integral part of process operation. The aim of process monitoring is to achieve early warning of changes in process operation that may result in low product yield or low quality production, or the identification of a disturbance that may result in the batch being terminated early as the product is out-of-specification. This results in much of the information latent within the data being ignored due to a lack of awareness of the necessary tools to extract the process features within the data.

In the first case study, the early detection and identification of process abnormalities during the manufacture of biochemicals intermediate is discussed; the formation of undesirable impurities during product manufacturing. The intermediate species studied is of particular importance since it forms a significant part of the final drug substance molecule. The objective of the study is to help identify those factors that have the greatest effect on impurity formation and to allow the establishment of acceptable operating procedures. Such protocols would potentially reduce laboratory analysis to only those batches that were close to the recommended operating limits, as opposed to testing all batches. It was conjectured that the problems that occurred during the reaction were most likely to be a result of a combination of factors rather being associated with just a single source. The industrial case study shows the benefits that can be gained from

(i) the application of multivariate statistical techniques performance monitoring to a key stage in a synthetic route for an active pharmaceutical ingredient (API).

In a second, simulation study, the use of batch-to-batch iterative control is used to ensure continued high quality product production. Here the impact off-line calculated control policies for batch fermentation processes are shown to be non-optimal when implemented on processes that exhibit variability due to model plant mismatch and/or the presence of unknown disturbances. In order to overcome these issues a batch to batch iterative learning control strategy is developed to modify the control actions for the next batch using the information obtained form current and previous batches. The control policy updating is calculated using a model linearized around a reference batch. In order to cope with process variations and disturbances, the reference batch can be taken as the immediate previous batch. After each batch, the newly obtained process operation data is added to the historical data base and an updated linearised model is re-identified. Since the control actions during different stages of a batch are usually correlated, it is proposed that the linearised model can be identified from partial least square regression. The proposed technique has been successfully applied to a fed-batch fermentation process.

Section 2 overviews multivariate statistical data analysis and modeling methodologies and Section 3 provides more details of principal component analysis (PCA) and its extension to multi-way PCA. Section 4 presents an industrial case study on using multivariate data analysis. Section 5 presents the partial least squares method for modeling and prediction. An iterative learning control strategy and its application to a fed-batch ferementation process are presented in Section 6. The last section provides a summary of nonlinear PCA, nonlinear PLS, and neural network approaches to the modeling, monitoring and control of pharmaceutical and biotechnological systems.

2 An Overview of Multivariate Statistical Data Analysis and Modeling Methodologies

A number of methodologies have been described in the literature for the maintenance of assured batch and fed-batch production. These include precise batch sequencing and automation and more recently multivariate statistical process control (e.g. [2, 3, 4, 5, 6, 7, 8, 9, 10]). More recently, advances in extending multivariate statistical performance (condition) monitoring algorithms and technologies have been made (e.g. [11, 12, 13, 14, 15, 16, 17, 18]). It is interesting to observe, however, that most applications of multivariate statistical process control have been applied to single recipe, single product or single grade process manufacturing. A major process performance monitoring challenge is to be able to address the issues associated with multi-recipe and multi-product modeling as well as modeling different unit processes on different manufacturing sites. A multi-group modeling approach [19] allows the modeling and monitoring of such flexible process manufacturing scenarios.

Nomikos and MacGregor [2, 3, 4] and Wold et al. [6] proposed on-line batch process monitoring schemes whose basis were the multivariate statistical projection based techniques of principal component analysis (PCA) and partial least squares (PLS). These methods are applicable to two-dimensional data sets and thus since batch data is inherently three dimensional (batch × variable × time), the first step is to reduce the dimensionality of the problem from three to two dimensions by rearranging the batch data. The two most common structures are (batch × (variable × time)) and ((batch × time) × variable). Alternative approaches that remove the need to reduce data dimensionality and that have been implemented for the monitoring of batch processes are the tri-linear approaches of parallel factor analysis (PARAFAC), PARAFAC2 and Tucker3 [20, 21, 22]. A number of comparative studies of the aforementioned monitoring methods have identified the advantages and limitations of the techniques in terms of process modeling and monitoring [20, 23, 24]. The batch process monitoring approaches mentioned above deal with within-batch variation and thus they monitor each batch independently during the batch run, or through-batch monitoring, which implies on-line monitoring. Lee and Dorsey [25] and Flores-Cerrillo and MacGregor [26] have also suggested monitoring schemes for detecting changes in batch-to-batch variations each the batch run, which relates to decisions made at the end of the batch, i.e. off-line analysis.

Batch and semi-batch processes typically exhibit significant dynamic behaviour with transient states, consequently measurements collected from these processes are usually auto- and cross-correlated. Thus the direct application of conventional PCA-based monitoring schemes is not possible as the raw batch data violates the underlying assumption that the process is asymptotically stationary. One approach proposed in the literature by Nomikos and MacGregor [2] to address this issue in batch process monitoring is to subtract the average time-series trajectory from the raw data and use the deviations from the mean trajectory as the inputs into the multi-way principal component analysis. By adopting this approach, the loading vectors contain information on the auto- and cross-correlation for all the variables for the duration of the batch. To implement this approach on-line, a key requirement is the need for the complete trajectory of the entire batch run since the score for an individual time point is a weighted sum of all the measurements collected throughout the batch run. Two different approaches to in-filling the unknown future measurements have been proposed: (1) zero deviation, or (2) current deviation. Even though these approaches have worked well in practice, both methods face two limitations. First, unknown future values from the current time to the end of the batch require to be estimated to attain the current score and residual [2, 27]. Secondly, the use of all the measurements throughout the duration of the batch assumes that the current condition of a process is affected by future operating conditions, which in practice is not necessarily valid. An alternative approach has been proposed whereby the score for an individual time point can be directly obtained by projecting only the available data onto the plane of a PCA model. Even though this method does not necessitate the estimation of future measurements, it has been shown to give poor monitoring performance especially in the early stage of a batch run [24].

Wold et al. [6] proposed an alternative approach that considered a different unfolding approach to the three dimensional batch data by preserving variable direction, i.e. (Batch × Time) × Variables, and does not require the estimation of future measurements. This method of data matrix unfolding explains the correlations between the variables centered by subtracting their mean over the entire batch; it does not explain their auto- and cross-correlation. As pointed out by Kourti [28], the variation describe by PCA in this method is not the type of variation of interest in on-line monitoring, i.e. dynamic behaviour of the deviations from the optimally desired time trajectory. Similar techniques have been applied to full scale fermentations (e.g. [11, 12, 29].

3 Principal Component Analysis

Principal Component Analysis (PCA) is a statistical tool that reduces the dimensionality of a problem by defining a new set of variables, each of which explains a decreasing amount of variability in the process data [30]. The new variables, termed principal components, are linear combinations of the original variables that are constrained to be orthogonal. The main areas of application of Principal Component Analysis in process analysis are data reduction, classification of variables, outlier detection, early warning of potential malfunctions and 'fingerprinting' for fault identification. Let $\mathbf{X} = \mathbf{x}_1, \mathbf{x}_2, \mathbf{x}_3, \ldots, \mathbf{x}_m$ be an m-dimensional data set describing either the process variables or the quality information. The first principal component is defined to be the linear combination of the columns of X which describe the greatest amount of variability, $\mathbf{t}_1 = \mathbf{X}\mathbf{p}_1$ subject to $|\mathbf{p}_1| = 1$ in the original data set. In m-dimensional space, $\mathbf{p}_1$ defines the direction of greatest variability, and $\mathbf{t}_1$ represents the projection of each object onto $\mathbf{p}_1$. The second principal component is the linear combination defined by $\mathbf{t}_2 = \mathbf{X}\mathbf{p}_2$ which describes the next greatest variance subject to $|\mathbf{p}_2| = 1$ and subject to the condition that it is orthogonal to the first principal component $\mathbf{t}_1$:

$$\mathbf{t}_2 = \mathbf{E}_1\mathbf{p}_2 \quad \text{where} \quad \mathbf{E}_1 = \left(\mathbf{X} - \mathbf{t}_1\mathbf{p}_1^{\mathrm{T}}\right) \tag{1}$$

This procedure is essentially repeated until m principal components are calculated. In effect, PCA decomposes the observation vector, X, as:

$$\mathbf{X} = \mathbf{T}\mathbf{P}^{\mathrm{T}} = \sum_{i=1}^{m} \mathbf{t}_i\mathbf{p}_i^{\mathrm{T}} \tag{2}$$

where **P** are the principal component loadings and **T** is the principal component scores matrix. The loadings vector provides information as to which variables contribute the most to individual principal components, i.e. they are the coefficients in the principal components model; whilst information on the clustering of the samples and the identification of transitions between different operating regimes is obtained from the scores vector, $\mathbf{t}_i$.

The main attribute of PCA, which enables the dimensionality of the problem to be reduced, is that if some of the process variables are collinear a smaller number of principal components than original variables are required to explain the variability in the data. Consequently it is desirable to exclude some of the components. In addition, the lower order components often describe the noise in the data. Typically, the number of principal components that provide an adequate description of the data is identified using cross-validation [31]. In practice two or three principal components are frequently sufficient for monitoring a process. The final model can thus be defined as:

$$\mathbf{X} = \mathbf{T}\mathbf{P}^{T} + \mathbf{E} = \sum_{i=1}^{k} \mathbf{t}_i\mathbf{p}_i^{T} + \mathbf{E} \tag{3}$$

The calculated principal components form the cornerstone of multivariate statistical process control charts.

3.1 Multi-way Principal Component Analysis

Batch data set consists of observations of each process variable at each time point throughout the batch production leading to a 2 dimensional data matrix, **X** for each batch. A multivariate approach of principal component analysis may be used to reduce the dimensionality of the process data set and provide a set of components to analyze visually, presenting the data in a format that may be used to detect process features and increase understanding of the fermentation. Data collected on a number of batches is typically arranged in a three-way matrix, batches x variables x time. To apply a bi-linear technique of multi-way principal component analysis the data matrix requires first to be unfolded to a two-dimensional array. Two approaches are used that unfold the data matrix in different ways leading to variations on the PCA analysis that draw different information from the data sets.

The first approach, MPCA was first introduced by Geladi et al. [32] and was later shown to be appropriate for the analysis of batch process data [2]. There are a number of ways that a three-way matrix can be decomposed, three routes are shown in Fig. 1, and the appropriate form depends on the ultimate objective of the analysis. If interest is in the behaviour at the end of the batch then the three-way data matrix given by **X** ($I \times J \times K$) is decomposed to give a two dimensional array where each row represents a single batch run, i, and comprises the individual process measurements at each time point Fig. 1 (Route 1). The second approach to multiple batch analysis is the case where the batches are stacked with each row corresponding to a particular variable for a specific time point, giving an $(I \times K) \times J$ matrix. The structure can be seen in Fig. 1 (Route 2). This vertical stacking approach where batches are not unfolded but kept as the original data matrix retains the variable trajectory for each batch as it evolves with time and is strictly not multiway PCA but essentially analogous to conventional PCA.

Now consider the case where there is available a database of I batches, where J process variables are measured over K time intervals (batches × variables × time), respectively, throughout the duration of the batch. This information can be arranged into a three-way data set **X** ($I \times J \times K$). There are three possible ways to unfold the array. In this study, the array was unfolded by partitioning the matrix into a series of two-dimensional slices and unfolding the matrix by the third variable. That is, each vertical slice ($I \times J$) is placed side by side, starting with the one corresponding to the first time interval [2]. This unfolding approach allows the analysis of the variability between batches, i.e. summarizing the information in the data with respect to both the variables and their time variation.

The objective of multi-way PCA is to decompose the three-way array **X** into R principal components comprising score vectors ($\mathbf{t}_r$) and loading matrices ($\mathbf{P}_r$), or unfolded as vectors ($\mathbf{p}_r$), plus a residual **E**, which is as small as possible, in a least squares sense :

$$\underline{\underline{X}} = \sum_{r=1}^{R} \underline{t_r} \otimes \underline{\underline{P_r}} + \underline{\underline{E}} \quad \text{or} \quad \underline{\underline{X}} = \sum_{r=1}^{R} \underline{t_r} \cdot \underline{p_r^T} + \underline{\underline{E}} \tag{4}$$

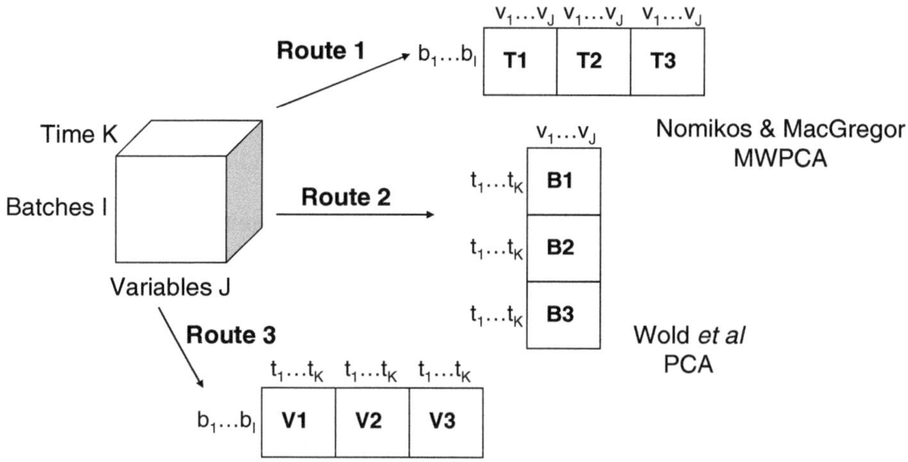

Fig. 1. Different Batch Unfolding Routes

This decomposition follows the principles of PCA, i.e. it separates the data block in an optimal way into two parts. The residual part (**E**) describes the noise associated with the data and the systematic part ($\sum_{r=1}^{R} \underline{t_r} \cdot \underline{p_r^T}$), which expresses the data set as one fraction ($\mathbf{t}_r$) related only to batches and a second fraction ($\mathbf{P}_r$) related to variables and their time variation. The loading matrix $\mathbf{P}_r(J \times K)$ in multi-way PCA contains most of the structural information about how the variable measurements deviate from their average trajectories under normal operation.

Mean-centering is usually applied to the data prior to performing a multi-way PCA. The mean of each column of the **X** data set is subtracted from each data element of this column. The way in which the **X** matrices are unfolded combined with mean-centering is very important since it results in the subtraction of the mean trajectory of each variable and, thereby, in the removal of the main non-linear component in the data. A PCA performed on this mean corrected data is, therefore, a study of the variation in the time trajectories of all the variables in all the batches about their mean trajectories. Furthermore, by scaling the variables in each column of **X** to unit variance, differences in the measurement units between variables can be taken into account and equal weight given to each variable at each time interval.

Performing a through-batch analysis without unfolding the variable trajectory of each batch enables the monitoring of the evolution of a new batch as analysis can be performed at each time point and the variable trajectory constantly updated. This approach also allows for batches of varying lengths to be analyzed in one data matrix without any adjustment to the observations. This is an important consideration for batch evolution where the time when an event occurs is often crucial to the batch performance. Identification of a batch being out of the control limits leads to further interrogation of the possible cause of the alarm signal. Contribution plots [33] are key to this by indicating the influence of each variable on the outlying observation.

The end-of-batch approach (Route 1 in Fig. 1) is applicable if each batch duration is constant across all batches to be analyzed. In most industrial batch processes this is not the case, with many batches over running or being terminated early. Work has been carried out into various methods for equalizing batch lengths. The three main methods include cutting the time series trajectories to a minimum length, dynamic time warping [34, 35] and the use of a surrogate variable (indicator variable) instead of time [36].

In this work, the most appropriate approach was that of multivariate dynamic time warping since it was important to capture process interaction both at the beginning of the Batch where the first additions takes place and at the end of the Batch where the second addition tales place. By cutting the batches to a minimum length, important information towards the end of some batches would have been lost whilst the surrogate variable approach was not possible since none of the variables exhibited the appropriate mathematical properties, i.e. monotonic increasing. Dynamic time warping (DTW) has previously been used as a method of pattern matching in speech recognition [34] and to the classification of faults in a continuous chemical process [35]. DTW distorts the test profile by stretching or compressing it to the length of a reference profile. For this application the mean trajectory was used as the reference trajectory with the other batches being aligned against this reference batch. Translating, expanding and contracting localized segments allowed the distance between the reference batch trajectories and those of the batch being synchronized to be minimized. This resulted in the time axis of both patterns being mapped onto a common axis.

4 Industrial Case Study Using Multivariate Data Analysis

4.1 Process Description

The raw process data comprised 29 batches, each containing between 300 and 500 samples. Five process variables were routinely logged at 30 second intervals – jacket outlet temperature (ºC), jacket inlet temperature (ºC), agitator speed, refrigerant flow and reactant temperature (ºC). Of the twenty-nine batches, the first nine were manufactured in a stainless steel vessel. During product development, this vessel was replaced by a Hastelloy reactor to provide an increase in cooling capacity and a reduction in batch duration. The remaining batches were manufactured in the Hastelloy reactor but the reactant capacity was increased for the last ten batches. However the larger cooling capacity of the Hastelloy vessel did not necessarily mean that the duration of these 'larger volume' batches would be longer than those produced with a lower volume of reactants.

Two measures of quality, yield (%) and impurity level (%) were made available. In this particular study, yield was not relevant to the analysis and was therefore excluded. According to product specifications, failed batches were classed as those that had an impurity level in excess of 0.14% although the desired level was less than or equal to 0.10%. Fig. 2 shows the variation in impurity levels. Examining the impurity readings for the two reactor types, it was apparent that the batches produced in the stainless steel reactor were unacceptable. For the Hastelloy reactor, all the batches were within the acceptable limits except for one of the batches, Batch 25,

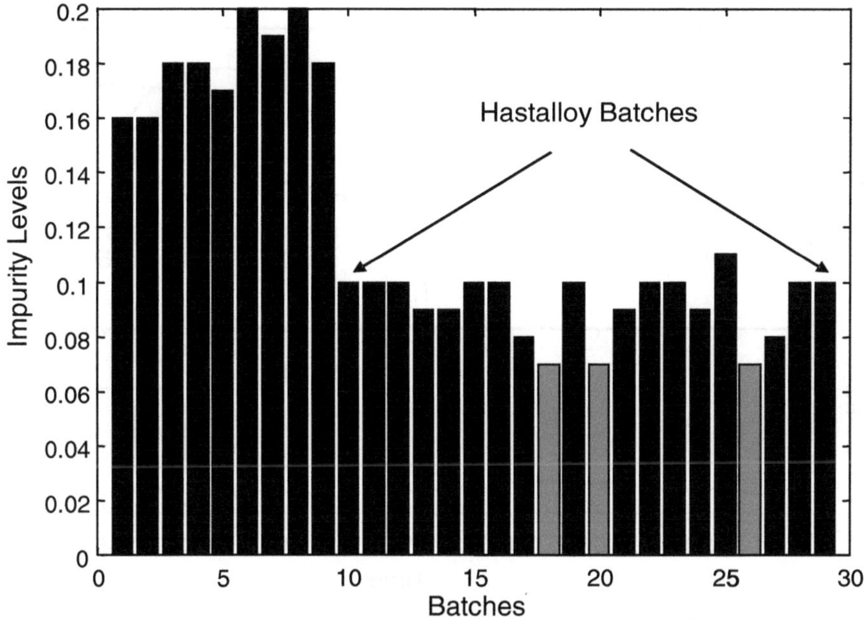

Fig. 2. Impurity levels of Stainless Steel and Hastelloy batches

which had an impurity level in excess of 0.10% However, it is interesting to observe that the minimum level of impurity recorded was 0.07% for Batches 18, 20 and 26. An interesting question is therefore to establish under what operational conditions, impurity levels of 0.07% might be guaranteed. This was explored as part of the study and is described in Section 4.3.

Before any detailed modeling or monitoring analysis can be performed, it is essential to visualize and interrogate the data collected, initially from all the batch runs and later focusing on only the Hastalloy batches where there were no impurities formed. This enables a clearer understanding of the data structure and how it reflects process behaviour. During this stage it was necessary to work closely with the production personnel and clarify any issues regarding the process, its operation and the data made available. The first step was to pre-screen the data for missing observations and outliers. Once these were identified, an appropriate in-filling algorithm was applied, e.g. zero or first order interpolation. It was then necessary to ensure that the model building data collected for each of the individual batches covered a similar period of operation. Finally the issue of the equalization of batch lengths was addressed. This is an essential pre-requisite if the data is to be analyzed using multi-way principal component analysis or partial least squares and is discussed later.

4.2 Batch Operations

The reactor is initially charged with a reactant slurry in a solution of ethyl acetate. Two additions are made during the reaction. From the time series plots of reactor vessel temperature and jacket outlet temperature for a particular Hastelloy Batch, Fig. 3 where

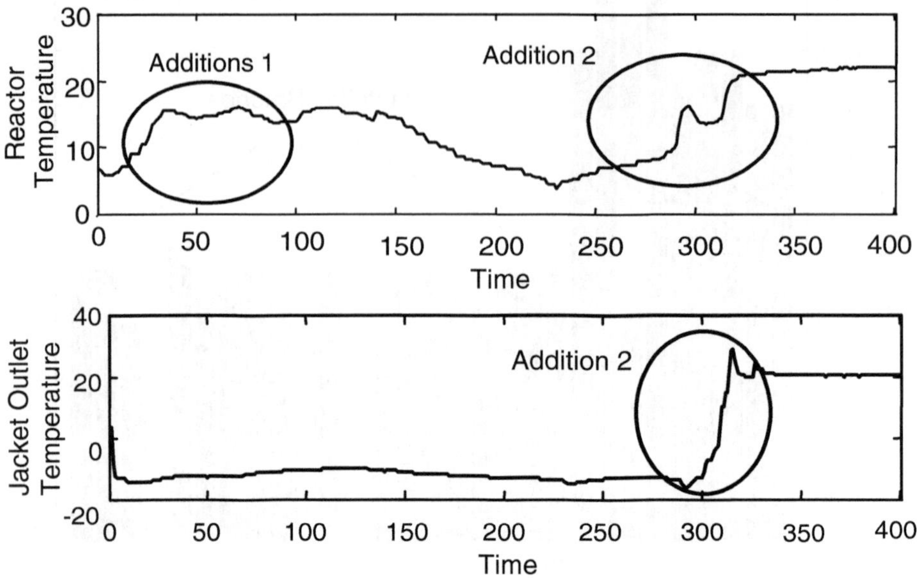

Fig. 3. Reactor vessel temperature with addition points

the key stages relating to the additions in the reaction can be observed. The first 'addition' takes place in two stages starting at around sample time 10, just prior to the sharp rise in temperature, due to the reaction exotherm, with a second part occurring around time 75 followed by a further temperature increase. During the addition and subsequent time period, the temperature must be kept below 17^0C. Thus the addition time is effectively determined by the cooling capacity of the vessel. Later in the batch, a second addition takes place. This can be observed as starting at around time point 270, where the second exotherm occurs. The temperature at this stage is required to be kept close to 21^0C. Once the second addition has been completed, a 4 hour stir commences. This is indicated in both time series plots by the temperature leveling-off from sample time 325 onwards. Reactor data logging commences during reactor preparation and charging, this results in data being collected prior to the first addition which is not of interest in this study. Thus to ensure that each set of data covered the same period of operation, the data collected during the initial reactor charging was truncated. Following the truncation of the batches, the data sets were still of unequal lengths with batch duration varying between 170 and 220 min. The main reason for this wide range in batch times was related to the differing vessel types in which the intermediate was manufactured, as well as to the additions that were manually controlled.

4.3 Multivariate Data Analysis Results

In this study multivariate Dynamic Time Warping (DTW) was used [35, 36]. DTW distorts the test profile by stretching or compressing it to the length of a reference profile. For this application the mean trajectory was used as the reference trajectory with the other batches being aligned against this reference batch. Translating, expanding and

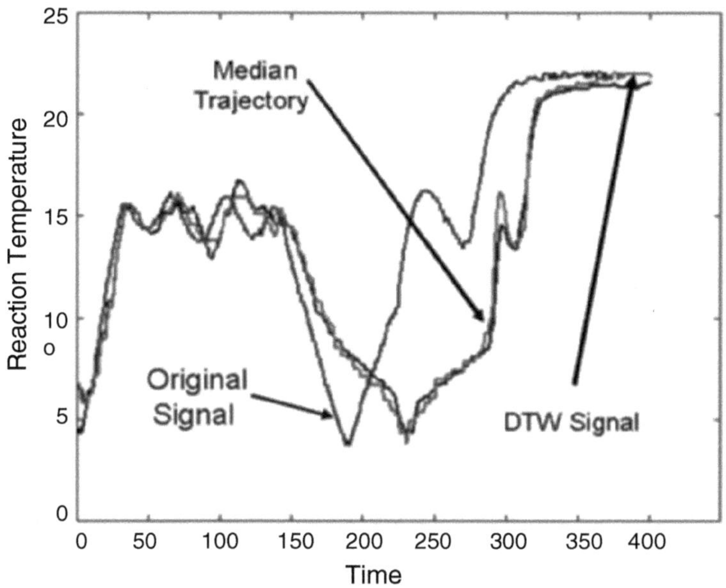

Fig. 4. Synchronisation of time trajectories for reactor temperature

contracting localized segments allows the distance between the reference batch trajectories and those of the batch being synchronized, to be minimized. This results in the time axis of both patterns being mapped onto a common axis. A total of 401 samples were selected for each of the 5 variables for the analysis.

Fig. 4 illustrates the synchronization for the variable, reactor temperature. The solid line is the truncated raw data and the dashed line is its time-warped equivalent. The aim of this particular study was to identify batches that exhibited different characteristics, eg low impurity batches 18, 20 and 26 and the high impurity batch 25, compared to those that demonstrated 'average' behavior, e.g. batches 10 to 17, 19, 21 to 24 and 27 to 29. This was achieved through the application of multi-way PCA. Thus interest was in batches which had either high or low levels of impurity.

The first step was to show operational personnel how multi-way PCA could visualize and identify the sources of the differences in performance of the different Batches. Multi-way PCA was carried out on the pre-screened data matrix comprising the 9 stainless steel and the 20 Hastelloy Batches. Six principal components, which explained approximately 80% of the total variability, were retained for subsequent analysis. The scores plot of principal component 1 and 2 shows three distinct groups within the data set, Fig. 5. The first group (A) includes Batches 1 to 9 and relates to the data associated with the Batches produced in the stainless steel vessel. As mentioned previously, the impurity levels for this vessel were higher than those for the batches produced in the Hastelloy vessel. In the subsequent more detailed analysis, the batches manufactured in the stainless steel vessel were removed from the study. The second group (B) contains Batches 10 to 26. These were manufactured in the Hastelloy reactor. This group includes batches that were produced with both small and larger batch reactant volumes. The third group (C) containing Batches 27, 28 and 29 were also produced in the

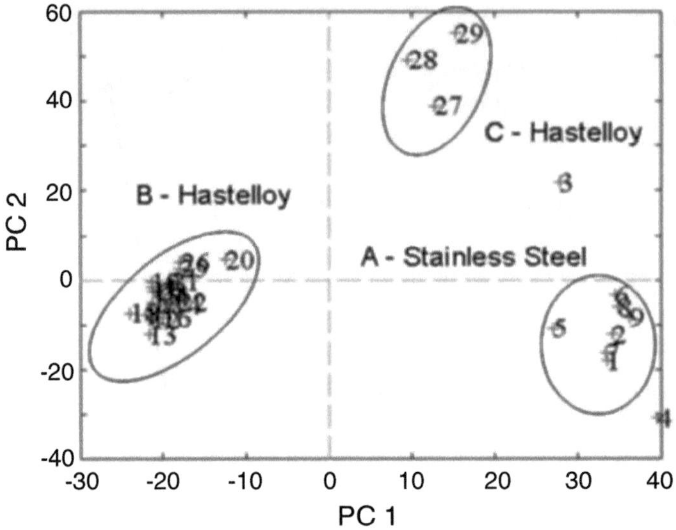

Fig. 5. Principal component plot for different reactors and reactor operations

Hastelloy vessel. Interrogating the data from these three batches revealed that the stirrer speed was higher than for the other Hastelloy batches. However, the impurity level for these batches was less than 0.10% indicating that for this particular process, stirrer speed did not appear to directly contribute to impurity levels.

Multi-way PCA was then applied to a multi-way PCA model built from the Hastelloy Batches from group B, Fig. 5. Focusing on the bivariate scores plot of principal component one and two plotted in Fig. 6a, the three batches generated at a higher speed 'o' (group C in Fig. 5), are observed to be well separated for the main cluster of 'acceptable' batches. Batch 25 is observed to fall outside the 99% action limit. It is interesting to observe that Batch 25 has the highest impurity level of all the Hastelloy Batches, 0.11%, compared with that of Batch 14 which has an impurity level of 0.09%. Figures 5 and 6a were able to demonstrate to operational personnel unfamiliar with multivariate statistical data analysis the power of data visualization.

Examining now the scores plot of principal components 5 and 6, Fig. 6b, these two Batches, 14 and 25, appear to be significantly 'different' from the other 'nominal' batches. Batch 14 (impurity 0.09%) lies within the action limit but apart form the rest of the batch scores, whilst Batch 25 (0.11% impurity) is seen to lie outside the action limit. To investigate the reason for Batch 25 being outside the action limit, a score contribution plot for this Batch is compared with that of a 'good', Batch 14 (impurity 0.09%), that lay near the origin of the PC1 versus PC2 scores plot in Figures 6a and 6b.

Contribution Plots, Figures 7a and 7b, provide information on changes in new observation variables relative to the average value calculated from the PCA model in that they represent the particular process variables that are unusual for a particular process condition. For example, consider the score relationship $t_1 = p_{11} * variable1 + p_{12} * variable2$; the combination ($p_{11} * variable1$) is the contribution of variable 1 to PC 1 and the combination (p_{12} * variable2) is the contribution of variable 2 to PC 1.

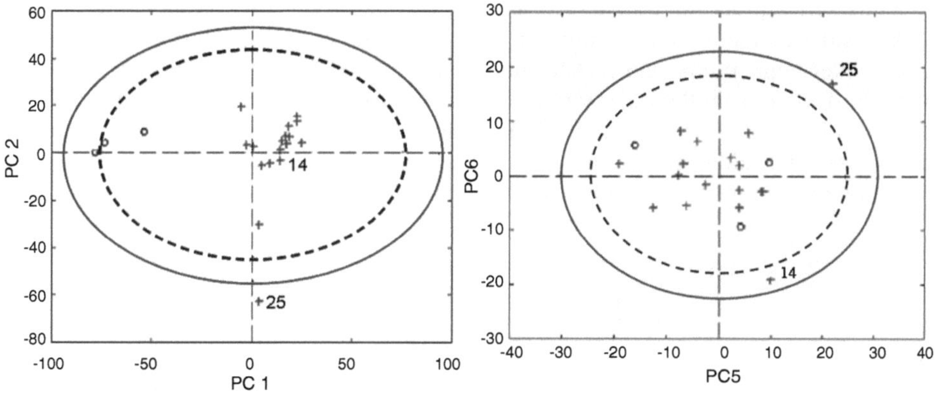

(a) Scores plot of Principal Component 1 versus Principal Component 2 showing higher speed batches (Blue circles) on the LHS and Batch 14 (0.09% impurity) and Batch 25 (0.11%)

(b) Scores plot of Principal Component 5 versus Principal Component 6 showing the closer clustering in the lower principal components and Batch 14 (0.09% impurity) and Batch 25 (0.11%)

Fig. 6.

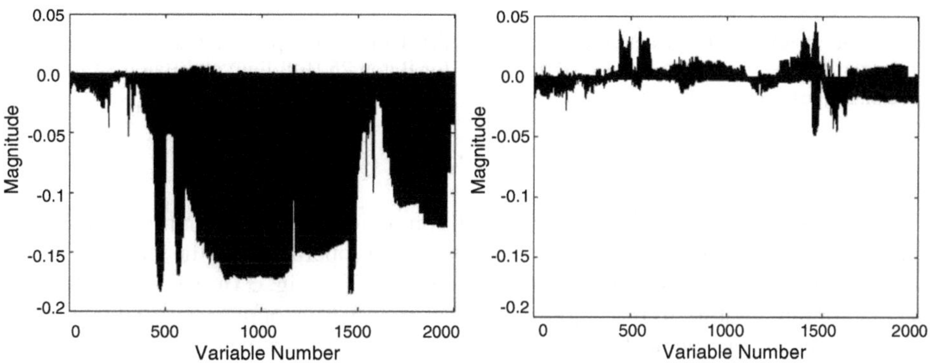

(a) Scores contribution plot of variables for Batch 25 PC 2

(b) Scores contribution plot of variables for Batch 14 for PC 2

Fig. 7.

In the multi-way contribution plots and resulting from the batch unfolding structure, the x-axis 'Variable Number', points 1 to 5 relate to actual process variables 1 to 5 at time point one, points 6 to 10 relate to actual process variables 1 to 5 at time point 2, etc. Contribution plots are described in more detail in [33].

Examining the score contribution plot for both Batches, Fig. 7a (Batch 25) and Fig. 7b (Batch 14), it can be seen that the magnitude of the scores are significantly greater for Batch 25. Examining the contribution plot for Batch 25 more closely at time point 200 (Variable Number 1000), the most influential process variables were identified as the jacket inlet temperature and the jacket outlet temperature, variables 2 and 1 respectively.

Time series plots for Batches 25 and 14, for the jacket inlet temperature and the jacket outlet temperature are shown in Figures 8a and 8b. It can be observed that the jacket inlet temperature and jacket outlet temperature of Batch 25 (0.11% impurity) starts to deviate from the trajectory of Batch 14 during the first addition period. Also, the reactor temperature for Batch 25, Fig. 8a is seen to be much less well controlled during the first addition. In contrast, for Batch 14, the reactor temperature is better controlled with a more constant jacket inlet temperature during the first addition.

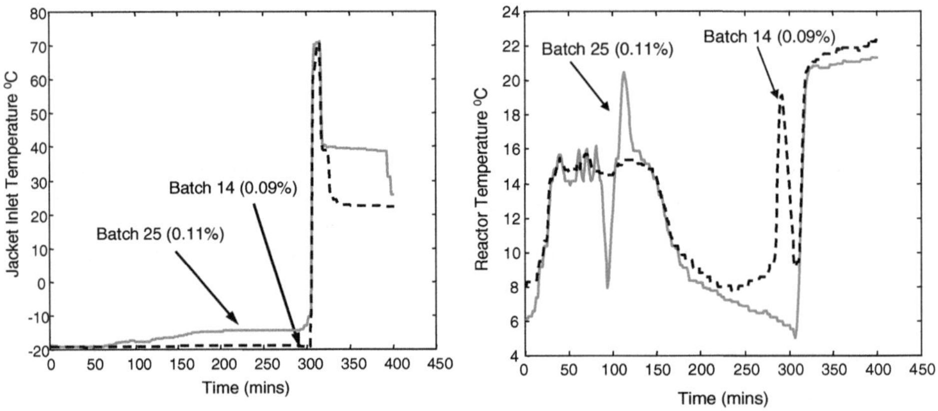

(a) Time series plot of jacket inlet temperature for Batch 25 (full line) and Batch 14 (dashed line)

(b) Time series plot of reactor temperature for Batch 25 (full line) and Batch 14 (dashed line)

Fig. 8.

Attention now focuses on the second addition period at around sample point 310. A characteristic of batch operation in this particular operation is the closing of the coolant control valve at the time of the second addition, Fig. 3 (lower plot), allowing the reactor temperature to rise rapidly to its desired 21ºC. Stopping the coolant flow in this way results in the jacket coolant temperature measurements being unreliable. Following a period of erratic coolant flow conditions, good control is seen to be regained in Batch 14 in contrast to that in Batch 25. This results in a clear difference in the responses of the jacket temperatures between the two batches during the subsequent four hour stir. Fig. 8b indicates that the impact on the reactor temperature for both Batch 25 and Batch 14 is similar with Batch 14 having a slightly higher temperature. It is clear that the main impact on impurity formation relates to the temperature control effects during the first addition, although the behaviour of the jacket inlet temperature is a cause for concern. Following discussions with plant personnel, it was concluded that issues relating to other plant coolant demands and coolant control issues were possibly contributing to increased impurity levels.

It was observed at the start of the study that three of the Hastellloy batches produced a final product with an impurity level of 0.07%. It is therefore instructive to examine the time series profiles of the jacket inlet temperature and the reactant temperature for two batches, Batch 19 with and impurity of 0.1% and Batch 20 with

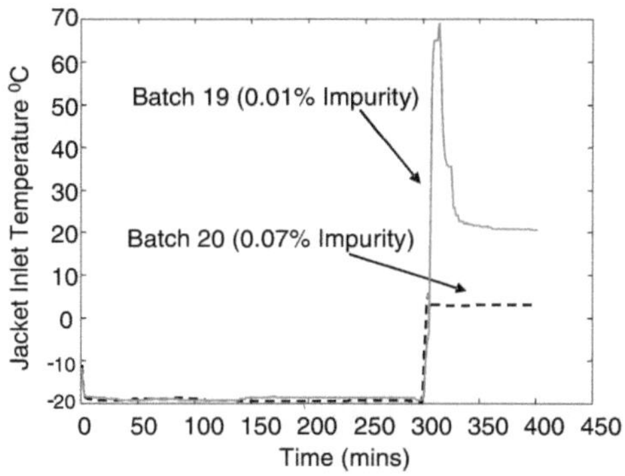

(a) Time series plot of jacket inlet temperature for Batch 20 (full line) and Batch 19 (dashed line).

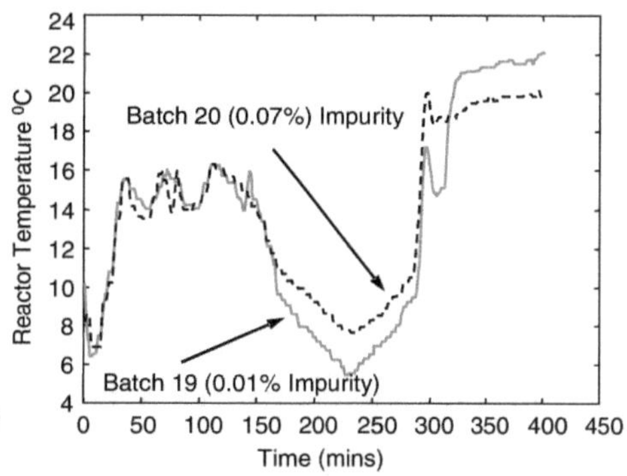

(b) Time series plot of reactor temperature for Batch 20 (full line) and Batch 19 (dashed line).

Fig. 9.

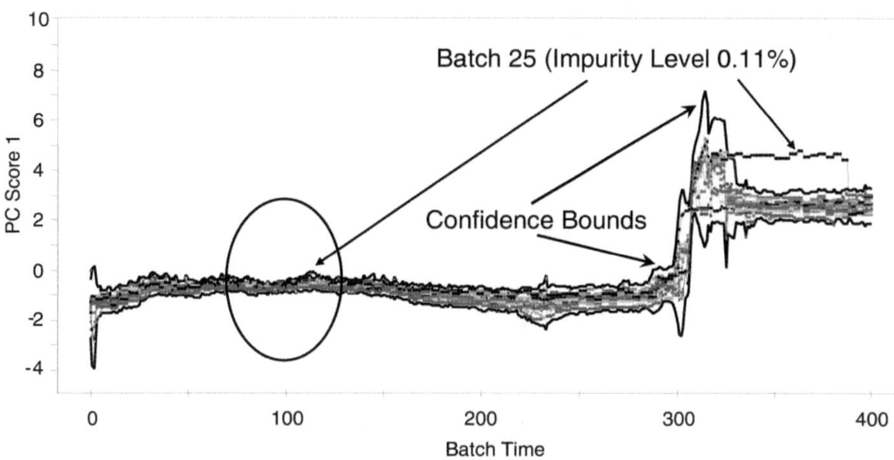

Fig. 10. Real Time through Batch Monitoring

an impurity of 0.07%, Figures 9a and 9b respectively. Clearly the superior control of the jacket coolant results in a much improved reactor temperature profile.

Finally, Fig. 10 demonstrates the power of being able to continuously monitor the performance (condition) of the batch trajectory during production. The time series scores plot shows a number of Hastelloy batch runs all lying within the action limits. A processing issue in Batch 25 can be observed during the first addition period by the batch trajectory breaching the action limit shortly after the first addition. Following the second addition and prior to the four hour stir, the batch again exhibits a different batch trajectory. Clearly, had through batch monitoring been in place at the time of the development runs the processing issue in Batch 25 would have been highlighted following the first addition leading to an early warning of what in this particular case was an impurity problem.

Finally, an interesting question that often arises is related to the ability to predict a process 'quality' output, in this case the impurity level. A multi-way Partial Least Squares model (was developed using the similar idea of unfolding the 3-dimensional data set to form a 2-dimensonal data matrix [4]. A good tutorial which describes the Projection to Latent Structures (or Partial Least Squares) approach is presented by Geladi and Kowalski [37].

5 Partial Least Squares Modeling and Prediction

PLS is a multivariate regression method that maximizes the correlation between the process variables (X) and the quality variables (Y) [37]. The objective of linear PLS is to project the data down onto a number of latent variables, say $\mathbf{t}_j$ and $\mathbf{u}_j$ and develop a regression model between the latent variables:

$$\mathbf{u}_j = b_j \mathbf{t}_j + \mathbf{e}_j \qquad (j = 1, \ldots, R) \tag{5}$$

where R is the number of latent variables defined, for example from cross-validation [31]. Let $\hat{\mathbf{u}}_j = \hat{b}_j \mathbf{t}_j$ be the prediction of $\mathbf{u}_j$ (Eq. 1). The matrices $\mathbf{X}$ and $\mathbf{Y}$ are decomposed as the sums of the outer products of the latent variables, $\mathbf{t}_j$, and the loadings, $\mathbf{p}_j$, and the prediction $\hat{\mathbf{u}}_j$ of $\mathbf{u}_j$ and the loadings $\mathbf{q}_j$, respectively:

$$\mathbf{X} = \sum_{j=1}^{R} \mathbf{t}_j \mathbf{p}_j^{\mathrm{T}} + \mathbf{E} \tag{6}$$

$$\mathbf{Y} = \sum_{j=1}^{R} \hat{\mathbf{u}}_j \mathbf{q}_j^{\mathrm{T}} + \mathbf{F} \tag{7}$$

where $\mathbf{E}$ and $\mathbf{F}$ are the residual matrices for the matrix decomposition of $\mathbf{X}$ and $\mathbf{Y}$.

However, when dealing with real and complex chemical and physical systems, linear PLS cannot reliably be used to model the underlying structure as it may exhibit significant non-linear characteristics. Several contributions have been made to integrate non-linear features within the linear PLS framework. Wold et al. [38] proposed a polynomial (quadratic) PLS algorithm that retains the framework of linear PLS but which modifies the relationship between the predictor and the response latent variables:

$$\mathbf{u}_j = c_{0j} + c_{1j} \mathbf{t}_j + c_{2j} \mathbf{t}_j^2 + \mathbf{e}_j \tag{8}$$

Baffi et al. [39] extended this approach to include a quadratic inner function.

To extend PLS to Multi-way PLS (MPLS), the case is considered where the data consists of I batches, where J process variables and L quality variables are measured over K and M time intervals, respectively, throughout the duration of the batch. As with multiway PCA, this information can be arranged into two three-way data matrices

$\mathbf{X}$ $(I\times J\times K)$ and $\mathbf{Y}$ $(I\times L\times M)$. In this study, the batch observation level unfolding approach of Wold et al. [6] is adopted. This enables the monitoring of a batch as it evolves as the batches are stacked without any unfolding of the variable trajectory. Here the data matrices $\mathbf{X}$ and $\mathbf{Y}$ are decomposed into a series of latent variables comprising score matrices and loading vectors, plus residual matrices $\mathbf{E}$ and $\mathbf{F}$:

$$\mathbf{X} = \sum_{r=1}^{R} \mathbf{T}_r \otimes \mathbf{p}_r + \mathbf{E} \quad \text{or} \quad \mathrm{X} = \sum_{r=1}^{R} \mathbf{t}_r \mathbf{p}_r^T + E \tag{9}$$

$$\mathbf{Y} = \sum_{r=1}^{R} \mathbf{T}_r \otimes \mathbf{q}_r + \mathbf{F} \quad \text{or} \quad \mathrm{Y} = \sum_{r=1}^{R} \mathbf{t}_r \mathbf{q}_r^T + F \tag{10}$$

These decompositions follow the principles of PLS with each score vector corresponding to the evolution of each of the batches with time.

5.1 Multivariate Modeling of Impurities

The data matrices necessary to undertake partial least squares modeling combined the addition times and the unfolded process data to infer the percentage impurity measured for each batch are shown in Fig. 11 and Fig. 12 respectively. The resulting PLS model used to predict impurity levels in the Hastelloy batches are shown in Fig. 13. The actual values (solid line) are observed to be close to the predicted values (dashed line) inferring that this particular model can be used to adequately predict the impurity levels of the Hastelloy batches in this particular case. It is stressed however that these predictions, although appearing to be good cannot be considered as being reliable or robust due to the minimal data available and only serve to indicate the potential of the PLS method.

Fig. 11. Construction of the PLS Data Matrices

This industrial case study highlights the potential and power of multivariate statistical techniques to assist in the identification and understanding of the reasons for the occurrence of different levels of impurity in the production of a pharmaceutical intermediate. In particular it has been shown that the reaction is sensitive to temperature changes during the first addition where it is essential that the temperature is well controlled within desired specifications as well as ensuring that the coolant supply

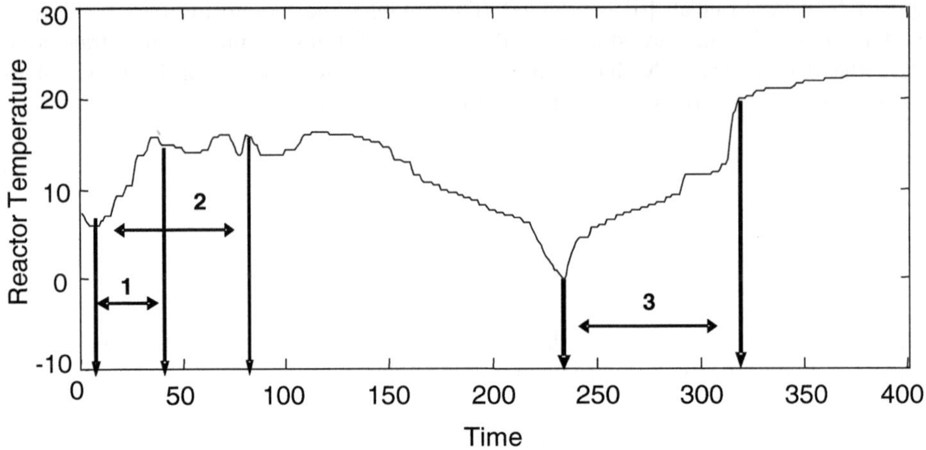

Fig. 12. Structure of the batch addition times

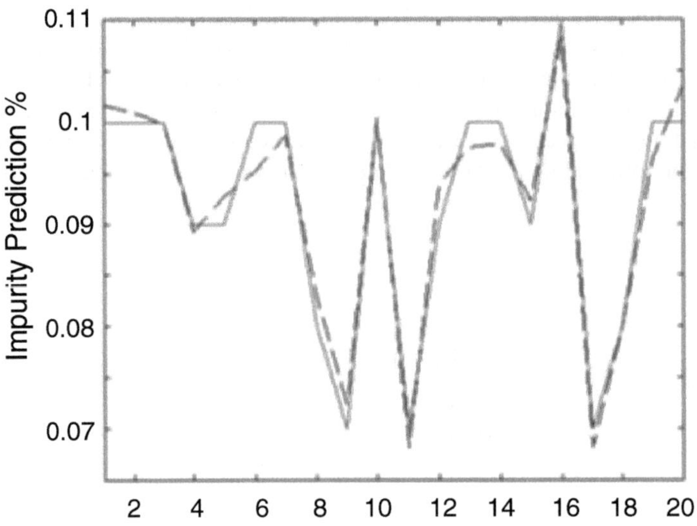

Fig. 13. Predicted versus Measured % Impurity

temperature is maintained at the desired level despite other plant utility demands. Finally the cluster attributed to increased stirrer speed confirmed that, for this particular reaction, speed did not appear to have an impact on impurity levels. This may be an issue that needs to be investigated further to determine if there are any real, but as yet unidentified mixing effects.

6 Iterative Learning Control of a Fed-Batch Fermentation Process

This section presents an iterative learning control strategy based on a linearised model that is identified from process operation data. The technique is demonstrated on a simulated fed-batch fermentation process.

6.1 Process Description

The fed-batch fermentation process is taken from [40]. It has been used for optimal control studies by a number of researchers [40, 41, 42]. The mechanistic model of the fed-batch ethanol fermentation process is described as follows

$$\frac{dX}{dt} = \mu X - \frac{X}{V}u \tag{11}$$

$$\frac{dS}{dt} = -10\mu X + \frac{(150 - S)}{V}u \tag{12}$$

$$\frac{dP}{dt} = \varphi X - \frac{P}{V}u \tag{13}$$

$$\frac{dV}{dt} = u \tag{14}$$

where

$$\mu = \frac{0.408S}{(1 + P/16)(0.22 + S)} \tag{15}$$

$$\varphi = \frac{S}{(1 + P/71.5)(0.44 + S)} \tag{16}$$

In Eq(11) to Eq(16), X is the cell mass concentration (g/L), S is the substrate concentration (g/L), P is the product concentration (g/L), V is the liquid volume of the reactor (L), u is the feed rate to the reactor (L/h), and t_f is the batch duration which is fixed as 65 h in this study. The initial condition is specified as $[X(0), S(0), P(0), V(0)]^T = [1, 150, 0, 10]^T$. The feed rate to the reactor is used for control and is constrained by

$$0 \le u \le 12 \tag{17}$$

In this fed-batch process, there is no outflow and, therefore, the feed rate must be chosen so that the reaction volume at the end of a batch does not exceed the physical volume of the reactor. The liquid volume of the reactor is limited by the vessel size

$$V(t_f) \le 200 \tag{18}$$

In this fed-batch process, the feed rate to the reactor is used for control. The control objective is to produce the maximum amount of the desired product, i.e. the yield of the reactor, which can be calculated as $P(t_f)V(t_f)$.

6.2 Batch-to-Batch Iterative Learning Control Based on Linearised Models

A typical feature of batch processes is that the same process unit is used in the manufacturing of different products and each product can be of a relatively small amount but with very high added value. The optimal operation of batch bio-reaction processes is very important in the face of increasing competition and increasing stringent regulations on product quality and waste disposal. To implement optimal control, accurate process models are usually required. Early research on batch process optimal control concentrated on first principle models. A difficulty in the optimal control of batch processes, such as batch bio-reactors, is that the first principle models are usually very complicated and difficult to obtain and implement for on-line control. Data based empirical modeling, such as PLS models and neural network models, can be a very useful alternative in this case. Tian et al. [43] used a hybrid stacked neural network model to calculate the optimal control actions for a batch polymerization process. In order to overcome the detrimental effect of model uncertainties, a reliable optimization control strategy was proposed by Zhang [44] where a model prediction confidence measure is incorporated in the optimization objective as an additional weighted term or as an extra objective in the multi-objective optimization framework.

The practical difficulties faced by the industrial implementation of optimal control strategy include the unavoidable model plant mismatches and the presence of unknown disturbances. Recently, iterative learning control (ILC) has been used in the run-to-run control of batch processes to directly update input trajectory [45, 46]. The basic idea of ILC is to update the control trajectory for a new batch run using the information from previous batch runs so that the output trajectory converges asymptotically to the desired reference trajectory. Refinement of control signals based on ILC can significantly enhance the performance of tracking control systems. Xiong and Zhang [47] presented a recurrent neural network based ILC scheme for batch processes where filtered recurrent neural network prediction errors from previous batches are added to the model predictions for the current batch and optimization is performed based on the updated predictions. Xiong and Zhang [48] proposed a batch to batch ILC strategy using linearised time variant perturbation models that are identified from process operational data. Here the linearised models were identified using multiple linear regression (MLR). In many batch processes, the control actions at different stages of a batch can be correlated since they are such determined to optimize the final product quality at the end of the batch. In such situations, models obtained using MLR may not be appropriate. In this study, it is proposed that the linearised models can be identified using PLS.

6.2.1 Linearised Models for Batch Processes

For the convenience of implementation, we consider batch processes where the batch run length (t_f) is fixed and consists of N sampling intervals (i.e. $N=t_f/h$, with h being the sampling time). Product quality variables (outputs), $y \in R^n$ ($n \geq 1$), can be obtained off-line by analyzing the samples taken during the batch run and the manipulated variable, $u \in R^m$ (m=1 in this work), can be measured at each sampling time on-line. The product quality and control trajectories are defined, respectively, as

$$\mathbf{Y}_k=[y_k^T(1), y_k^T(2),\ldots, y_k^T(N)]^T \tag{19}$$

$$\mathbf{U}_k=[u_k(0), u_k(1),\ldots,u_k(N\text{-}1)]^T \tag{20}$$

where the subscript k denotes the batch index. It is assumed here that the desired reference trajectories of product quality can be set reasonably and are defined as

$$\mathbf{Y}_d=[y_d^T(1), y_d^T(2),\ldots, y_d^T(N)]^T \tag{21}$$

A batch operation is typically modeled with a dynamic model, but it would be convenient to consider a static function relating the control sequence to the product quality sequences over the whole batch duration [46]. Considering the causality, the product quality variables at time t, $y_k(t)$, is a static non-linear function of all control actions up to time t, $\mathbf{U}_k(t)=[u_k(0), u_k(1),\ldots,u_k(t\text{-}1)]^T$, i.e.

$$y_k(t)=f_t(\mathbf{U}_k(t)) + v_k(t), \qquad t=1, 2, ..., N, \qquad y_k(0)=y_0 \tag{22}$$

where $f_t(.)$ represents the non-linear function between $\mathbf{U}_k(t)$ and $y_k(t)$ and $v_k(t)$ is the measurement noise at time t. Eq(22) can be rewritten in matrix form as:

$$\mathbf{Y}_k= \mathbf{F}(\mathbf{U}_k) + \mathbf{v}_k \tag{23}$$

where $\mathbf{F}(\cdot)$ represents the non-linear static functions between $\mathbf{U}_k(t)$ and $y_k(t)$ at different sampling times and $\mathbf{v}_k=[v_k^T(0), v_k^T(1),\ldots, v_k^T(N\text{-}1)]^T$ is a vector of measurement noises.

Let the nominal control trajectory and its corresponding product quality trajectory from a reference batch be defined as

$$\mathbf{U}_s=[u_s(0), u_s(1),\ldots,u_s(N\text{-}1)]^T \tag{24}$$

$$\mathbf{Y}_s=[y_s^T(1), y_s^T(2),\ldots, y_s^T(N)]^T \tag{25}$$

where the subscript s denotes the nominal batch and $y_s(0) = y_0$. Then the *perturbation variables* of the control and product quality variables are defined, respectively, as

$$\overline{\mathbf{U}}_k = \mathbf{U}_k - \mathbf{U}_s \tag{26}$$

$$\overline{\mathbf{Y}}_k = \mathbf{Y}_k - \mathbf{Y}_s \tag{27}$$

where $\overline{y}_k(0) = 0$.

Linearising the non-linear batch process model described by Eq(23) with respect to $\mathbf{U}_s$ around the nominal trajectories ($\mathbf{U}_s$, $\mathbf{Y}_s$), the following can be obtained.

$$\mathbf{Y}_k = \mathbf{Y}_s + \left.\frac{\partial \mathbf{F}(\mathbf{U}_k)}{\partial \mathbf{U}_k}\right|_{\mathbf{U}_s} (\mathbf{U}_k - \mathbf{U}_s) + \mathbf{w}_k + \mathbf{v}_k \tag{28}$$

where $\mathbf{w}_k=[w_k^T(1), w_k^T(2), \ldots, w_k^T(N)]^T$ is a sequence of model errors due to the linearisation (i.e., due to negalecting the higher oder terms) and $\mathbf{v}_k$ represents the effects of noise and unmeasured disturbances. Define the linearised model $\mathbf{G}_s$ as

$$\mathbf{G}_s = \left.\frac{\partial \mathbf{F}(\mathbf{U}_k)}{\partial \mathbf{U}_k}\right|_{\mathbf{U}_s} \tag{29}$$

The structure of $\mathbf{G}_s$ is restricted to the following lower-block-triangular form due to the causality.

$$\mathbf{G}_s = \begin{bmatrix} g_{10} & 0 & \cdots & 0 \\ g_{20} & g_{21} & \cdots & 0 \\ \vdots & \vdots & \ddots & \vdots \\ g_{N0} & g_{N1} & \cdots & g_{N(N-1)} \end{bmatrix} \tag{30}$$

The linearised model can be identified from historical process operation data using MLR [48]. To cope with process drift, the linearised model can be re-identified after each batch with data from the most recent batch added to the historical process data. Furthermore, the control trajectory and quality variable trajectory from the most recent batch can be used as the reference trajectories.

In many batch processes, the control policies are typically determined to optimise the product quality at the end of a batch. Therefore, the control actions during different stages of a batch are usually correlated. In such cases, appropriate linearised model may not be obtained from MLR. To overcome the colinearity in the regression variables, PLS regression [37] can be used to obtain the linearised models.

6.2.2 Batch to Batch Control

Since the prime objective in batch process operation is on enhancing the end-of-batch product quality, here we only consider final product quality control. For the kth batch, the actual product quality at the end of a batch can be written as the model prediction plus an error term

$$y_k(t_f) = \hat{y}_k(t_f) + e_k \tag{31}$$

where $y_k(t_f)$ and $\hat{y}_k(t_f)$ are, respectively, the actual and model predicted final product quality, and e_k is model prediction error.

The model prediction for the (k+1)th batch can be represented using the first order Taylor series expansion based on the kth batch:

$$\begin{aligned} \hat{y}_{k+1}(t_f) &= \hat{y}_k(t_f) + \left.\frac{\partial f}{\partial u_1}\right|_{U_k} (u_1^{k+1} - u_1^k) + \cdots + \left.\frac{\partial f}{\partial u_N}\right|_{U_k} (u_N^{k+1} - u_N^k) \\ &= \hat{y}_k(t_f) + \mathbf{G}^T \Delta\mathbf{U}^{k+1} \end{aligned} \tag{32}$$

where

$$\Delta\mathbf{U}^{k+1} = [\Delta u_1^{k+1}\ \Delta u_2^{k+1}\ \ldots\ \Delta u_N^{k+1}]^T$$

and

$$\mathbf{G}^T = \left[\left.\frac{\partial f}{\partial u_1}\right|_{U_k} \left.\frac{\partial f}{\partial u_2}\right|_{U_k} \cdots \left.\frac{\partial f}{\partial u_N}\right|_{U_k} \right]^T .$$

Suppose that the model prediction errors are the same for the kth batch and the (k+1)th batch, then the optimal control of the (k+1)th batch can be represented as:

$$\min_{\Delta U^{k+1}} J = \left\| \hat{y}_k(t_f) + \mathbf{G}^T \Delta \mathbf{U}^{k+1} + e_k - y_d \right\|_Q^2 + \left\| \Delta \mathbf{U}^{k+1} \right\|_R^2 \tag{33}$$

where $\mathbf{Q}$ is the weighting for product quality control errors and $\mathbf{R}$ is the weighting for control action.

Let $\frac{\partial J}{\partial \Delta \mathbf{U}^{k+1}} = 0$, then the optimal control updating can be expressed as:

$$\Delta \mathbf{U}^{k+1} = (\mathbf{GQG}^T + \mathbf{R})^{-1} \mathbf{GQ}(y_d - \hat{y}_k(t_f) - e_k) \tag{34}$$

$$\mathbf{U}^{k+1} = \mathbf{U}^k + \Delta \mathbf{U}^{k+1} \tag{35}$$

Using Eq(34) and Eq(35), control actions for the (k+1)th batch are obtained as modifications to those for the kth batch using an updated model $\mathbf{G}$.

6.3 Fed Batch Fermentation Application Results

In this fed-batch process, the batch time is 65 hours. The batch length is divided into 10 equal intervals and, within each interval the fed rate is kept constant. Thus, the feeding profile (control profile) for a batch is a vector containing 10 elements. The weighting matrices Q and R were selected as 1 and $15I$ respectively. The desired final produce concentration is selected as 103.5 g/l. It is noted that this desired value is slightly higher than that would be achievable in a normal batch run. This is to ensure that the batch to batch iterative learning control strategy will always try to achieve the highest product concentration possible. It is assumed that 30 batches of historical process operation data are available to build linearised models. Of the 30 batches, 20 batches were used as training data and 10 batches were used as testing data.

6.3.1 Results with a Fixed Model

With a fixed linearised model, the reference batch was taken as the 30th (the last) batch in the historical data. For PLS modelling, the number of latent variables kept in the model was determined through cross validation. A number of latent variables were tried and the one giving the least sum of squared errors is selected. In this case, the number of latent variables used is 4.

Fig. 14 shows the performance of batch to batch iterative learning control under a fixed model. At the 11th batch, the parameter μ was increased by 3% to simulate a disturbance. During the first 10 batches, the product concentration under the batch to batch control with the PLS linearised model converges quickly and approaches the desired value. In contrast, although the product concentration improves significantly

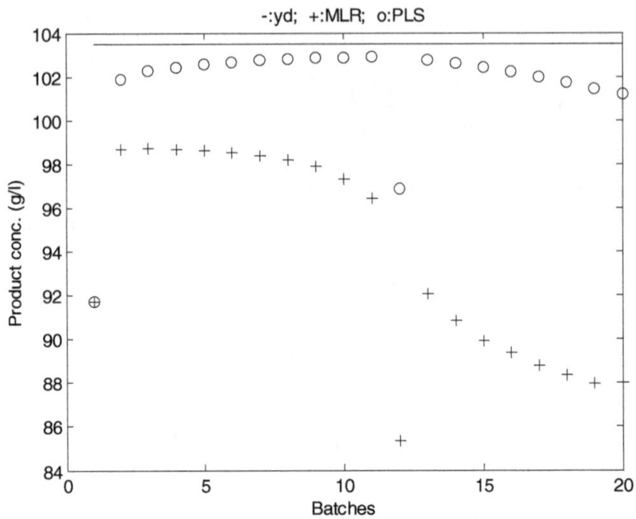

Fig. 14. Product concentration under batch to batch control with a fixed model

in the second batch under the batch to batch control with the MLR linearised model, it starts getting worse from the third batch. This could be due to that the MLR model is not appropriate. The regression matrix has a condition number of 648 indicating that the control actions (feed rates) during different stages of a batch are correlated. Thus, MLR may not give an appropriate model.

With the unknown disturbance presented at the 11th batch, both control strategies fail to track the desired product concentration. The performance of the batch to batch control under the MLR model is much worse than that under the PLS model. This can be due to the fact that the fixed MLR and PLS models are no longer valid with the presence of the disturbance.

6.3.2 Results with Updated Models

Since the control performance under a fixed model is not satisfactory, models should be updated after each batch. After each batch run, data from this batch are added to the historical process operation database. The process operation database is randomly partitioned into a training data set (60%) and a testing data set (40%). With the updated models, the linearised model is re-identified using the immediate previous batch as the reference batch. For PLS modelling, once again, the number of latent variables kept in the model was determined through cross validation. A number of latent variables were tried and the one giving the least sum of squared errors on the testing data set was selected. Fig. 15 shows the number of latent variables retained in the model after each batch. It can be seen that the number of latent variables varies from batch to batch. At the 11th batch, the parameter μ was increased by 3% to simulate a disturbance.

Fig. 16 shows the performance of batch to batch iterative learning control under updated models. During the first 10 batches, the product concentration under the

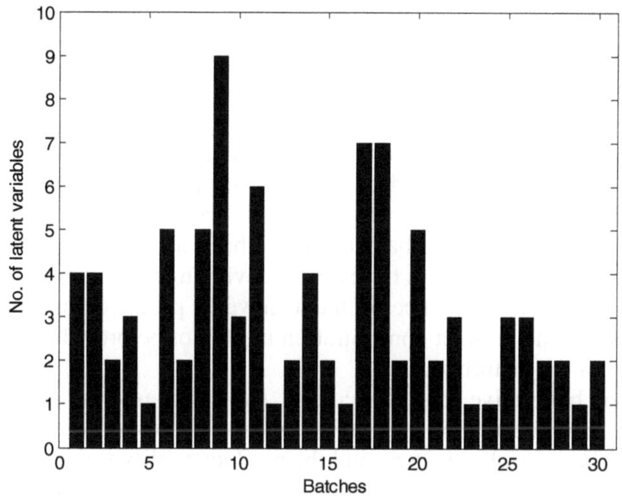

Fig. 15. Numbers of latent variables kept in the updated PLS models

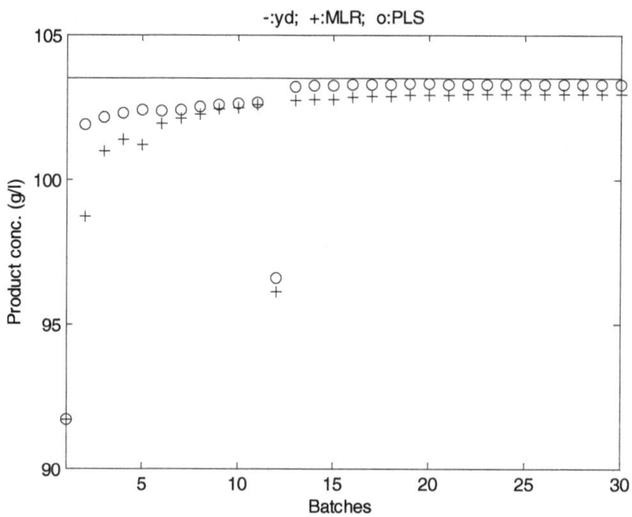

Fig. 16. Product concentration under batch to batch control with updated models

batch to batch control based on the updated PLS models converges faster and to better values than that under the batch to batch control based on the updated MLR models.

At the 11th batch, due to the presence of disturbance, the control policy from the 10th batch is no longer appropriate and product concentrations under both control strategies were lowered to around 96 g/l. With batch to batch control based on updated models, the product concentrations under both control strategies significantly improved in subsequent batches. However, with the updated PLS models, the product concentration improved faster and converged to better value than that with the updated MLR models.

To further test the performance of the batch to batch control strategy under disturbances, the parameter μ was increased by 3% during batches 11 and 20. Fig. 17 shows the performance of batch to batch iterative learning control under updated models. It can be seen that the product concentration improves during the first 10 batches. With the presence of disturbance in batches 11 to 20, product concentration drops significantly. With batch to batch control based on updated MLR models, product compositions during these batches do not have an improving trend. However, those based on updated PLS models have an improving trend. It should be realized that this disturbance has drastic influence on the achievable product concentration. Due to operating constraints, the product concentration might not be brought back to the same level as that without the disturbance.

A batch to batch iterative learning control strategy based on linearised process models is proposed and applied to a fed-batch fermentation process. By considering the deviations of the control trajectory and the product quality trajectory from those in a reference trajectory, a linearised model can be established and identified from historical process operational data. Partial least square regression can be used to address the correlation between the control actions at different stages of a batch. For the linearised model to better represent the current batch the immediate previous batch can be taken as the reference batch and the linearised model is updated from batch to batch. Based on the linearised model, an optimal control trajectory updating law can be derived analytically. Application to a simulated fed-batch bio-reactor demonstrates that the proposed batch to batch iterative learning control strategy can effectively address the problems of model plant mismatches and the presence of unknown disturbances.

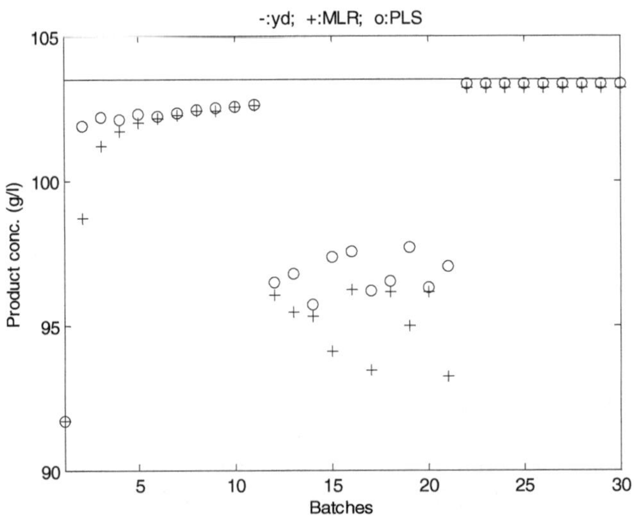

Fig. 17. Product concentration under batch to batch control with updated models (disturbances present in batches 11 to 20)

7 Non-linear PCA, PLS and Neural Network Approaches

Pharmaceutical and biotechnological systems are not only extremely complex but also tend to be highly nonlinear, highly multivariate and time varying. The monitoring control and supervision of such highly non-linear and time variant processes using non-linear methodologies and technologies, rather than liner approaches, is therefore of considerable interest as a means of achieving improved productivity and reduced process variability. Artificial intelligence methodologies, including artificial neural networks, knowledge based systems, case-based-reasoning, etc have been shown to contribute significantly to the achievement of these objectives. Such approaches are not competitive with the approaches and methodologies discussed above but rather should be viewed alongside and possible collaborative with multivariate statistical methodologies. For example in software sensors and non-linear multivariate statistical process control using PCA and PLS and their non-linear variants e.g. [49, 50], artificial neural network approaches e.g. [51], or Knowledge based systems approaches. Bringing together all these approaches in a robust platform alongside fundamental process knowledge could offer improvement in bioprocess monitoring, control, supervision and optimisation by speeding up product and process development and scale-up and reducing overall process variability and potential human error.

Acknowledgements

The authors gratefully acknowledge the permission of the *Journal of Biotechnology* for allowing the reproduction of selected figures and related text from Martin and Morris, 2002, *Journal of Biotechnology* and the contributions from Zhihua Xiong and the UK Department for Innovation, Universities and Skills under the UK/China Fellowship for Excellence program and the EU Bio-production Project (No. 026515-2) for their contributions to case study 2.

References

1. Clementschitsch, F., Bayer, K.: Microbial Cell Factories (published online, May 22), doi:10.1186/1475-2859-5-19
2. Nomikos, P., MacGregor, J.F.: Monitoring batch processes using multi-way principal components. AIChE Journal 40(8), 1361–1375 (1994)
3. Nomikos, P., MacGregor, J.F.: Multivariate SPC Charts for monitoring batch processes. Technometrics 37(1), 41–59 (1995)
4. Nomikos, P., MacGregor, J.F.: Multi-way partial least squares in monitoring batch processes. Chemometrics and Intelligence Laboratory Systems 30, 97–108 (1995)
5. Wold, S., Geladi, P., Esbensen, K., Ohman, J.: Multi-way principal components and PLS analysis. Journal of Chemometrics 1, 41–56 (1987)
6. Wold, S., Kettaneh, N., Friden, H., Holmberg, A.: Modelling and Diagnostics of Batch Processes and Analogous Kinetic Experiments. Chemometrics and Intelligent Laboratory Systems 44, 331–340 (1998)
7. Gregersen, L., Jorgensen, S.B.: Supervision of fed-batch fermentations. Chemical Engineering Journal 75(1), 69–76 (1999)

8. Lennox, B., Hiden, H.G., Montague, G.A., Kornfield, G., Goulding, P.R.: Application of Multivariate Statistical Process Control to Batch Operations. Computers and Chemical Engineering 24(2-7), 291–296 (2000)
9. Martin, E.B., Morris, A.J.: Impurity Detection in Batch Pharmaceutical Production. In: 8th International Conference on Computer Applications in Biotechnology, Modelling and Control of Biotechnology Processes, Quebec City, Canada, June 2001, pp. 37–46 (2001)
10. Martin, E.B., Morris, A.J.: Enhanced bio-manufacturing through advanced multivariate statistical technologies. Journal of Biotechnology 99, 223–235 (2002)
11. Fletcher, N.M., Martin, E.B., Morris, A.J., Quinn, H., Hinge, B.: The Monitoring of an Industrial Fed-batch Fermentation Process. In: CAB-8, Quebec, Canada, pp. 149–154 (2001)
12. Fletcher, N.M., Martin, E.B., Morris, A.J.: Comparison of Dynamic Approaches to Batch Modelling. In: Proc. 12th European Symposium of Computer Aided Process Engineering, pp. 487–492 (2002)
13. McPherson, L.A., Martin, E.B., Morris, A.J.: Super model-based techniques for batch performance monitoring, ESCAPE-12. In: European Symposium on Computer Aided Process Engineering, vol. 12, pp. 523–528 (2002)
14. Martin, E.B., Bettoni, A., Morris, A.J.: Manufacturing Performance Monitoring of a Mixed Batch Continuous Process through Resampling. Journal of Quality Technology 34(2), 171–186 (2002)
15. Martin, E.B., Morris, A.J., Lane, S.: Monitoring Process Manufacturing Performance. IEEE Control Systems Magazine 22(5), 26–39 (2002)
16. Mercer, E.C.W., Martin, E.B., Morris, A.J.: State-space residual based monitoring. In: ESCAPE-12. European Symposium on Computer Aided Process Engineering, vol. 12, pp. 727–732 (2002)
17. Simoglou, A., Martin, E.B., Morris, A.J.: Statistical Performance Monitoring of Dynamic Multivariate Processes using State Space Modelling. Computers Chem. Engng. 26, 909–920 (2002)
18. Alabi, I., Morris, S., Martin, A.J., Martin, E.B.: On-line Dynamic Process Monitoring using Wavelet-based Generic Dissimilarity Measures. Trans. IChemE Part A Chemical Engineering Research and Design 83(A6), 698–705 (2005)
19. Lane, S., Martin, E.B., Kooijmans, R., Morris, A.J.: Performance monitoring of a multi-product semi-batch process. Journal of Process Control 11, 1–11 (2001)
20. Louwerse, D.J., Smilde, A.K.: Multivariate statistical process control of batch processes based on three-way models. Chemical Engineering Science 55, 1225–1235 (1999)
21. Wise, B.M., Gallagher, N.B., Martin, E.B.: Application of PARAFAC2 to fault detection and diagnosis in semiconductor etch. Journal of Chemometrics 15, 285–298 (2001)
22. Meng, X., Morris, A.J., Martin, E.B.: On-line monitoring of batch processes using a PARAFAC representation. Journal of Chemometrics 17, 65–81 (2003)
23. Westerhuis, J.A., Kourti, T., MacGregor, J.F.: Comparing alternative approaches for multivariate statistical analysis of batch process data. Journal of Chemometrics 13, 397–413 (1999)
24. Van Sprang, E.N.M., Ramaker, H.-J., Westerhuis, J.A., Gurden, S.P., Smilde, A.K.: Critical evaluation of approaches for on-line batch process monitoring. Chemical Engineering Science 57, 3979–3991 (2002)
25. Lee, J.H., Dorsey, A.W.: Monitoring of batch processes through state-space models. AIChE Journal 50, 1198–1210 (2004)
26. Flores-Cerrillo, J., MacGregor, J.F.: Multivariate monitoring of batch processes using batch-to-batch information. AIChE Journal 50, 1219–1228 (2004)

27. García-Munoz, S., Kourti, T., MacGregor, J.F.: Model predictive monitoring for batch processes. Industrial and Engineering Chemistry Research 43(18), 5929–5941 (2004)
28. Kourti, T.: Multivariate dynamic data modeling for analysis and statistical process control of batch processes, start-ups and grade transitions. Journal of Chemometrics 17, 93–109 (2003)
29. Fletcher, N.M., Morris, A.J., Montague, G., Martin, E.B.: Local Dynamic Partial Least Squares Approaches for the Modelling of Batch Processes. Can. J. Chem. Eng. 86, 960–970 (2008)
30. Jolliffe, I.T.: Principal Component Analysis. Springer, New York (1986)
31. Wold, S.: Cross-validatory estimation of the number of components in factor and principal components model. Technometrics 20(4), 397–405 (1978)
32. Geladi, P., Wold, S.: Local principal component models, rank maps and contextuality for curve resolution and multi-way calibration inference. Chemom. Intell. Lab. Syst. 2, 273–281 (1987)
33. Miller, P., Swanson, S.E., Heckler, C.F.: A Missing Link in Multivariate Quality Control. Int. J. Appl. Math. Comput. Sci. 8, 775–792 (1998)
34. Gollmer, K., Posten, C.: Supervision of bioprocesses using a ynamdic time warping algorithm. Control Engineering Practice 40, 1287–1295 (1996)
35. Kassidas, A., MacGregor, J.F., Taylor, P.A.: Synchronization of Batch Trajectories Using Dynamic Time Warping. AIChE Journal 44(8), 1361–1375 (1998)
36. Rothwell, S.G., Martin, E.B., Morris, A.J.: Comparison of Methods for Dealing with Uneven Length Batches with Application to MPCA Monitoring of a Batch Process. In: Proceedings 7th Int. Conf. On Computer Applications in Biotechnology (CAB7), Osaka, Japan (1998)
37. Geladi, P., Kowalski, B.R.: Partial Least Squares Regression: A Tutorial. Analytica Chimica Acta 185, 1–17 (1986)
38. Wold, S., Kettaneh-Wold, N., Skagerberg, B.: Nonlinear PLS modeling. Chemometrics and Intelligent Laboratory Systems 7(1-2), 53–65 (1989)
39. Baffi, G., Martin, E.B., Morris, A.J.: Non-linear Projection to Latent Structures Revisited: the Quadratic Approach. Computers and Chemical Engineering 23, 395–411 (1999)
40. Hong, J.: Optimal substrate feeding policy for fed batch fermentation with substrate and product inhibition kinetics. Biotechnol. Bioeng. 27, 1421–1431 (1986)
41. Chen, C.T., Hwang, C.: Optimal control computation for differential-algebraic process systems with general constraints. Chem. Engng Commun. 97, 9–26 (1990)
42. Luus, R.: Application of dynamic programming to differential - algebraic process systems. Computers Chem. Engng. 17, 373–377 (1993)
43. Tian, Y., Zhang, J., Morris, A.J.: Modeling and optimal control of a batch polymerization reactor using a hybrid stacked recurrent neural network model. Ind. Eng. Chem. Res. 40, 4525–4535 (2001)
44. Zhang, J.: A reliable neural network model based optimal control strategy for a batch polymerisation reactor. Ind. Eng. Chem. Res. 43, 1030–1038 (2004)
45. Gao, F., Yang, Y., Shao, C.: Robust iterative learning control with applications to injection molding process. Chemical Engineering Science 56, 7025–7034 (2001)
46. Lee, S., Chin, I.S., Lee, H.J., Lee, J.H.: Model predictive control technique combined with iterative learning control for batch processes. AIChE J. 45, 2175–2187 (1999)
47. Xiong, Z., Zhang, J.: Batch-to-batch iterative optimisation control based on recurrent neural network models. Journal of Process Control 15(1), 11–21 (2005)

48. Xiong, Z., Zhang, J.: Product quality trajectory tracking in batch processes using iterative learning control based on time-varying perturbation models. Ind. Eng. Chem. Res. 42(26), 6802–6814 (2003)
49. Martin, E.B., Morris, A.J., Zhang, J.: Process performance monitoring using multivariate statistical process control. IEE Proc. -Control Theory Appl. 143(2), 132–144 (1996)
50. Jia, F., Martin, E.B., Morris, A.J.: Non-linear principal components analysis with application to process fault detection. International Journal of Systems Science 31(1), 1473–1487 (2000)
51. Glassey, J., Ignova, M., Ward, A.C., Montague, G.A., Morris, A.J.: Bioprocess supervision: neural networks and knowledge based systems. Journal of Biotechnology 52, 201–205 (1997)

11

Modelling of Biotechnological Processes – An Approach Based on Artificial Neural Networks

Eduardo Valente[1,2], Miguel Rocha[1], Eugénio C. Ferreira[3], and Isabel Rocha[3]

[1] Department of Informatics / CCTC – University of Minho
Campus de Gualtar, 4710-057 Braga, Portugal
mrocha@di.uminho.pt
[2] Department of Informatics Engineering, ESTCB
Avenida do Empresário, 6000-767 Castelo Branco, Portugal
eduardo@est.ipcb.pt
[3] IBB – Institute for Biotechnology and Bioengineering
Centre of Biological Engineering – University of Minho
Campus de Gualtar, 4710-057 Braga, Portugal
{ecferreira,irocha}@deb.uminho.pt

Abstract. In this chapter we describe a software tool for modelling fermentation processes, the FerMoANN, which allows researchers in biology and biotechnology areas to access the potential of Artificial Neural Networks (ANNs) for this task. The FerMoANN is tested and validated using two fermentation processes, an *Escherichia coli* recombinant protein production and the production of a secreted protein with *Saccharomyces cerevisiae* in fed-batch reactors. The application to these two case studies, tested for different configurations of feedforward ANNs, illustrate the usefulness of these structures, when trained according to a supervised learning paradigm.

1 Introduction

This chapter discusses a software tool, the FerMoANN, that intends to be a useful instrument in research on the modelling, control and optimization of fermentation processes, whose aim is to allow an increase on process productivity.

Most biological processes are difficult to describe mathematically and, as a consequence, it is difficult to predict their behaviour. One of the most difficult aspects of its analysis, especially when working in real time, is the dynamic behaviour of biochemical reactions, since most of the times the kinetic equations associated have parameters that are difficult to estimate. Moreover, problems with sterility and lack of reliable measuring techniques are pointed as factors for not having measurements of key process variables [22].

Therefore, it becomes necessary to use non-deterministic models that could be implemented based on data and that are not tied to real-time measurements.

The Artificial Neural Networks (ANNs) are one of such modelling tools that have been extensively used for representing complex phenomena, since they are universal non-linear function approximators and their application does not require any knowledge about the structure of the model [20]. ANNs are therefore suitable for modelling fermentation processes.

M. do Carmo Nicoletti, L.C. Jain (Eds.): Comp. Intel. Tech. for Bio. Mod., SCI 218, pp. 311–332.
springerlink.com

This work integrates the ANNs in FerMoANN, with an intuitive design and interface, aiming at becoming a powerful instrument for researchers in the field of Biology / Biotechnology to study fermentation processes. A similar tool has not been found available, so its specifications were made from scratch, based on suggestions given by researchers from the biological area.

In this work, in order to test the performance of the developed tool, two case studies are considered, with different degrees of complexity. These cases are fermentation processes of the yeast *Saccharomyces cerevisiae* (*S. cerevisiae*) and of the bacterium *Escherichia coli* (*E. coli*).

2 Fermentation Processes

In general, fermentation can be defined as a bioprocess that generates a product through the mass culture of a microorganism [30] in a bioreactor.

There are several factors that can influence a fermentation process and among them there is the quantity and composition of the substrate that are provided to the cell culture. The substrate is what supports the life of a particular microorganism and could include nutrients like glucose, oxygen, and light or nitrogen source. Other important parameters affecting cellular growth are the physical and chemical conditions of the environment (temperature, pH, pressure, etc.) and protein induction [16].

This work focuses on the influence of the substrate administration on microorganism cultures, making possible to test any feed profile and see their influence on the productivity of the culture.

2.1 Cultivation Techniques

There are three main cultivation techniques, which are chosen considering the type of the desired product and the characteristics of the microorganism. These modes of operation are continuous, batch and fed-batch cultures:

Batch. In this cultivation mode, an initial amount of substrate is provided, and the process runs with the generation of biomass and products without the addition of more nutrients, until all the substrate is consumed.

Continuous. In this cultivation mode, the substrate is continuously provided to the bioreactor. Also, a part of the fermentation broth is removed in quantities that allow the culture continuity without reaching the maximum capacity of the bioreactor volume. If the rate of substrate addition is constant, a steady state can be reached inside the bioreactor with constant values for biomass, substrate and product concentrations.

Fed-batch. This technique is based on the continuous or sequenced batch supply of one or more nutrients, without removal of the bioreactor broth. The process continues until the desired volume is reached or until the maximum capacity of the container is achieved. The fed-batch cultures have some advantages, such as prevention of substrate inhibition, high cell density cultivation, etc [34].

It is in this last type of culture where simulation and optimization tools could be of greater value for researchers, since they produce data that could be used in fermentation process control, influencing their behaviour in run-time. A vast set of studies could be found, that use fed-batch bioreactors for process control and optimization, namely those in [2], [3], [16] or [21].

However, some of the main difficulties in the design of on-line optimization and control for biological processes lie in the insufficiency of cheap and reliable hardware and software sensors capable of providing direct and on-line measurements of the biological variables. Many times an off-line optimization is made for the determination of the operating conditions before the cultivation is carried out and the optimized operating conditions are not modified during the cultivation [34]. Even with recent developments in hardware and software sensors, only some of the state variables involved, that are critical for efficient control, are available for on-line measurement. For example, usually the dissolved oxygen concentration and gaseous flow rates are available for on-line measurement while the biomass, products and substrates concentrations are often available via off-line analysis. Given the low frequency in biomass, substrate and product measurements in most fermentation processes, it is very difficult to estimate on-line important variables like specific growth, consumption and production rates, the so-called kinetic variables [14]. The specific cell growth rate is one of the most important process variables characterizing the state of microorganisms during fermentations mainly because the biosynthesis of many products of interest is often related with the values assumed by this variable.

The developed tool was designed to aid in the effort of estimating kinetic variables in fermentation processes from other available measurements. For evaluating the software performance, data were gathered from two separate fermentation processes. These case studies will be briefly described in the next two sections.

2.2 The *Saccharomyces cerevisiae* Fermentation Process

This process is based in the production of a secreted protein with post-translational modifications, using the yeast *Saccharomyces cerevisiae* (SEY2102) as the host organism and SUC2-s2 as secreted protein. This process was studied by Park and Ramirez (PR) in a fed-batch bioreactor [21]. The analogy between the products secreted by this type of cells and mammalian cells, the lack of pathogenicity with humans and other characteristics makes this culture suitable for the large production of useful human proteins.

The equations that describe the PR process are as follows [19] (the detailed explanation of each equation is beyond the objectives of this work):

$$\frac{dx_1}{dt} = \frac{4.75A(x_2 - x_1)}{0.12 + A} - \frac{ux_1}{x_5} \tag{1}$$

$$\frac{dx_2}{dt} = \frac{x_3 x_4 \exp(-5x_4)}{0.1 + x_4} - \frac{ux_2}{x_5} \tag{2}$$

$$\frac{dx_3}{dt} = \left(A - \frac{u}{x_5}\right)x_3 \tag{3}$$

$$\frac{dx_4}{dt} = -7.3Ax_3 - \frac{u(x_4 - 20)}{x_5} \tag{4}$$

$$\frac{dx_5}{dt} = u \tag{5}$$

$$A = \frac{21.87x_4}{(x_4 + 0.4)(x_4 + 62.5)} \tag{6}$$

x_1: Concentration of secreted protein (units / L)
x_2: Concentration of total protein (units / L)
x_3: Concentration of cells (g / L)
x_4: Concentration of substrate (g / L)
x_5: Bioreactor Volume (L)
u: Feed rate (L / h)
A: Specific growth rate (h^{-1})

In this experiment, the aim is to find the best operating conditions that maximize the production of the secreted protein, which is intrinsically related with the yeast specific grow rate (A). In equation (6) we can observe that the state variable x_4 (substrate concentration) is the essential element for determining A. So, this equation will be used to produce data for the learning process of the ANN model.

2.3 The *Escherichia coli* Fermentation Process

The bacterium *Escherichia coli* (*Ecoli*) has been one of the most commonly used recombinant protein production systems, because of its ability to grow rapidly and at high densities on inexpensive substrates like glucose. Furthermore, it is one of the best characterized systems regarding molecular genetics, physiology and expression systems. An important problem in this culture lies in the tight control that is necessary during the fermentation process, since at particular conditions (excessive or low glucose concentration) recombinant protein production can be significantly reduced [25], due to the effect of acetic acid or cellular starvation. The main optimization objective in this process is to control the feeding trajectory in a manner that concentration of glucose maintains at optimum levels for cell growth and subsequent high protein production.

The detailed mathematical model describing this process can be found in [26]. The substrate specific consumption rate is given by the following equation:

$$q_s = q_{S,\max} \frac{S}{S + K_S} \frac{K_{i,S}}{K_{i,S} + A} \tag{7}$$

q_S: Specific substrate uptake rate (h^{-1})
$q_{S,max}$: Maximum specific uptake rate (h^{-1})
K_S: Affinity constant for glucose uptake (g / kg)
$K_{i,S}$: Inhibition constant of acetate on glucose uptake (g / kg)
A: Acetate concentration (g/kg)

The specific substrate uptake rate is related with the specific growth rate by:

$$q_s = \frac{\mu}{k_{XS}} \tag{8}$$

μ: Specific growth rate (h^{-1})
k_{XS}: Global yield coefficient (g/g)

The model differential equations obtained from mass balances to the main state variables are the following, in a matrix format:

$$\frac{d}{dt}\begin{bmatrix} X \\ S \\ A \\ O \\ C_T \end{bmatrix} = \begin{bmatrix} 1 & 1 & 1 \\ -k_1 & -k_2 & 0 \\ 0 & k_3 & -k_4 \\ -k_5 & -k_6 & -k_7 \\ k_8 & k_9 & k_{10} \end{bmatrix} \begin{bmatrix} \mu_1 \\ \mu_2 \\ \mu_3 \end{bmatrix} X - D \begin{bmatrix} X \\ S \\ A \\ O \\ C_T \end{bmatrix} + \frac{F_{in,S}}{W} \begin{bmatrix} 0 \\ S_{in} \\ 0 \\ 0 \\ 0 \end{bmatrix} + \begin{bmatrix} 0 \\ 0 \\ 0 \\ OTR \\ -CTR \end{bmatrix} \tag{9}$$

X: Biomass concentration (g/kg)
S: Substrate concentration (g/kg)
O: Dissolved oxygen concentration (g/kg)
C_T: Total dissolved carbon dioxide concentration (g/kg)
k_x: Yield parameters
D: Dilution rate (h^{-1})
$F_{in,S}$: Liquid mass flow of the stream containing glucose fed into the bioreactor (kg h^{-1})
W: Weight of liquid inside the bioreactor (kg)
S_{in}: Substrate concentration in the inflow (g kg^{-1})
OTR: Oxygen transfer rate from the gaseous to the liquid phase (g/kg/h)
CTR: Carbon dioxide transfer rate from the liquid to the gaseous phase (g/kg/h)

In the *Ecoli* process, the parameters of interest are the specific growth rates (μ_1, μ_2, μ_3) and the specific substrate uptake rate (q_s), which represent the dynamic behaviour of this process. These are the parameters that are difficult to measure in real-time on the culture and for whom an ANN model could be of great interest. For the ANN validation, the mathematic equations 7 and 8 will be used, together with the heuristics from [26], since they determine the kinetic coefficients from other measurable parameters like X, S, A, O, and C_T.

2.4 Levels of Complexity

The main differences in the complexity of the two case studies arise from:

- Number of the input and output variables;
- Continuity and complexity of the mathematical functions to approximate by the ANN.

The PR kinetic behaviour could be described by equation (6), which means that with only one input variable (x_4: substrate concentration) it is possible to achieve the values of the only kinetic coefficient relevant in this process (A: Specific growth rate).

The *Ecoli* process kinetic behaviour, expressed by equation (9) shows that five variables (X, S, A, O, C_T) can have influence in the values of the three kinetic variables (μ_1, μ_2, μ_3).

The number of input and output variables defines the number of neurons at the input layer and at the output layer, respectively, in an ANN. More inputs and more outputs mean more processing units and more connections for the Artificial Neural Network, which increases significantly the processing burden.

The PR equation does not have discontinuity points, and the variations on the kinetic variables are relatively smooth.

The *Ecoli* process has considerable changes in state variables, from a low initial to a very high biomass and product concentration [24]. Also, the switches from different metabolic states cause discontinuities in the values that are very difficult for an ANN to learn.

3 Artificial Neural Networks

Artificial Neural Networks (ANNs) are based on an analogy with biological brains. Their way of operation provides them with features that are not present in traditional computing machines, such as massive parallelism, distributed computing and representation, adaptability, among many others [12].

The building blocks of ANNs are the artificial neurons (Fig. 1). The weights define the interaction of each neuron with its neighbours. For each neuron, any signal that arrives is multiplied by the weight of the corresponding connection. These values are added with an offset, or bias, to reach the activation of the neuron. The activation function applied over this value grants the non linearity of ANN's behaviour.

ANNs are mainly characterized by the topology of the connections between neurons, also known as architecture, the method of adjusting its connection weights (training algorithm) and the activation function [5], [7]. The neurons in an ANN are typically disposed in layers, with a layer for input neurons that receives the signal from the environment and has no processing associated. There is also an output layer, containing

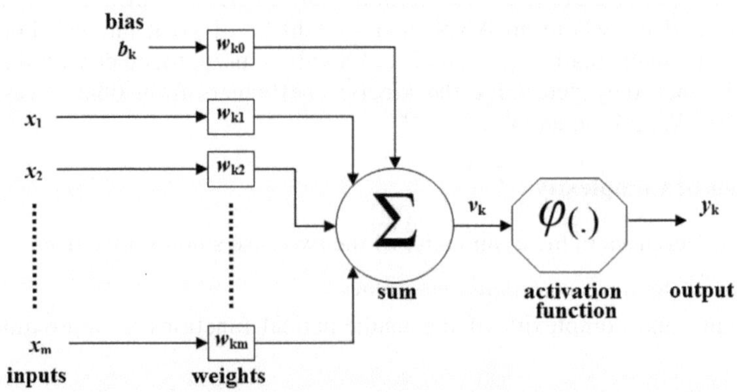

Fig. 1. Artificial neuron structure

neurons with a normal operation, whose outputs are sent to the environment. Internally, between these two layers, other layers of neurons can exist, which are called intermediate or hidden layers. In the same layer, usually all neurons have the same activation function and the same structure of links to other layers of neurons [6].

According to its architecture, an ANN can be classified as a single layer network (it actually has two layers, the input layer and output layer) or multilayer (input layer, one or more hidden layers and the output layer). ANNs can also be classified according to the direction of the internal information flow. If the information flows layer by layer, without cycles, this is a unidirectional or feed-forward network. Otherwise, it is classified as a recursive network. The choice of which model of information flow should be used to represent a process depends on the dynamics and complexity of the problem under analysis [17], [18].

A multilayer ANN has, in principle, a higher ability to solve more complex problems. The greater the number of neurons in each layer and the number of layers, the higher is the degree of complexity of the problems that the network can solve. However, the excessive complexity of a network is often unnecessary and carries an extra processing burden, probably causing overfitting problems. The optimal number of hidden neurons is the smallest number that yields an acceptable prediction error [10]. Therefore, there are mechanisms for constructing and pruning networks, with the aim of building them with only the necessary size for a good performance in learning and generalization of the problem.

In this work, ANN training is done according to the paradigm of supervised learning. There is a data set of known input and output vectors with interrelated data, which the ANN will have to assimilate, through the successive application of these training cases. The training process is repeated while the response of the network does not coincide or not satisfactorily approaches the desired outputs [8].

To ensure the effectiveness of the model, ANNs with three layers (input, hidden and output) were used, since they are able to represent any continuous functional mapping [1], [10]. The number of nodes at the hidden layer could be achieved by practical methods, testing several hypotheses until finding the number that demonstrates best results.

The training algorithm used was *Backpropagation* [10], which is based on the known delta rule, with roots in the optimization of filters [32]. This algorithm is composed of two phases. In the first phase, a stimulus is applied to the network and the error difference between the obtained and the expected output value is registered. In the second phase, that error is propagated backwards through the network, causing an adjustment of the connection weights, thus implementing the learning of the network.

This algorithm has proven to be very effective in most cases but it may, in certain circumstances, show some drawbacks. For example, in learning, where the lowest possible error of training is sought, there may be stagnating points when a local minimum is found. This value is the lowest error in a certain neighbourhood but not the overall minimum. This and other problems that lead the network learning to virtually stagnate can be solved, or at least mitigated, through the use of some heuristics.

One of these heuristics is used by the Resilient Back Propagation (RProp) [23], which is considered the best learning algorithm for networks with various topologies. The basic idea of this algorithm is to update the weights of the network using only the rate of learning and the sign of the partial derivative of the error function of each weight. This process provides convergence speed, robustness and is easy to implement. Experiences made comparing the use of *Rprop* with the use of the *Backpropagation* algorithm show that the first method performs better in 98% of the cases. If, however, an appropriate pre-processing of the data is made, the performance is almost equal [11], [28].

The learning process could lead to several models presenting all a good learning performance. To choose the best of these models, some selection methods could be used to estimate their generalization capabilities, like cross validation [13]. By training an ANN with a particular data set, there is a risk that the network specializes and adapts its behaviour specifically to the solution of the cases used for training. This leads to loss of generalization and, hence, the loss of capacity to solve new problems not considered in the training data set. To identify the most specialized models, the ANN needs to be validated by applying new cases that are not included in the training set and the responses to those cases should be determined.

4 State of the Art

Nowadays, it is already possible to find a vast set of studies in the area of bioprocesses modelling. For describing fermentation processes, usually three types of models are used: white box, grey box and black box models. These designations classify the models according to the degree of visibility that an observer has on the internal structure of the respective systems. The choice of a particular type generally depends on the existing amount of *a priori* knowledge about the system to be modelled [17,18], the nature of the problem to be solved and on the process variables that could be measured in real-time of the biological experiments.

Traditionally, the whole process is described in a mechanistic way, being the physic behaviour of the process externally visible. Because of this visibility, these models are called white box models. These models are mainly composed by differential equations with kinetics based on the Monod or Monod-derived equations, which describe the cell growth limited by substrate concentrations and often contemplate inhibition by different compounds [5]. These models are empirical and highly non-linear, requiring the determination of both model structure and parameters for each particular case [6, 9, 33]. One of the major applications of these models is process optimization.

The white box approach is used in the work of Veloso *et al.* [31], where the equations of mass balance, a complex kinetic model and an Extended Kalman Observer were used for the on-line determination of biomass concentration. Park and Ramirez [21] proposed an optimal control policy for the optimization of a secreted heterologous protein production process, described by a white-box model. They based their work the Pontryagin's Minimum Principle for determining the necessary conditions for system optimization. Lee and Ramirez [16] established optimal control strategies for maximizing the production of induced foreign protein by recombinant bacteria, by deterministic algorithms that calculate exactly the periods where the cell specific growth rate and protein production are maximized. Some process optimizations require more advanced techniques. For example, Genetic and Evolutionary Algorithms had been used in several works [24, 25, 26, 27]. Mendes *et al.* [19] presented an experimental comparison between Particle Swarm Optimization, Differential Evolution and real-valued Evolutionary Algorithms for the optimization of fed-batch fermentation processes described by white-box models.

On the other hand, Artificial Neural Networks have proven to be of great value, due to their ability to model nonlinear systems trough input-output mapping. ANNs can be constructed from process experimental data, being suitable to use as black box models of biological processes. Lee and Park [15] used an ANN model as a software sensor for monitoring and controlling wastewater treatment process. The problem is that existing measures does not give on-line measurements of nutrient dynamics.

The use of a single ANN to optimize processes does not always brings the best results, even at the level of process function approximation, or at the level of processing burden when the system to approach is too complex and has a great number of inputs and outputs. A possible solution is to divide the problem into several parts, which could be implemented by a set of ANNs or by ANNs combined with other solutions (hybrid model). Chen *et al.* [3] presented an optimization of a fed-batch bioreactor using two cascade recurrent neural networks for process modelling which predict the biomass production based on the feed rate profile. A Genetic Algorithm is then combined with the extended recurrent neural networks model to determine the feed rate profile that maximizes the final quantity of biomass for the fed-batch fermentation process. Another example is the work of Koprinkova *et al.* [14] that focuses on the time delay between the changes in substrate level and the adjustment of the growth rate at the appropriate level in chemostat cultivations. They proposed different feed-forward ANNs with delay elements for both specific growth rate and specific consumption rate estimation as a solution for modelling this behaviour.

When white box and black box components are joined together, a grey box model is obtained. For example, Zuo and Wu [34] proposed a method to semi real-time optimization and control of a fed-batch fermentation system using hybrid neural networks for process modelling and genetic algorithms for determining optimal feeding rates. The hybrid model joins the knowledge-based part (differential equations, from mass balance of the system) and the black box part (ANN, for dynamic behaviour of the specified system). Peres *et al.* [22] presented a method of expressing and combining different types of knowledge, consisting in a gating system that joins knowledge modules that could be white, black or gray box models. By this way, it is possible to combine the capacity of the three types of models to achieve a better prediction. A similar method is applied by Simutis *et al.* [29], which also highlighted the importance of starting with a clear definition of the problem and a corresponding quantitative objective function.

Given all this work in the area, it is important to establish what is intended for the current work stating what is the problem to solve and what is the difference between this and other solutions for related problems:

1. The objective is to optimize a fed-batch bioreactor, increasing product formation. A similar objective could be found in the work of Chaudhuri and Modak [2], but this method should be applicable to any strain of microorganisms;
2. The control variable is the feed profile;
3. Real feed profiles exist, considered like near optimal solutions for maximum cell growth and product formation, obtained by optimization methods;
4. Given 1, 2 and 3, the aim of this chapter is to present a computational tool that allows the use of ANNs in the modelling of these processes.

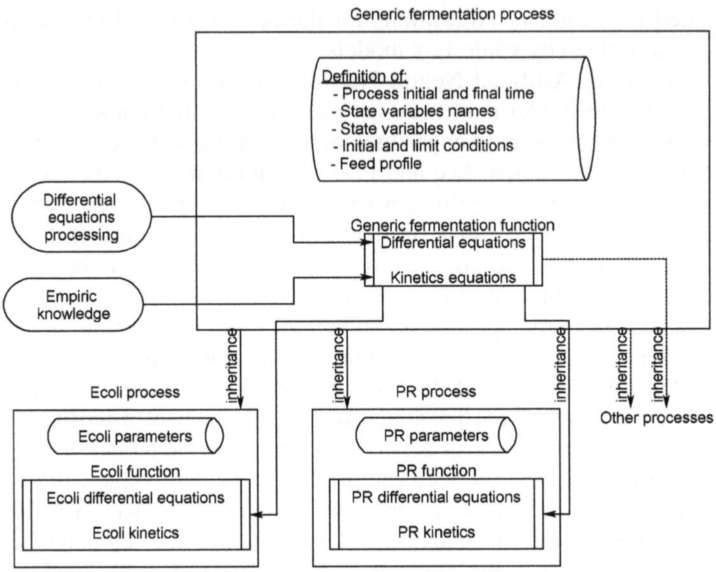

Fig. 2. Schema of the deterministic (white box) model for fermentation processes

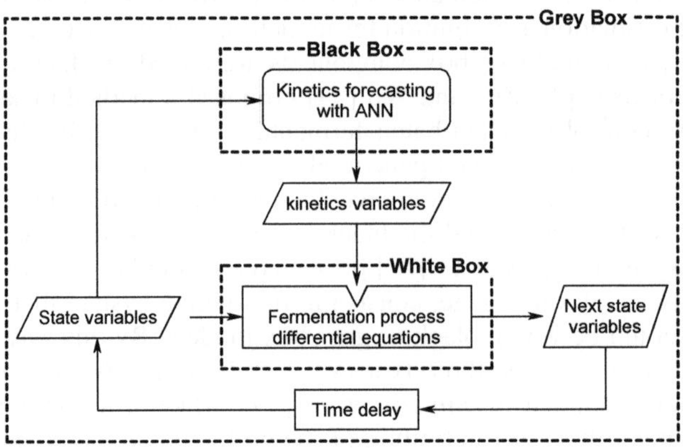

Fig. 3. A schematic representation of the generic grey box model for the simulation of fermentation processes

The starting point was the existing knowledge about the processes, in the form of white-box models composed of ordinary differential equations that describe the fermentation processes behaviour (Fig. 2).

To establish the optimum feed profile it is necessary to have a reliable model for the prediction of process state and kinetic variables. These last ones are not possible to measure directly in the culture experiment, and so they are obtained through other variables. Moreover, the structure of the kinetic equations is often obtained with empirical methods that are dependent of the culture strain in study. To avoid this limitation, an

ANN is used to simulate the kinetic behaviour for each process, being the number of inputs and outputs defined by existing process data.

Establishing a structure that could be applied to any fermentation process, the objectives for process optimization becomes too general to be resumed in an objective function like the one proposed by Simutis *et al.* [29]. Instead of achieving an optimal feed profile for a particular process, the objective was to give biologists a tool with which it is possible to construct a reliable model for any fermentation process. In this tool, it should be possible to introduce feed profiles and to test their contribute in process yield maximization.

In this way, the existing mechanistic knowledge is combined with a component implemented with a black box. The result is a hybrid grey box (Fig. 3).

5 The FerMoANN Tool

The FerMoANN computational tool proposed in this work should allow:

- To produce process data off-line, that can be used to test the grey box model without having to collect data from the real experiment, decreasing developing time;
- To select one of the available processes and respective white-box models. It should also be possible to automatically add new modules for different processes;
- To evaluate several ANN configurations, selecting the one that better approximates the process kinetic behavior, i.e the one that shows lower Root Mean Square Error in their learning and validation. In this evaluation, it should be possible to choose the training parameters (e.g. training algorithm), inputs and outputs, validation method, etc.;
- To use trained ANNs to model a given process;
- To visually check the modeling results, and allow testing diverse feed profiles with the aim of finding the one that maximizes the process yield.

The general guidelines established for the code design were:

- Modularity, that was achieved with the division of the software in functional blocks, which can operate independently one of each other, ensuring also better scalability and easing the process of incorporating new software modules;
- Generalization, that was ensured giving the application the ability to model any fermentation process, not merely a set of pre-defined processes;
- Functionality, that was obtained from the previous two principles and ensuring a programme with a simple and user friendly interface that can easily be understood.

The use of ANNs reliefs the researcher from defining mathematically the whole process, leaving the kinetic component as a black box, which is obtained through the model inference from data coming from the microorganism's culture.

The FerMoANN was built using mainly three software packages written in Java by the authors (*ann*, *bio* and *annFerm*):

- The *ann* package provides the necessary features for a generic application using feedforward ANNs, following the paradigm of supervised learning [4, 27]. In this module, it is possible to set up the implementation of the network at various levels, such as its structure (characterization of the layers), the adjustment of weights (definition of the initial values and adjustment during training), the definition of the data to be presented to train the ANN, the selection of the best training algorithm for the case study, the type of model validation to use, etc.

- The *bio* package refers to fermentation processes modeling by white box where kinetic models are based on Monod-type equations [26]. With this module the behaviour of various pre-established cultures can be simulated offline. Every culture is associated with a fermentation white box model with its own characteristics where, in addition to other parameters, a set of differential equations and the kinetic equations are included [23, 25]. The fermentation processes here considered are the production of recombinant proteins with *S. cerevisiae* and *E. coli* (see section 2) and the overall schema is given in Fig. 2.
- The *annFerm* package provides the bridge between *ann* and *bio* packages. It is in this package that the fermentation processes models are adapted to allow estimation of kinetic coefficients by artificial neural networks. This adaptation results in a grey box model (Fig. 3).

5.1 Development

The FerMoANN was split into two modules with different interfaces. One module is for white box modelling, that is used to produce data for the other module, the grey box model, thus accelerating the development of the application. Each interface contains a set of sections that provide specific functionalities. In the panels, text areas have been included to display information to the user when necessary.

At the programming level, there was a separation between the part concerning the visual elements of the interface and the part responsible for carrying out the main functions of the program (core functionality). All classes for the interface elements have "Intrfc" as a suffix for better distinction of classes that belong to the core.

The white box interface includes several sections for the generation of feed profiles, for the generation of training data, for combination of examples, for the introduction of noise and also for drawing relevant graphs (Fig. 4):

Feed profiles generation. The feeding profiles, i.e. the amount of substrate to feed to the culture at each time, represent the inputs of the simulation. Through this panel multiple profiles can be created with constant, linear, random, saw, wave or sinus functions. Using different feeds to create data allows a broader spectrum of training examples.

White box process simulation. This simulation uses Monod-type models for generating the kinetics variables, which are then introduced into the ODE models. For each feeding profile, a distinct simulation can be conducted resulting in the values for state and kinetic variables over time. The data from this panel is used for ANN training and a comparison between data that is obtained by the model and with ANNs can be performed.

Mixing examples. A number of files with simulation results can be handled, and used to create new files through composition, cutting, interpolation or sampling data. These

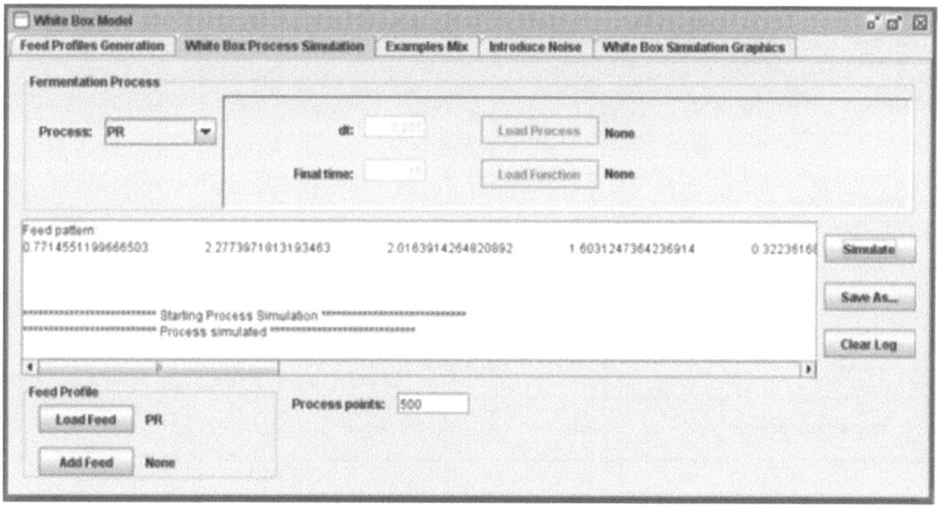

Fig. 4. White box interface functionalities

can create more comprehensive sets of examples or decrease the size of some collections without losing generality.

Adding noise. The introduction of noise on data serves to create data sets that increase the generalization abilities of the ANN model. This can be useful when the set of available data is not rich enough to provide a good training process for the ANN.

White box simulation graphs. The graphs allow for a better visualization of the data generated by the white-box mathematical model. The preview is done in two separate graphs for the state and kinetics variables. It also allows visual comparisons between results of different simulations.

The grey box interface is intended to determine the best ANN configuration that fits the problem under study and the application of those ANNs to simulate the process. Here a number of functionalities can be found, namely the ANN's evaluation, the ANN's training, the process simulation using a trained ANN and the evaluation and training graphs (Fig. 5):

ANN's evaluation. This tool can assess various configurations of ANNs, varying the input and output variables, the number of nodes in the hidden layer (all ANNs are feed-forward with one hidden layer, completely connected), the training algorithm and the number of training epochs. Each model is evaluated by its training error and generalization error in a validation set. It is thus possible to determine which parameter combinations will lead to more accurate models.

ANN's training. An ANN with a given configuration (typically the one that resulted from the above study) can be trained. It then can be used to predict the kinetic variables, integrated within the grey box model.

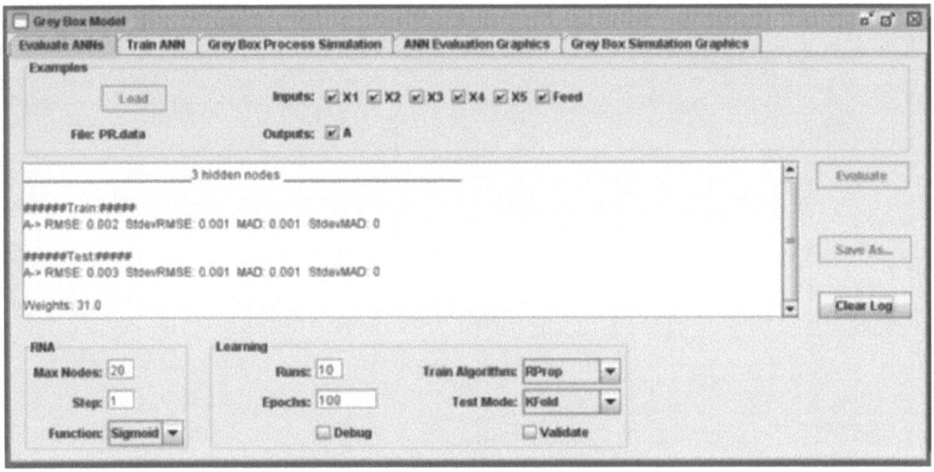

Fig. 5. Grey box interface functionalities

Grey box process simulation. This section enables the simulation of the fermentation process. Trained ANNs calculate the kinetic variables from the values of the state variables in a particular time point and differential equations represent the mass-balance equations that determine the time evolution of the values of the state variables.

ANN evaluation graphs. These graphs show the errors of the evaluation of ANN configurations. They provide information on the settings, data and variables that were used in the training. The graphs allow a better interpretation of the error curves (both in training and validation sets).

Grey box simulation graphs. These graphs show data resulting from grey box model simulations. They show the time evolution for each state and kinetic variable. It is also possible to compare two simulations and thus examine, for example, the difference between the curves obtained with a grey box model and the ones obtained with white box models.

5.2 Experiments and Results

For a demonstration of the software capabilities, two case studies were used: PR and *Ecoli* processes, already described in section 2.

The first step of the modelling process was the feeding profiles generation that allow a wide spectrum of examples to be achieved. The profiles range from optimized ones (using Evolutionary Algorithms [24]), random profiles, constant profiles, etc. These files were then used to simulate the process with white box models, thereby obtaining one data set for each profile. Fig. 6 shows the results for the PR case study with two distinct profiles.

The state variables (x_1, x_2 …) are those involved somehow in the process, and for these variables, it is possible to obtain culture measures at different time points. The

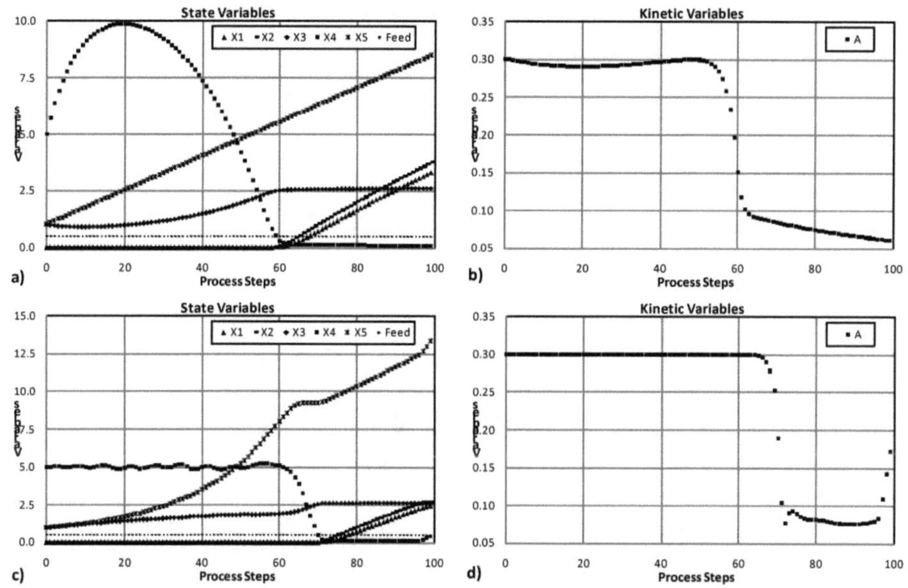

Fig. 6. Example of white box modelling simulation for PR process. a) and b) Constant feed profile; c) and d) Feed profile optimized using Evolutionary Algorithms.

blocks on the left show the graphs for each state variable. In the blocks on the right, the kinetic variable (A) curve is shown, which represents the dynamics of the process. In Fig. 6, only the cases of PR process simulation with optimized and constant feed profiles are shown, but it is also possible to use other feed profiles (e.g. saw, sinusoidal, random and ramp functions). The process steps are the simulation points used in the interpolation (100 for PR). In Fig. 7, the graphs for the *Ecoli* case study are shown.

22 different feed profiles were used for the PR process with values between 0 and 3 L/h (given that this was assumed as the maximum capacity of the feeding device), considering 4 random profiles, 4 linear, 3 constant, and 11 optimized. With each of those 22 profiles the simulation of the PR process was performed.

Using the data composition interface it was possible to group all simulations forming a single global training set for the ANN. In principle, the PR process will not be too difficult to simulate, because of its simplicity. So, from the global data set a sample of 150 values was retrieved, which make the ANN training data set.

For the *Ecoli* case study, 21 different feed profiles were used: 5 random, 5 linear, 4 constant and 7 optimized. Being a more complex process, 500 examples were extracted from the entire data set for ANN evaluation.

If all the configuration parameters are considered, the universe of possibilities for different network configurations is enormous. To limit this universe it was stipulated that certain parameters remain unchanged, such as training algorithm (RProp), the activation function (sigmoid), the number of epochs (1000) and the validation mode (KFold). This configuration grants a sufficient base for a good ANN training.

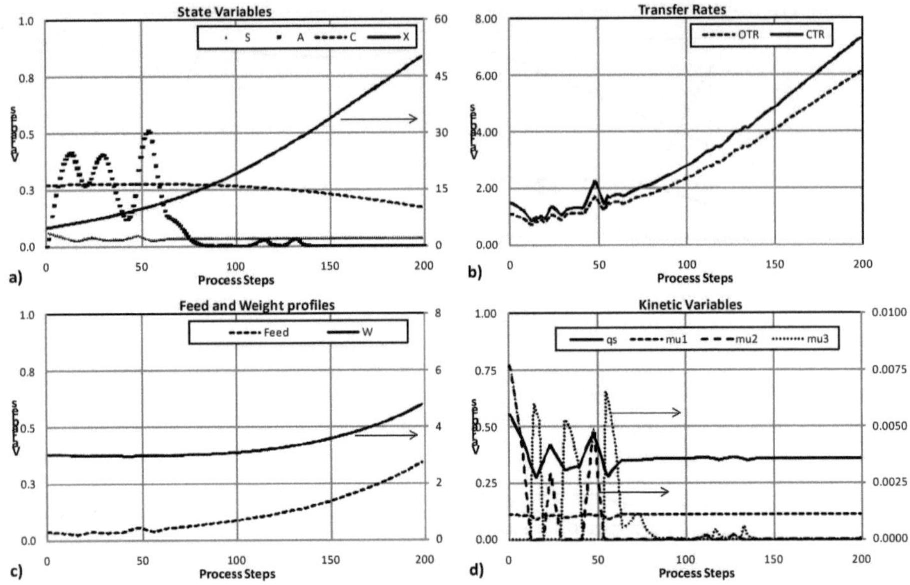

Fig. 7. Example of white box simulation for *Ecoli* process with optimized feed profile

For each evaluation, the parameters changed were the state variables used as ANN inputs and the number of nodes in the ANN hidden layer. An ANN with 20 nodes in hidden layer should be able to simulate very complex cases, so every value between 0 and 20 nodes was tested. To achieve the best combination of input variables (state variables) several tests took place ranging from the inclusion of all variables in the model to the inclusion of only one of them. Supposedly, the more input variables are included, the more the network is capable to determine the outputs (kinetic variables), even for processes with complex characteristics.

For the PR process, the first test was made with each state variable separately. In Fig. 8 we show two examples of evaluation results. This figure shows the error over the training and test sets during the evaluation. The measures of error used were the Root Mean Square Error (RMSE) and Mean Absolute Deviation (MAD). The results for both measures have been proportionately identical, which leads to assume that conclusions can be drawn using only RMSE results.

In the horizontal axis the number of hidden nodes for each ANN structure evaluated is shown. The training error shows how the ANN assimilated the process in question. The test set error shows the adaptation of each network, after trained, to new examples from the same biological process. The best solution is the one that shows lower values of error in both training and test sets. For example, in Fig. 8 it is possible to see that process simulation with the input variable x_4 presents lower values of error than when the input variable x_5 is used. It may also be noted that, using x_4, a network with only one node in the inner layer is sufficient to simulate the process, since the increase in the number of nodes does not bring significant gains in reducing the errors.

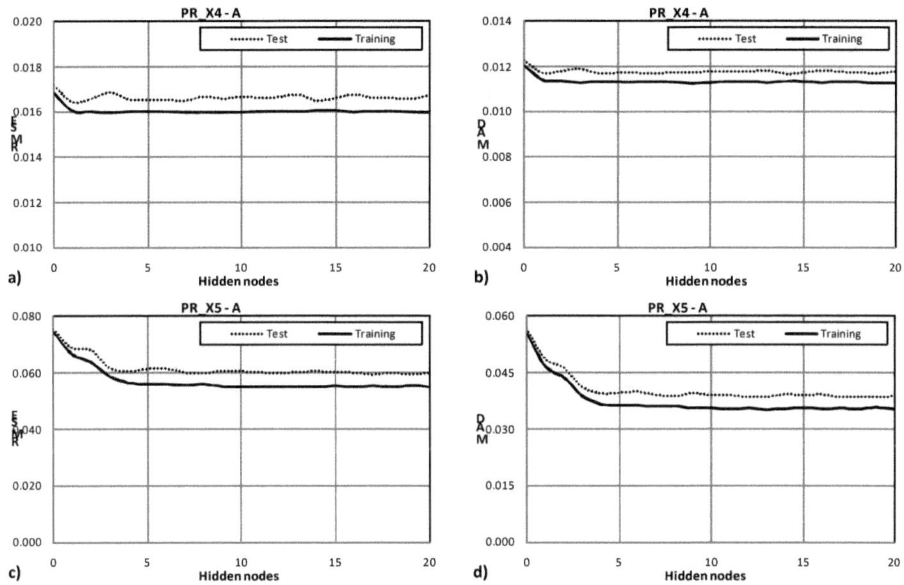

Fig. 8. Result of evaluation for the PR process only with x_4 (a)(b) and only with x_5 (c)(d) state variables as inputs

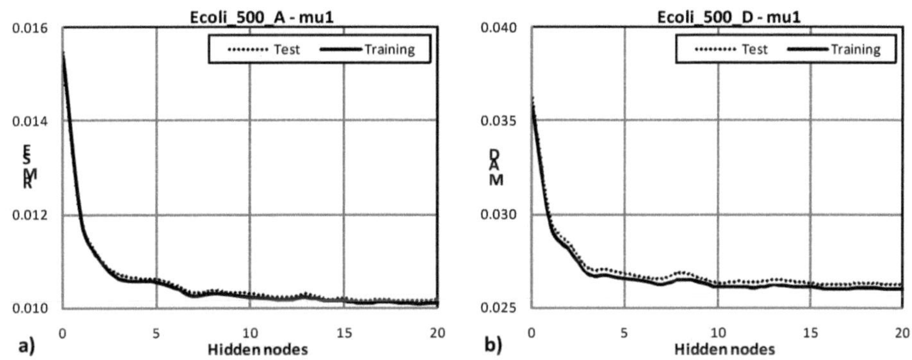

Fig. 9. Evaluation of *mu1* forecasting with the state variable A (a) and with state variable D (b) as inputs

In the *Ecoli* process, the amount of kinetic variables increases the complexity of the selection method. All possibilities considered for the PR case are now multiplied by 4 (the number of *Ecoli* kinetic variables). For an easier interface implementation, kinetic variables were named *mu1*, *mu2* and *mu3*. For example, regarding *mu1*, if we want to test the influence of each state variable in its prediction we would obtain 10 graphs (the number of *Ecoli* state variables). Consider, for instance, the RMSE graphs for prediction of *mu1*, when only the A or the D state variables are used as inputs, as shown in Fig. 9.

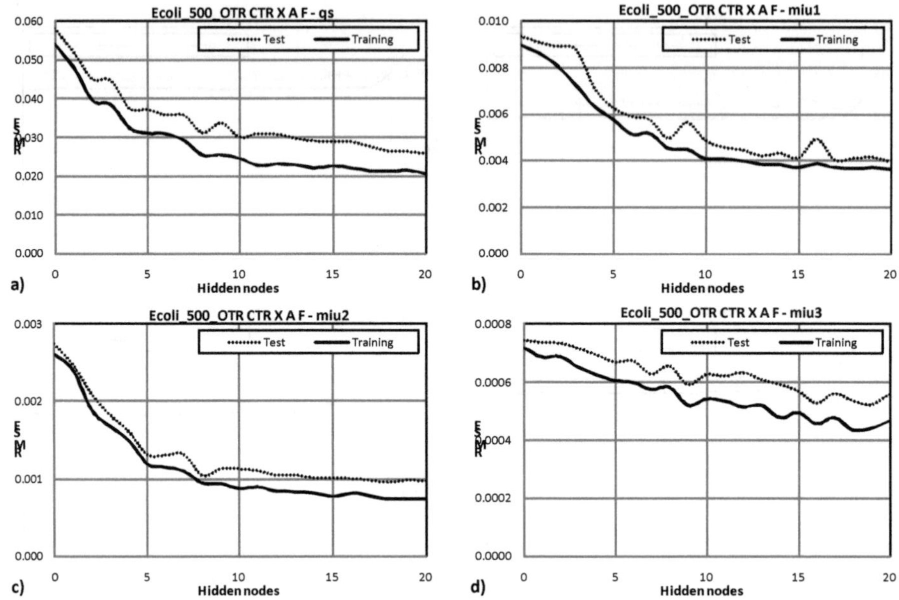

Fig. 10. Evaluation of *Ecoli* kinetic variables prediction with the state variables OTR, CTR, X, A and F as inputs

Fig. 11. Trained ANN. Inputs: state variables; Outputs: kinetic variables; Weight Matrix: nodes connections weights; Activation function: activation function for each non-input node. Layers: number of layers and number of nodes by layer.

It is possible to create any combination of input variables. For example, in the *Ecoli* case study, the variables OTR, CTR, X, A and F can be chosen to estimate the 4 kinetic variables, and the result is shown in Fig. 10.

With the possibility of changing the state variables in the simulation, the researcher can confirm whether the variables that initially were thought to be the most suitable for the prediction are in fact those with better results. This analysis may even lead to the proposal of new hypotheses and a different view of the processes.

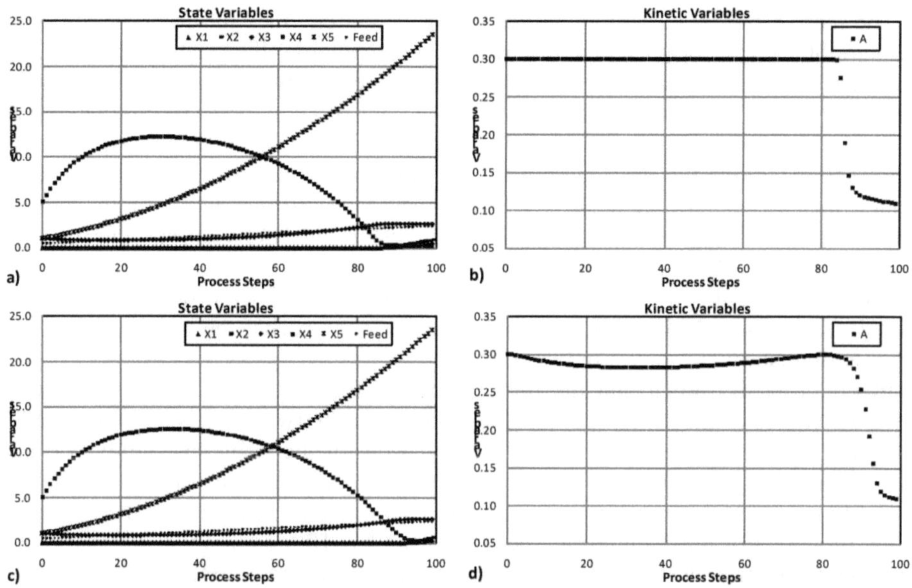

Fig. 12. PR process simulation using a linear feed profile: with grey box (above); with white box (below)

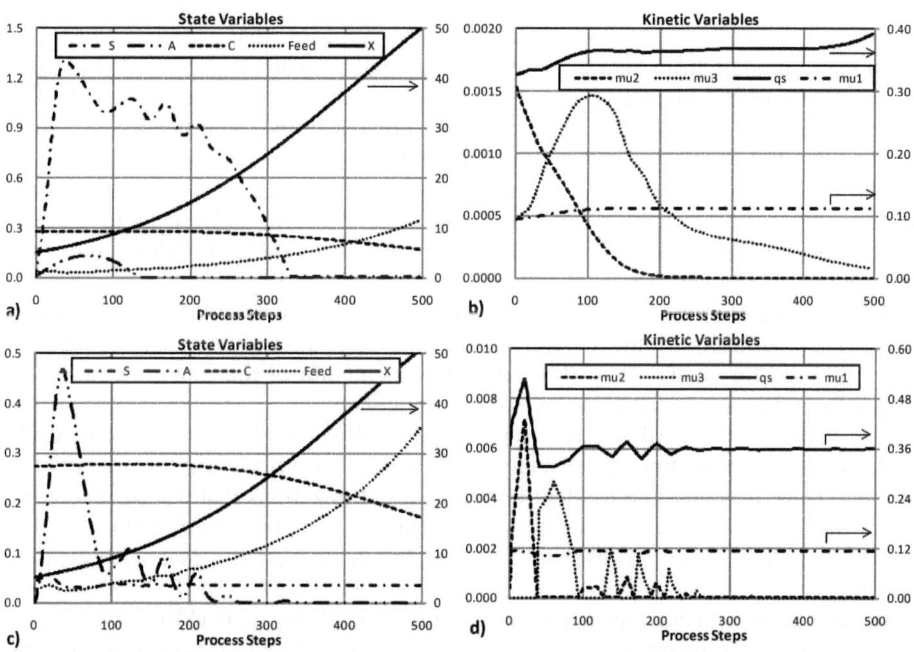

Fig. 13. *Ecoli* process simulation using an optimized feed profile. With grey box (above); with white box (below).

The more different combinations are tested the better, since the combination of input variables does not present linear results. For example, the junction of two state variables may result in a worst prediction than the estimation made with only one of them separately. Similarly, a variable that does not produce great results when used alone, or in certain combinations, could produce excellent results when used in other combinations. Once the best combination of input variables and the best number of hidden nodes is determined, an ANN is trained to be further used for the process simulation. For example, in the PR case study, using only the state variable x_4 and 6 hidden nodes, the ANN would be the one given in Fig. 11.

The ANNs are then trained and used to simulate the process, now with a grey box model. Using the ANN in Fig. 11, and a linear feed profile, the results for the PR process simulation are shown in Fig. 12.

In Fig. 12 it is shown that the performance of the model was quite satisfactory, since the curves of the kinetics and state variables of the two models are very similar, even using only 150 training cases. The simulation for the *Ecoli* case study was also satisfactory (Fig. 13), although it is evident that the ANNs have some difficulties to model functions with abrupt variations, as for example the variable q_S at the beginning of the process. For this simulation, the ANN used has 10 hidden nodes, it was trained with the input variables OTR, CTR, X, A and F and using an optimized feed profile.

6 Conclusions

In the work described in this chapter a software tool, the FerMoANN, was developed to aid in the modelling of fed-batch fermentation processes that can be used to create ANN based models from data coming from real experiments. It is possible to shape a process without the need to know the mathematical description of its mechanism, since learning is made directly from examples created from experimental data, showing the potential of the application of ANNs in this type of problems.

The application makes possible to introduce new fermentation processes at runtime, enabling the application for a multitude of cases. It not only seeks to provide a solution to a specific case, but serves as a platform for a layman in ANNs to evaluate and use multiple models for each culture.

This is possible since user interfaces are implemented in a simple and intuitive way. The modularity of the FerMoANN makes it scalable; the main modules can easily be used for future work. The core implementation of the functionalities is detached from the user interface, to make code understanding and re-use easier.

The implementation of several graph displays allows making visual comparisons between results, providing easier analysis of the factors influencing the course of the process. Thus, it is possible to identify potential optimizations that can be made, such as to change the feed profile. These optimizations can be tested within the tool, thereby avoiding some real experiments that bring, necessarily, increased costs for research.

In the future, a major aim is to improve the capabilities of FerMoANN with new functionalities. An important concern is to create an interface to make the introduction of new processes easier, without having to write Java code. A generic objective function that could use the model to automatically define an optimization curve for some parameters (like feed rate profile) could also be implemented.

References

1. Bishop, C.M.: Neural Networks for Pattern Recognition. Oxford University Press, USA (1995)
2. Chaudhuri, B., Modak, J.M.: Optimization of Fed-Batch Bioreactor Using Neural Network Model. Bioprocess Engineering 19, 71–79 (1998)
3. Chen, L., Nguang, S.K., Chen, X.D., Li, X.M.: Modelling and Optimization of Fed-Batch Fermentation Processes Using Dynamic Neural Networks and Genetic Algorithms. Biochemical Engineering Journal 22, 51–61 (2004)
4. Cortez, P.A.: Nature Inspired Models for Time Series Prediction. PhD Thesis. University of Minho (2002)
5. Coulman, G.A., Stieber, R.W., Gerhardt, P.: Dialysis Continuous Process for Ammonium-Lactate Fermentation of Whey: Mathematical Model and Computer Simulation. Appl. Env. Microbiol. 34(6), 725–732 (1977)
6. Fan, G., Wolkowicz, G.S.K.: Analysis of a Model of Nutrient Driven Self-Cycling Fermentation Allowing Unimodal Response Functions. Discrete and Continuous Dynamical Systems 8(4), 801–831 (2007)
7. Fausett, L.: Fundamentals of Neural Networks: Architectures, Algorithms, and Applications. Prentice-Hall, US (1994)
8. Fyfe, C.: Artificial Neural Networks and Information Theory. University of Paisley (2000)
9. Georgiev, T., Ratkov, A., Kristeva, J., Ivanova, V., Dimitrova, A., Manchev, A., Ratkov, B.: Improved Mathematical Model of L-Lysine Fermentation Process Based on Manure. In: 24th Int. Conf. Information Technology Interfaces IT/ 2002, Cavtat, Croatia, pp. 24–27 (2002)
10. Haykyn, S.: Neural Networks – A Comprehensive Foundation, 2nd edn. Prentice-Hall, New Jersey (1999)
11. Igel, C., Toussaint, M., Weishui, W.: Rprop Using the Natural Gradient. In: Trends and Applications in Constructive Approximation. International Series of Numerical Mathematics, vol. 151, pp. 259–272. Birkhäuser Verlag (2005)
12. Jain, A.K., Mao, J., Mohiuddin, K.M.: Artificial Neural Networks: A Tutorial. IEEE Computer 29(3), 56–63 (1996)
13. Kohavi, R.: A Study of Cross-validation and Bootstrap for Accuracy Estimation and Model Selection. In: International Joint Conference on Artificial Intelligence, pp. 1137–1143 (1995)
14. Koprinkova, P., Petrova, M., Patarinska, T., Bliznakova, M.: Neural Network Modelling of Fermentation Processes: Specific Kinetic Rate Models. Cybernetics and Systems: An International Journal 29, 303–317 (1998)
15. Lee, D.S., Park, J.M.: Neural Network Modelling for On-line Estimation of Nutrient Dynamics in a Sequentially-operated Batch Reactor. Journal of Biotechnology 75, 229–239 (1999)
16. Lee, J., Ramirez, W.F.: Optimal Fed-Batch Control of Induced Foreign Protein Production by Recombinant Bacteria. AIChE Journal 40(5), 899–907 (1994)
17. Levišauskas, D., Tekorius, T.: Model-Based Optimization of Fed-Batch Fermentation Processes Using Predetermined Type Feed-Rate Time Profiles. A Comparative Study. Information Technology and Control, Kaunas, Technologija 34(3), 231–236 (2005)
18. Mandic, D.P., Chambers, J.A.: Recurrent Neural Networks for Prediction: Learning Algorithms, Architectures and Stability. John Wiley & Sons, Chichester (2001)

19. Mendes, R., Rocha, M., Rocha, I., Ferreira, E.C.: A Comparison of Algorithms for the Optimization of Fermentation Processes. In: Proceedings of the 2006 IEEE Conference on Evolutionary Computation, vol. 7, pp. 7371–7378. IEEE Press, Los Alamitos (2006)
20. Oliveira, R.: Combining First Principles Modelling and Artificial Neural Networks: A General Framework. Computers and Chemical Engineering 28, 755–766 (2004)
21. Park, S., Ramirez, W.F.: Optimal Production of Secreted Protein in Fed-Batch Reactors. AIChE Journal 34(9), 1550–1558 (1988)
22. Peres, J., Oliveira, R., Azevedo, S.F.: Knowledge Based Modular Networks for Process Modelling and Control. Computers and Chemical Engineering 25, 783–791 (2001)
23. Riedmiller, M.: Rprop – Description and Implementation Details. Technical Report. Citeseer (1994)
24. Rocha, M., Neves, J., Rocha, I., Ferreira, E.C.: Evolutionary Algorithms for Optimal Control in Fed-Batch Fermentation Processes. In: Raidl, G.R., Cagnoni, S., Branke, J., Corne, D.W., Drechsler, R., Jin, Y., Johnson, C.G., Machado, P., Marchiori, E., Rothlauf, F., Smith, G.D., Squillero, G. (eds.) EvoWorkshops 2004. LNCS, vol. 3005, pp. 84–93. Springer, Heidelberg (2004)
25. Rocha, M., Neves, J., Veloso, A., Ferreira, E.C., Rocha, I.: Evolutionary Algorithms for Static and Dynamic Optimization of Fed-Batch Fermentation Processes. In: Ribeiro, B., Albrecht, R.F., Dobnikar, A., Pearson, D.W., Steele, N.C. (eds.) Adaptive and Natural Computing Algorithms, pp. 288–291. Springer, Heidelberg (2005)
26. Rocha, I.: Model-Based Strategies for Computer-Aided Operation of a Recombinant E. coli Fermentation. Phd Thesis, University of Minho (2003)
27. Rocha, M.: Optimization of Connectionists Learning Models With Genetic and Evolutionary Computation. Phd Thesis, University of Minho (2004)
28. Rojas, R.: Neural Networks – A Systematic Introduction. Springer, Heidelberg (1996)
29. Simutis, R., Oliveira, R., Manikowski, M., Azevedo, S., Lübbert, A.: How to Increase the Performance of Models for Process Optimization and Control. Journal of Biotechnology 59, 73–89 (1997)
30. Stanbury, P.F., Whitaker, A.: Principles of Fermentation Technology. Pergamon Press, Oxford (1984)
31. Veloso, A.C., Rocha, I., Ferreira, E.C.: On-Line Estimation of Biomass in an E. coli Fed-Batch Fermentation. In: Proceedings of the 2nd Mercosur Congress on Chemical Engineering – ENPROMER and 4th Mercosur Congress on Process Systems Engineering, Rio de Janeiro, Brazil (2005)
32. Widrow, B., Sterns, S.D.: Adaptive Signal Processing. Prentice-Hall, New York (1985)
33. Zheng, Y., Gu, T.: Analytical Solutions to a Model for the Startup Period of Fixed-Bed Reactors. Chemical Engineering Science 51(15), 3773–3779 (1996)
34. Zuo, K., Wu, W.T.: Semi-realtime Optimization and Control of a Fed-Batch Fermentation System. Computers and Chemical Engineering 24, 1105–1109 (2000)

Appendix

Journals

Advanced Engineering Informatics, Elsevier, The Netherlands.
http://www.elsevier.com/wps/find/journaldescription.cws_home/622240/description

AI Magazine, AAAI Press, USA.
http://www.aaai.org/Magazine/magazine.php

Biochemical Engineering Journal, Elsevier, The Netherlands.
http://www.elsevier.com/wps/find/journaldescription.cws_home/600804/description

Bioprocess and Biosystems Engineering, Springer, Berlin/Heidelberg.
http://www.springerlink.com/content/100401/

Biotechnology Letters, Springer, The Netherlands.
http://www.springer.com/life+sci/microbiology/journal/10529

Computers & Chemical Engineering, Elsevier, The Netherlands.
http://www.elsevier.com/wps/find/journaldescription.cws_home/349/description

Current Opinion in Biotechnology, Elsevier, The Netherlands.
http://www.elsevier.com/wps/find/journaldescription.cws_home/601293/description

Fuzzy Optimization and Decision Making, Springer, USA.
http://www.springer.com/math/journal/10700

IEEE Intelligent Systems, IEEE Press, USA.
http://www.computer.org/portal/site/intelligent/

IEEE Transactions on Neural Networks, IEEE Press, USA.
http://ieee-cis.org/pubs/tnn/

IEEE Transactions on Evolutionary Computing, IEEE Press, USA.
http://www.ieee-cis.org/pubs/tec/

IEEE Transactions on Fuzzy Systems, IEEE Press, USA.
http://ieeexplore.ieee.org/xpl/RecentIssue.jsp?punumber=91

IEEE Computational Intelligence Magazine, IEEE Press, USA.
http://ieeexplore.ieee.org/xpl/RecentIssue.jsp?punumber=10207

Intelligent Decision Technologies: An International Journal, IOS Press, The Netherlands.
http://www.iospress.nl/loadtop/load.php?isbn=18724981

International Journal of Hybrid Intelligent Systems, IOS Press, The Netherlands.
http://ijhis.hybridsystem.com/

International Journal of Intelligent and Fuzzy Systems, IOS Press, The Netherlands.
http://www.iospress.nl/loadtop/load.php?isbn=10641246

International Journal of Knowledge-Based and Intelligent Engineering Systems, IOS Press, The Netherlands.
http://www.iospress.nl/loadtop/load.php?isbn=13272314

Journal of Bioscience and Bioengineering, Elsevier, The Netherlands.
http://www.elsevier.com/wps/find/journaldescription.cws_home/505516/description#description

Journal of Biotechnology, Elsevier, The Netherlands.
http://www.elsevier.com/wps/find/journaldescription.cws_home/505515/description

Machine Learning, Springer, USA.
http://www.springer.com/computer/artificial/journal/10994

Neural Computing and Applications, Springer, London.
http://www.springer.com/computer/mathematics/journal/521

Neurocomputing, Elsevier, The Netherlands.
http://www.elsevier.com/locate/neucom

Trends in Biotechnology, Cell Press, Elsevier, The Netherlands.
http://www.trends.com/tibtech/default.htm

Special Issue of Journals

Jain, L. C.; Lim, C .P.; Nguyen, N. T. (Guest Editors) (2008) Recent Advances in Intelligent Paradigms Fusion and Their Applications, International Journal of Hybrid Intelligent Systems, vol. 5, no. 3.

Lim, C. P.; Jain, L. C.; Nguyen, N. T.; Balas, V. (Guest Editors) (2008) Advances in Computational Intelligence Paradigms and Applications, An International Journal on Fuzzy Optimization and Decision Making, Kluwer Academic Publisher vol. 7, no. 3.

Nguyen, N. T.; Lim, C. P.; Jain, L. C.; Balas, V. E. (Guest Editors) (2008) Theoretical Advances and Applications of Intelligent Paradigms, Journal of Intelligent and Fuzzy Systems, IOS Press, vol. 19, no. 6.

Ghosh, A.; Seiffert, U.; Jain, L. C. (2007) Evolutionary Computation in Bioinformatics, Journal of Intelligent and Fuzzy Systems, IOS Press, The Netherlands, vol. 18, no. 6.

Palade, V.; Jain, L. C. (2005) Practical Applications of Neural Networks, Journal of Neural Computing and Applications, Springer, Germany, vol. 14, no. 2.

Abraham, A.; Jain, L. C. (2004) Optimal Knowledge Mining, Journal of Fuzzy Optimization and Decision Making, Kluwer Academic Publishers, vol. 3, no. 2.

Palade, V.; Jain, L. C. (2004) Engineering Applications of Computational Intelligence, Journal of Intelligent and Fuzzy systems, IOS Press, vol. 15, no. 3.

Alahakoon, D.; Abraham, A.; Jain, L. C. (2004) Neural Networks for Enhanced Intelligence, Neural Computing and Applications, Springer, UK, vol. 13, no. 2.

Abraham, A.; Jonkar, I.; Barakova, E.; Jain, R.; Jain, L. C. (2004) Special issue on Hybrid Neurocomputing, Neurocomputing, Elsevier, The Netherlands, vol. 13, no. 2.

Abraham, A.; Jain, L. C. (2003) Knowledge Engineering in an Intelligent Environment, Journal of Intelligent and Fuzzy Systems, IOS Press, The Netherlands, vol. 14, no. 3.

Jain, L. C. (1999) Fusion of Neural Nets, Fuzzy Systems and Genetic Algorithms in Industrial Applications, IEEE Transactions on Industrial Electronics, USA, vol. 46, no. 6.

Conference Proceedings

Lovrek, I.; Howlett, R. J.; Jain, L. C. (Editors) (2008) Knowledge-Based and Intelligent Information and Engineering Systems, Lecture Notes in Artificial Intelligence, vols. 5177, 5178 and 5179, KES 2008, Springer-Verlag, Germany.

Apolloni, B.; Howlett, R. J.; Jain, L. C. (Editors) (2007) Knowledge-Based Intelligent Information and Engineering Systems, Lecture Notes in Artificial Intelligence, vols. 4692, 4693 and 4694, KES 2007, Springer-Verlag, Germany.

Howlett, R. P.; Gabrys, B.; Jain, L. C. (Editors) (2006) Knowledge-Based Intelligent Information and Engineering Systems, Lecture Notes in Artificial Intelligence, vols. 4251, 4252 and 4253, KES 2006, Springer-Verlag, Germany.

Khosla, R.; Howlett, R.P.; Jain, L. C. (Editors) (2005) Knowledge-Based Intelligent Information and Engineering Systems, Lecture Notes in Artificial Intelligence, vols. 3681, 3682, 3683 and 3684, KES 2005, Springer-Verlag, Germany.

Negoita, M.; Howlett, R. P.; Jain, L. C. (Editors) (2004) Knowledge-Based Intelligent Information and Engineering Systems, Lecture Notes in Artificial Intelligence, vols. 3213, 3214 and 3215, KES 2004, Springer-Verlag, Germany.

Palade, V.; Howlett, R. P.; Jain, L. C. (Editors) (2003) Knowledge-Based Intelligent Information and Engineering Systems, Lecture Notes in Artificial Intelligence, vols. 2773 and 2774, Springer-Verlag, Germany.

Damiani, E.; Howlett, R. P.; Jain, L. C.; Ichalkaranje, N. (Editors) (2002) Proceedings of the Fifth International Conference on Knowledge-Based Intelligent Engineering Systems (KES'2002), vols. 1 and 2, IOS Press, The Netherlands.

Baba, N.; Jain, L. C.; Howlett, R. P. (Editors) (2001) Proceedings of the Fifth International Conference on Knowledge-Based Intelligent Engineering Systems (KES'2001), vols. 1 and 2, IOS Press, The Netherlands.

Howlett, R. P.; Jain, L. C. (Editors) (2000) Proceedings of the Fourth International Conference on Knowledge-Based Intelligent Engineering Systems (KES'2000), vols. 1 and 2, IEEE Press, USA.

Book Series

Advanced Intelligence and Knowledge Processing, Springer-Verlag, Germany

Nguyen, N. T. (2008) Advanced Methods for Inconsistent Knowledge Management, Springer-Verlag, London.

Wang, J. T. L. *et al.* (Editors) (2005) Data Mining in Bioinformatics, Springer-Verlag, London.

Grana, M. *et al.* (Editors) (2005) Information Processing with Evolutionary Algorithms, Springer-Verlag, London.

Fyfe, C. (2005) Hebbian Learning and Negative Feedback Networks, Springer-Verlag, London.

Husmeier, D. *et. al.* (Editors) (2005) Probabilistic Modelling in Bioinformatics and Medical Informatics, Springer-Verlag, London.

Tan, K. C. *et al.* (2005) Multiobjective Evolutionary Algorithms and Applications, Springer-Verlag, London.

Bandyopadhyay, S. *et. al.* (Editors) (2005) Advanced Methods for Knowledge Discovery from Complex Data, Springer-Verlag, London.

Abraham, A.; Jain, L. C.; Goldberg, R. (2005) Evolutionary Multiobjective Optimization, Springer-Verlag, London.

Gomez-Perez *et al.* (2004) Ontological Engineering, Springer-Verlag, London.

Zhang, S. *et. al.* (2004) Knowledge Discovery in Multiple Databases, Springer-Verlag, London.

Vazirgiannis, M. *et al.* (2003) Uncertainty Handling and Quality Assessment in Data Mining, Springer-Verlag, London, 2003.

Advanced Information Processing, Springer-Verlag, Germany

Harris, C.; Hong, X.; Gan, Q. (2002) Adaptive Modelling, Estimation and Fusion from Data, Springer-Verlag, Germany.

Wang, L.; Fu, X. (2005) Data Mining with Computational Intelligence, Springer-Verlag.

Computational Intelligence and its Applications Series, IGI Publishing, USA

Fulcher, J. (Editor) (2006) Advances in Applied Intelligences, IGI Publishing.

Knowledge-Based Intelligent Engineering Systems Series, IOS Press, The Netherlands

Abraham, A.; Koppen, M.; Franke, K. (Editors) (2003) Design and Applications of Hybrid Intelligent Systems, vol. 104, IOS Press, The Netherlands.

Turchetti, C. (2004) Stochastic Models of Neural Networks, vol. 102, IOS Press, The Netherlands.

Wang, K. (2003) Intelligent Condition Monitoring and Diagnosis Systems, vol. 93, IOS Press, The Netherlands.

Abraham, A. *et al.* (Editors) (2002) Soft Computing Systems, vol. 87, IOS Press, The Netherlands.

Motoda, H. (2002) Active Mining, vol. 79, IOS Press, The Netherlands.

The CRC Press International Series on Computational Intelligence, The CRC Press, USA

Jain, L. C.; Fanelli, A. M. (Editors) (2000) Recent Advances in Artificial Neural Networks: Design and Applications, CRC Press, USA.

Medsker, L.; Jain, L. C. (Editors) (2000) Recurrent Neural Networks: Design and Applications, CRC Press, USA.

Jain, L.C. (Editor) (2000) Evolution of Engineering and Information Systems, CRC Press, USA.

Dumitrescu, D.; Lazzerini, B.; Jain, L. C.; Dumitrescu, A. (2000) Evolutionary Computation, CRC Press, USA.

Dumitrescu, D.; Lazzerini, B.; Jain, L. C. (2000) Fuzzy Sets and their Applications to Clustering and Training, CRC Press, USA.

Jain, L. C.; De Silva, C. W. (Editors) (1999) Intelligent Adaptive Control, CRC Press, USA.

Jain, L. C.; Martin, N. M. (Editors) (1999) Fusion of Neural Networks, Fuzzy Logic and Evolutionary Computing and their Applications, CRC Press, USA.

Jain, L. C.; Vemuri, R. (Editors) (1998) Industrial Applications of Neural Networks, CRC Press, USA.

Jain, L. C.; Johnson, R. P.; Takefuji, Y.; Zadeh, L. A. (Editors) (1998) Knowledge-Based Intelligent Techniques in Industry, CRC Press, USA.

International Series on Natural and Artificial Intelligence, Advanced Knowledge International (AKI), Australia

Lee, R. S. T. (2005) Advanced Paradigms in Artificial Intelligence, Advanced Knowledge International, Australia.

Jain, R. *et al.* (Editors) (2003) Innovations in Knowledge Engineering, Advanced Knowledge International, Australia.

Graziella, T.; Jain, L. C. (2003) Innovations in Decision Support Systems, Advanced Knowledge International, Australia.

Series on Innovative Intelligence, World Scientific, Singapore

Zaknich, A. (2003) Neural Nets for Intelligent Signal Processing, vol. 4, World Scientific Publishing Company, Singapore.

Hirose, A. (2003) Complex Valued Neural Networks, vol. 5, World Scientific Publishing Company, Singapore.

Books

Tolk, A.; Jain, L. C. (Editors) (2009) Complex Systems in Knowledge-based Environments, Springer-Verlag, Germany.

Jain, L. C.; Sato, M.; Virvou, M.; Tsihrintzis, G.; Balas, V.; Abeynayake, C. (Editors) (2008) Computational Intelligence Paradigms: Volume 1 – Innovative Applications, Springer-Verlag.

Virvou, M.; Jain, L. C. (Editors) (2008) Intelligent Interactive Systems in Knowledge-Based Environments, Springer-Verlag.

Holmes, D.; Jain, L. C. (Editors) (2008) Innovations in Bayesian Networks, Springer-Verlag, Germany.

Jain, L. C.; Palade, V.; Srinivasan, D. (Editors) (2007) Advances in Evolutionary Computing for System Design, Springer-Verlag.

Chahl, J. S.; Jain, L. C., Mizutani, A.; Sato-Ilic, M. (Editors) (2007) Innovations in Intelligent Machines 1, Springer-Verlag.

Sato, M.; Jain, L. C. (2006) Innovations in Fuzzy Clustering, Springer-Verlag.

Palade, V.; Bocaniala, C. D.; Jain, L. C. (Editors) (2006) Computational Intelligence in Fault Diagnosis, Springer-Verlag.

Holmes, D.; Jain, L. C. (Editors) (2006) Innovations in Machine Learning, Springer-Verlag.

Seiffert, U.; Jain, L. C.; Schweizer, P. (Editors) (2005) Bioinformatics Using Computational Intelligence Paradigms, Springer-Verlag.

Ghosh, A.; Jain, L. C. (Editors) (2005) Evolutionary Computation in Data Mining, Springer-Verlag, Germany.

Pal, N.; Jain, L. C. (Editors) (2005) Advanced Techniques in Knowledge Discovery and Data Mining, Springer-Verlag, London.

Abraham, A.; Jain, L. C.; van der Zwaag, B. (Editors) (2004) Innovations in Intelligent Systems, Springer-Verlag, Germany.

Fulcher, J.; Jain, L. C. (Editors) (2004) Applied Intelligent Systems, Springer-Verlag, Germany.

Abraham, A.; Jain, L. C.; Kacprzyk, J. (Editors) (2003) Recent Advances in Intelligent Paradigms and Applications, Springer-Verlag, Germany.

Seiffert, U.; Jain, L. C. (Editors) (2002) Self-Organising Neural Networks, Springer-Verlag, Germany.

Jain, L. C.; Kacprzyk, J. (Editors) (2002) New Learning Paradigms in Soft Computing, Springer-Verlag, Germany.

Jain, L. C.; De Wilde, P. (Editors) (2001) Practical Applications of Computational Intelligence Techniques, Kluwer Academic Publishers, USA.

Howlett, R. J.; Jain, L.C. (Editors) (2001) Radial Basis Function Networks 1, Springer-Verlag, Germany.

Howlett, R. J.; Jain, L. C. (Editors) (2001) Radial Basis Function Networks 2, Springer-Verlag, Germany, 2001.

Jain, L. C.; Lazzerini, B.; Halici, U. (Editors) (2000) Innovations in ART Neural Networks, Springer-Verlag, Germany.

Jain, L. C. (Editor) (1997) Soft Computing Techniques in Knowledge-Based Intelligent Engineering Systems, Springer-Verlag, Germany.

Sato, M.; Sato, S.; Jain, L. C. (1997) Fuzzy Clustering Models and Applications, Springer-Verlag, Germany.

Jain, L. C.; Jain, R. K. (Editors) (1997) Hybrid Intelligent Engineering Systems, World Scientific Publishing Company, Singapore.

Vonk, E.; Jain, L. C.; Johnson, R. P. (1997) Automatic Generation of Neural Networks Architecture Using Evolutionary Computing, World Scientific Publishing Company, Singapore.

Van Rooij, A. J. F.; Jain, L. C.; Jain, L. C. (1996) Neural Network Training Using Genetic Algorithms, World Scientific Publishing Company, Singapore.

Book Chapters

Tolk, A.; Jain, L. C. (2009) An Introduction to Complex Systems in Knowledge-based Environments, vol. 168, SCI Series, Springer-Verlag, pp. 1–5.

Pedrycz, W.; Ichalkaranje, N.; Phillips-Wren, G.; Jain, L. C. (2008) Introduction to Computational Intelligence for Decision Making, Springer-Verlag, Chapter 3, pp. 75–93.

Virvou, M.; Jain, L. C. (2008) Intelligent Interactive Systems in Knowledge-Based Environments: An Introduction, Springer-Verlag, Chapter 1, pp. 1–8.

Jain, L. C.; Lim, C. P. (2008) An Introduction to Computational Intelligence Paradigms, Springer-Verlag, Chapter 1, pp. 1–15.

Jain, L .C. *et al.* (2007) Intelligent Machines: An Introduction, Springer-Verlag, Chapter 1, pp. 1–9.

Jain, L. C. *et al.* (2007) Introduction to Evolutionary Computing in System Design, Springer-Verlag, Chapter 1, pp. 1–9.

Jain, L. C. *et al.* (2007) Evolutionary Neuro-Fuzzy Systems and Applications, Springer-Verlag, Chapter 1, pp. 11–45.

Tran, C.; Abraham, A.; Jain, L. C. (2006) Soft Computing Paradigms and Regression Trees in Decision Support Systems, in Advances in Applied Artificial Intelligence, Idea Group Publishing, Chapter 1, pp. 1–28.

Abraham, A.; Jain, L. C. (2005) Evolutionary Multiobjective Optimization, Springer-Verlag, Chapter 1, pp. 1–6.

Sisman-Yilmaz, N .A.; Alpaslan, F.; Jain, L. C. (2004) Fuzzy Multivariate Auto-Regression Method and its Application, in Applied Intelligent Systems, Springer, pp. 281–300.

Jain, L. C.; Konar, A. (2001) An Introduction to Computational Intelligence Paradigms, in Practical Applications of Computational Intelligence Techniques, Springer, pp. 1–38.

Filippidis, A.; Russo, M.; Jain, L. C. (2000) Novel Extension of ART2 in Surface Landmine Detection, Springer-Verlag, Chapter 1, pp. 1–25.

Author Index

Editors

Maria do Carmo Nicoletti holds an undergrad in Mathematics (UNESP,1973), a master in Computer Science (ICMSC-USP,1977), a M.Sc by research in Computer Science (Oxford University, 1980) and a Ph.D. in Computational Physics (IFSC-USP, 1994). She did her post-doctorate studies in Artificial Intelligence at the UNSW, Sydney, Australia (1998–2000). During 1981–1988 she worked as a software analyst in tasks related to software development as well as analysis and evaluation of programs and systems. Since 1988 she has worked at the CS Department of Universidade Federal de S. Carlos, S. Carlos, SP, Brazil and currently is an associate professor in charge of undergrad and graduate courses in CS. Her main area of research is Artificial Intelligence, with an emphasis on methods and techniques related to machine learning and knowledge representation. She is also interested in the use of intelligent computational techniques for controlling biochemical systems as well as for modeling biomolecular processes.

Professor Lakhmi C. Jain is a Director/Founder of the Knowledge-Based Intelligent Engineering Systems (KES) Centre, located in the University of South Australia. He is a fellow of the Institution of Engineers Australia.

His interests focus on the artificial intelligence paradigms and their applications in complex systems, art-science fusion, virtual systems, e-education, e-healthcare, unmanned air vehicles and intelligent agents.

MIX
Papier aus verantwortungsvollen Quellen
Paper from responsible sources
FSC® C105338

If you have any concerns about our products,
you can contact us on
ProductSafety@springernature.com

In case Publisher is established outside the EU,
the EU authorized representative is:
Springer Nature Customer Service Center GmbH
Europaplatz 3, 69115 Heidelberg, Germany

Printed by Libri Plureos GmbH
in Hamburg, Germany